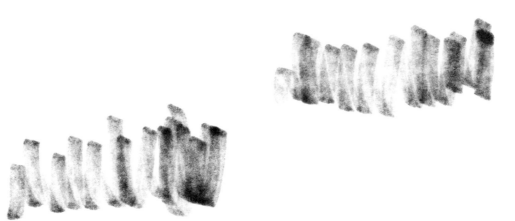

Mapping
Our World

GIS Lessons for Educators

Lyn Malone • Anita M. Palmer • Christine L. Voigt

ESRI PRESS
REDLANDS, CALIFORNIA

ESRI
    Mapping Our World: GIS Lessons for Educators
    ISBN 1-58948-022-8

First printing March 2002.

Printed in the United States of America.

Published by ESRI, 380 New York Street, Redlands, California 92373-8100.

Books from ESRI Press are available to resellers worldwide through Independent Publishers Group (IPG). For information on volume discounts, or to place an order, call IPG at 1-800-888-4741 in the United States, or at 312-337-0747 outside the United States.

To our families, especially our husbands, Pat, Roger, and Brian—

Thank you for your support, guidance, patience, and love.

To Joe Ferguson and Ann Judge, devoted teachers and eager students

who lost their lives on September 11, 2001—

Your passion and dedication to geographic education lives on.

Contents

Preface

This book is a milestone in GIS history. It doesn't describe the latest developments in software design, or the newest applications, but it does advance understanding of the potential scope and power of GIS in one of the most significant ways possible: these are the first comprehensive lessons in the theory and practice of GIS to be designed for middle- and high-school students.

This is not to say that *Mapping Our World: GIS Lessons for Educators* is the "new geography," a replacement of the "old geography" that you are now teaching. This is a supplement, an enhancement. GIS and *Mapping Our World* are tools. The study of geography is the thing itself.

The more important point, however, is this: these tools are the same ones being used at NASA, the United States Geological Survey, the National Oceanic and Atmospheric Administration, the Environmental Protection Agency, and many other governmental agencies and businesses. This isn't software that's been watered down, or made up for special educational purposes. It, too, is the real thing.

The fuel that drives GIS, and the benefits or rewards of using it, are in one sense the same thing: critical thinking—or more specifically, spatial reasoning. Without it, GIS is just software and data. With it, it almost becomes a perpetual-motion machine: new ideas and avenues of inquiry follow each other with astonishing speed and precision. It's not a matter of numbers being crunched more quickly—it's a way of solving problems.

At the same time, it teaches basic skills—skills increasingly necessary as knowledge of the world becomes digital: how to organize and analyze large data sets; input, store, and output data; see maps as numbers and numbers as maps—it encourages an independent and critical way of thinking, a way of seeing the world that moves easily from the grandest macroscopic and global perspectives to the finest and subtlest details.

This is the vital, human spectrum. Everything is connected. What we do here and now—with good, bad, or indifferent intentions—can have profound, even startling consequences 50 years or 50 minutes from now on the other side of the globe. There is an incredible amount of information that accompanies every problem to be solved and every idea we come up with to solve it. GIS is an integrative tool. And *Mapping Our World* is the beginning of the process.

Acknowledgments

We would like to thank all those who helped make this book possible:

At ESRI: The K–12 team of George Dailey, Charlie Fitzpatrick, and Angela Lee, whose vision, leadership, and hard work made this book a reality; Kim Zanelli English, who directed the project and edited lesson content; Amaree Israngkura, who designed the cover; Laura Feaster, who edited lesson content; Gary Amdahl, who edited the entire book; William Garoutte, who designed the Teacher Resource CD; Scott McNair, who put together the ArcView Software & Data CD; Peter Schreiber and Debbie Wright, who handled myriad legal issues; Thomas Purk, who provided support and resources for the Apple® Macintosh® platform; Rich Turner, Anita Friedman, and Kathleen Willison, responsible for the ArcView® software that comes with this book; Deane Kensok, for timely help with Internet issues; Milton Ospina and Ann Johnson of the ESRI Higher Education team; Kelly Ablard-Hickman, who helped with a variety of administrative details; Christian Harder, manager of ESRI Press; Judy Boyd, manager of Educational Products; Nick Frunzi, director of Educational Services; and Jack Dangermond, ESRI founder and president, who made publication of this book a priority for his company.

All the lessons in this book were reviewed and tested by classroom teachers. We thank them for taking the time to work through these lessons with their students to ensure their accuracy and relevance: Brad Baker of Bishop Dunne Catholic School, Dallas, Texas; Gerry Bell of Port Colborne High School, Port Colborne, Ontario; Shanna Hurt of Arapahoe High School, Littleton, Colorado; Marsha MacLean of the Redlands Unified School District, Redlands, California; Bart Manson and Joe Myszkowski of Red River High School, Grand Forks, North Dakota; Cathy Pleau of V. J. Gallagher Middle School, Smithfield, Rhode Island; Cynthia J. Ryan of Barrington Middle School, Barrington, Rhode Island; Herb Thompson of Greenspun Junior High School, Henderson, Nevada; and Patricia Walls of Taylor County Middle School, Grafton, West Virginia.

We also thank Joseph Kerski, geographer with the United States Geological Survey, who reviewed the book for geographical accuracy.

Thanks also to Roger Palmer of Red River High School, Grand Forks, North Dakota, who spent countless hours to ensure that the book aligns with the national science and technology standards.

We also say thanks to Jim Trelstad-Porter, Director of International Student Advising at Augsberg College, who interpreted Spanish data and correspondence with the Instituto Nacional de Estadistica, Geografia e Informatica in Mexico.

Data acknowledgments and permissions

We would like to thank the following data publishers for their contributions to this work:

Color shaded relief image is provided courtesy of WorldSat International, Inc. Copyright © 2001. All rights reserved (www.worldsat.com). It is used herein with permission.

Temperature, precipitation, relief, land-use, and agriculture are ArcAtlas™ data layers. They are the intellectual property of Data+ and ESRI and are used herein with permission. Copyright © 1996 Data+ and ESRI. All rights reserved.

CountryWatch databases and data from U.S. and international sources provided courtesy of CountryWatch.com, Inc. Copyright © CountryWatch.com, Inc. (www.countrywatch.com) and is used herein with permission. All rights reserved.

2000 earthquake data provided courtesy of Council of the National Seismic System (CNSS) members networks and is used herein with permission.

Climate data provided courtesy of Worldclimate.com (www.worldclimate.com) and is used herein with permission.

Climate, religion, and language map information provided courtesy of National Geographic Maps and used herein with permission.

Historical city data from *Four Thousand Years of Urban Growth: An Historical Census* copyright © 1987, provided courtesy of Edwin Mellen Press and is used herein with permission.

World city population estimate data (2000) provided by the Population Division of the United Nations (ST/ESA/Ser.R/50) and is used herein with permission.

Middle East data, obtained from USGS Digital Atlas of the Middle East, is courtesy of the National Imagery and Mapping Agency and is used herein with permission.

Yemen border map data constructed from research provided courtesy of the British Yemeni Society and is used herein with permission.

Economic and trade data provided courtesy of U.S. Census Bureau and Instituto Nacional de Estadistica, Geografia e Informatica, Mexico, and is used herein with permission.

South Pole photograph obtained from the National Oceanic and Atmospheric Administration (NOAA) Corps.–South Pole Station, courtesy of NOAA.

Image files of flooded Earth created from bathymetry data courtesy of NOAA.

Precipitation data, hurricane tracks, and satellite images for Hurricane Mitch obtained from the United States Geological Survey (USGS) Digital Atlas of Central America, courtesy of NOAA.

Central America data obtained from USGS Digital Atlas of Central America courtesy of National Imagery and Mapping Agency and is used herein with permission.

AVHRR sensor images of Antarctica created from NOAA satellites and USGS DEMs obtained from USGS Terra Web (TerraWeb.wr.usgs.gov).

Satellite image of Larsen Ice Shelf from Landsat 7 satellite courtesy of the National Aeronautics and Space Administration (NASA).

World Wildlife Fund ecoregions data provided courtesy of World Wildlife Fund and is used herein with permission.

Introduction

Dear Teacher,

Welcome to *Mapping Our World: GIS Lessons for Educators,* a book of ideas, resources, exercises, and data, that we believe will prove a valuable supplement to your World Geography course. It's not meant to take the place of your current textbook or curriculum, but to enhance it, to expand it into the world of high-speed computing, vast databases, the World Wide Web, and the supermaps of geographic information systems.

Mapping Our World contains not only lessons, student handouts, assessments, and rubrics, but a one-year site license of ArcView 3.x for your school, as well. In short, you not only have everything you need to create GIS projects, you will be using the same tools that professional planners, emergency and disaster response personnel, government agencies around the world, and businesses of every description are using. In addition to the software and data CD you also have a Teacher Resource CD that will help make the integration of GIS technology and your curriculum as easy as possible.

Because we are all teachers, we have developed this book from a teacher's perspective: we tried to think of things that we wished other curriculum materials had. For example, the physical shape of this book is a little different. It is one inch wider than the standard 8-½-by-11 size, to make it easier for you to make copies of the student handouts. The binding on the book is also unique: it allows the book to lie as flat as possible when making copies, or when you just want it to stay open on your desk and don't have a brick handy. Throughout the lesson you will see graphical cues to remind your students of questions they should answer, or key items to take special notice of. The Teacher Resource CD is auto-run (it starts as soon as you put it in your machine) and has an easy-to-use graphic interface (all you have to do is point and click).

The lessons are organized both thematically and regionally. For each theme, there are regional, global, and advanced lessons. If you like, you could go through and work only the regional lessons, or only the global lessons. You could work through module 7 first and go backward to module 2. The point is that there is no special order. Each global and regional lesson was written with the idea that someone brand new to using GIS could pick up the investigation and get started right away. Each is complete and assumes no prior knowledge. We have provided a matrix of the National Geography Standards so that you can look up a lesson by standard.

Please take some time to glance through the items in the How to Use this Book section. There you will find information on the best and easiest ways to bring *Mapping Our World* into your classroom.

We hope you find this to be a useful tool for your classroom!

Sincerely,
The Mapping Our World Team

What Is GIS?

GIS is geographic information system

A *geographic information system* (GIS) uses computers and software to leverage the fundamental principle of geography—that location is important in people's lives. GIS helps locate new businesses and track environmental degradation. It helps route garbage trucks and manage road paving. It helps marketers find new prospects, and helps farmers grow healthier, larger crops. In education, GIS provides students the power of technology to solve geographic problems.

GIS takes the numbers and words from the rows and columns in databases and spreadsheets and puts them on a map. Placing your data on a map highlights where you have lots of customers if you own a store, or lots of leaks in your water system if you run a water company. It allows you to view, understand, question, interpret, and visualize your data in ways simply not possible in rows and columns. In the classroom, students may look at data on precipitation to study climate patterns, or analyze imports and exports for global trading partners.

And, with data on a map, you can ask more questions. You can ask "where?" and "why?" and "how?"—all with the location information on hand. And you make better decisions with the knowledge that geography and spatial analysis are included.

With the vast sources of information available today, GIS is a key tool in determining what it all means. With so much information tied to a location, GIS helps find patterns we might not see without a map.

GIS can make thematic maps (maps coded by value) to help illustrate patterns. To explore highway accidents we might first make a map of where each accident occurred. We could explore further by coding accidents by time of day. We might use one color to locate those that occur at night and a second color for those that occur during the day, and then we might see a more complex pattern.

Mapping the locations of school-age children can help reveal where day care is needed. Mapping crime incidents helps reveal where there may be a need for increased police patrols. Mapping customers' home and work locations can help banks locate ATM machines to provide better service.

GIS helps us look for patterns in both the human-made and natural realms and understand our world.

GIS is about solving problems

Sometimes we need to create new patterns or reshape existing ones. Planners of all kinds—city planners, environmental planners, school district officials—do this every day. Their job is to lay out a framework so growth can occur in a managed way and benefit as many people as possible while respecting our natural resources.

Schools plan bus-routing schedules to efficiently pick up and drop off students across the city. Each bus needs a pattern of how to pick up each student. GIS provides tools to create those patterns, in this case *routes,* to solve the problem.

In the military, leaders need to understand terrain to make decisions about how and where to deploy their troops, equipment, and expertise. They need to know what areas to avoid and which are safe. GIS provides tools to help get personnel and materials where they need to be safely so they can do their job.

In forestry, existing and future trees need to be cared for to ensure a regular supply for the world's building needs. GIS provides tools to help determine where to cut today and where to seed tomorrow.

In many areas of business, such as manufacturing and banking, organizations must meet government regulations regarding pollution and interstate trade. GIS provides tools to help companies comply with local, state, and national regulations.

In floods and hurricanes, emergency response teams save lives and property. GIS provides tools to help locate shelters, distribute food and medicine, and evacuate those in need.

In telecommunications, when phone service is out, it means part of the network may be disconnected. GIS provides tools to help find out what part of the network is affected. With that information at hand, workers in the field can get everyone talking again.

GIS is part of your world

If you flipped on a light switch today, chances are a GIS helped make sure the electricity was there to light up the room. If you drive down a highway today, chances are a GIS managed the signs and streets along the way. If you received a delivery today, chances are a GIS helped the driver find the way to your house. If you bought fresh vegetables today, chances are a GIS helped manage the land and calculate the fertilizer needed for the crop. If you looked at a map on the Internet, chances are a GIS had a hand in that, too.

GIS is a technology at work today to make your world better.

How to Use this Book

The modules   Whether you're a newcomer to GIS or an experienced GIS teacher, module 1 is the best place to begin. The one-lesson design of this module introduces students to the concept of GIS, basic ArcView skills, and the steps of geographic inquiry.

Once you've finished module 1, you and your students are free to explore the content and lessons in *any* order you choose. Each module and each lesson can be taught independently of the others, and you can tailor the material to suit the specific needs of your class and curriculum.

Each module illustrates an important theme or concept of geographic knowledge. These concepts were derived from the *National Geography Standards—Geography for Life,* published in 1994. The following is a list of modules 2–7 and a brief description of the concepts they address:

Module 2: Physical Geography I—Landforms and Physical Processes

Powerful forces originating deep in the earth shape the landforms that characterize its surface.

Module 3: Physical Geography II—Ecosystems, Climate, and Vegetation

Four major determinants of climate are latitude, elevation, landforms, and proximity to the ocean.

Module 4: Human Geography I—Population Patterns and Processes

Many factors play parts in the distribution and migration of human populations.

Module 5: Human Geography II—Political Geography

Cultural concerns and conflicts continually reshape the political makeup of the world.

Module 6: Human Geography III—Economic Geography

Economic development, modernization, and trade illustrate the interrelatedness of the global community.

Module 7: Human/Environment Interaction

Physical processes influence patterns of human activity, just as human activities have an effect on the environment.

The lessons Modules 2–7 approach a geographic theme from three perspectives: a global perspective, a regional case study, and an advanced investigation. Like the modules, each lesson can be used in isolation; they do not need to be completed in any particular order.

 Global perspective explores a geographic concept from a worldwide point of view. The focus is on how a particular concept affects human society on a global level. These lessons are designed for the beginning GIS student and provide detailed step-by-step instructions in the GIS Investigation.

 Regional case study targets the same geographic concept from a regional perspective. Students examine one world region and use examples and data specific to that area. These lessons also are designed for beginning GIS students and provide detailed step-by-step instructions.

 Advanced investigation addresses the same geographic concept, but provides the students an opportunity to research and input data from outside sources such as the Internet or other GIS data sets. This lesson is designed for experienced GIS students. These lessons typically have the students creating and saving their own ArcView projects, using the more advanced tools in the software, requiring significant self-motivation and critical thinking skills.

Using the teacher notes **Lesson overview:** A short summary gives you a snapshot of the lesson.

Materials and time: Time and materials needed for each lesson are included for your planning.

Standards: A list of the National Geography Standards (as published in *Geography for Life: The National Geography Standards,* 1994) covered in this lesson for middle school and high school students is included here.

Objectives: Specific learning objectives for each lesson give you more detailed information about the lesson.

GIS skills and tools: Important GIS skills and tools that are taught and used in this lesson are summarized here.

Geographic inquiry graphic: These are the five steps of geographic inquiry that are incorporated into each lesson. For additional information on the geographic inquiry method see the section titled "Geographic Inquiry: Thinking Geographically."

Lesson introduction: This section incorporates practical information—when to pass out specific handouts, for example—along with lists of questions for a class discussion introducing the topic.

Student activity: The primary student activity in each lesson is the GIS Investigation performed at the computer. This section offers Teacher Tips and information specific to the GIS Investigation.

Assessment: A summary of what your students will do to demonstrate their understanding of concepts and the proficiency of their skills upon completion of each GIS Investigation. Complete student assessment handouts are located at the end of the lesson, along with their evaluation rubrics.

Extensions: Consult these lists of ideas if you wish to expand the lesson for your class or individual students.

Rubric-based assessment

The lessons in *Mapping Our World* allow and encourage your students to explore a variety of geographic concepts and topics. A single letter or number grade won't be an accurate representation of the depth or completeness of their understanding of all concepts they've dealt with. The rubrics included with each lesson will allow you to evaluate your student's performance in a number of different ways. A learner may show mastery of one particular concept, but perform another task at the introductory level. The rubrics will also help you provide specific feedback to your students, showing them exactly where they need additional assistance or practice.

Exemplary: The student has gone above and beyond a particular standard. He or she has a strong understanding of the concept and has the ability to mentor other students.

Mastery: This is the target level for all students. Performance at this level shows that they have a good understanding of the concept illustrated in the standard.

Introductory: The student has limited understanding of the standard. Or, the product he or she produced shows little evidence of meeting the standard.

Does not meet requirements: The student does not show any foundational knowledge of the standard and the products they produce show no evidence of their understanding.

Suggested ways to use the evaluation rubrics:

- Distribute a copy of the rubric to students when you return their evaluated work. Circle or highlight the student's level of achievement for each standard. This provides the greatest amount of feedback for the student on each particular standard. Use the back of the form to make additional comments.
- Use the rubric as a form of student self-evaluation. Give students an unscored copy of the rubric and ask them to evaluate their own work. In the case of group projects, you can also let each group member evaluate the performance of the other members of the group.

Windows and Macintosh platforms

All of the GIS Investigations written in this book have been tested on and function within the Microsoft® Windows® and Apple Macintosh environments. The ArcView 3.x software included with this book looks and operates the same on both platforms with few exceptions.

Because Windows is the recommended platform for all ESRI® software products, all graphics and instructions in the book use the Windows platform. If you are using a Macintosh computer, please see the Macintosh Technical Guide located at the back of the book for a description of differences in the operation of ArcView on the Windows and Macintosh platforms.

Using the CDs   *ArcView Software and Data CD*

This CD contains a one-year licensed copy of ArcView 3.2a for Windows platforms and ArcView 3.0a for the Macintosh platform, as well as the data and projects used in the GIS Investigations. Refer to the installation guides in the back of the book for detailed instructions on how to install the data and the software. If you do not feel comfortable installing programs on your computers, please be sure to ask your campus technology specialist for assistance. The data and software on this CD need to be installed on your computer and all computers that the students will use to complete the GIS Investigations.

Teacher Resource CD

This CD is designed to put a wide variety of multimedia GIS resources and other helpful information at your fingertips. If you are new to GIS, you should visit the Getting Started section of the CD immediately. Keep this CD handy—you will want to refer to it often.

Where to begin   Before using these materials with your class, we recommend that you review and complete the following ordered list:

1 Install software, data, and projects on your computer and student computers.
2 Complete the Getting Started section found on the Teacher Resource CD.
3 Work through module 1 activity by yourself.
4 Introduce students to GIS.
5 Make copies and work through module 1 with your students.
6 Explore other lessons in modules 2–7.

Geographic Inquiry: Thinking Geographically

Geography is the study of the world and all that is in it: its peoples, its land, air, and water, its plants and animals, and all the connections among its various parts. When you are investigating the world and its events you are dealing with geography. As you move through space in your everyday life you are observing and interacting with geography and making geographic decisions based on those encounters. You may not be aware of it but you are involved in geographic inquiry. This mode of thinking is not unlike other research-oriented approaches, such as the scientific method; however, it has one big difference: space. Knowing where something is, how its location influences its characteristics, and how its location influences relationships with other phenomena are the foundation of geographic thinking. This mode of investigation asks you to see the world and all that is in it in spatial terms. Like other research methods, it also asks you to explore, analyze, and act upon the things you find.

Geographic inquiry is at the core of *Mapping Our World: GIS Lessons for Educators*. Throughout this book, you will use this method to explore many issues, using GIS as your tool kit (see pages xv–xvi for more discussion on what a GIS is). The same steps will be with you in each activity, and this method can be used to investigate any subject, at any scale. It is also important to recognize that this is the same method used by professionals around the world working to address social, economic, political, environmental, and a range of scientific issues. They, like you, have geography and GIS as key organizers.

So, what are the steps of geographic inquiry?

1 Ask geographic questions
2 Acquire geographic resources
3 Explore geographic data
4 Analyze geographic information
5 Act upon geographic knowledge

Let's clarify them.

Ask geographic questions Think about a topic or place, and identify something interesting or significant about it. Spin that observation into the form of a question, such as "Why do these particular trees show signs of stress?" or "How do the types of businesses change as we move along this street?" or "What does it matter if that whole area is cleared of trees?" By turning the interesting observation into a question, you can focus the exploration. Good geographic questions range from the simple "Where are things?" to "How do things change between here and there?" to deeper questions, such as "Why does this thing change between here and there?" or "What is the result of this thing changing between here and there?" Thus, you might be tempted to ask "Where do songbirds nest?" or "Why is there drought in this region while that region is flooded?" or "What is the result of refugees moving from this land across the border to that place?" A good question sets up the exploration.

Acquire geographic resources

Once you have a question, you can think about the information needed to answer it. Here, it's helpful to consider at least three aspects of the issue: geography, time, and subject.

What's the geographic focus of your research? In studying a country in relation to others, your inquiry might require country-level data, and you would need data for the country of interest as well as for neighboring countries. Defining the geographic focus helps you define the scale (global, regional, local) of your inquiry, and helps you define the extent (a city, a country, a continent, the globe) of your inquiry.

For what period of time do you need the data? Answering questions of today would sensibly mean using information about the present or a very recent time. However, those questions could gain greater clarity by including a historical or future perspective. Alternatively, a question focused on past events requires historical rather than contemporary data.

For what subject(s) and specific topics do you need data? This may sound silly but it is very useful to take time to consider the topical aspects of the data needed. Population may be the general nature of your study but international migration may be your actual focus. Learn to dissect your data needs. The more specific you can be in defining your focus, the less likely you are to get lost in piles of unrelated and unnecessary data.

Often, you can find the necessary geographic data quite easily, in readily available packages or downloadable from the Internet. Sometimes you have to produce the data yourself, or convert data from one form into a more appropriate form. In the early days of GIS, almost all data had to be produced independently. These days, the explosion of technology and rise of the Internet has made it much easier to acquire information. This explosion of data means you may find material in a wide range of formats, at multiple scales, and with variable quality. After tracking down what is readily accessible, and recording any source information about your data, you need to look at what is still missing and decide if you can answer your question. Even if you are missing some desired data, you may still be able to answer your initial question, or a variation of it, by exploring your resources carefully.

Explore geographic data

Turn the data into maps, tables, and charts. Maps are especially valuable, because they give you a powerful view of patterns, or how things change over space. Maps also allow you to integrate different kinds of data from different sources—pictures (aerial photos, satellite images) and features (roads, rivers, borders)—in layer after layer. Explore this data in a variety of combinations. Look at individual items and what is around them. Explore how spatial phenomena relate to things around them—for instance, mountains and streams, cities and coastlines or rivers, agriculture and deforestation. Be creative. Observe.

For any one set of data, there are many ways to twist and turn it. By integrating maps with tables, charts, and other representations, some patterns may begin to appear, patterns that might spur you to refine your original question, or to seek out one more set of data. Such refinement at this stage is common, and sensible. For example, when first exploring regional rainfall patterns, you might not have anticipated that you would need the locations of mountain ranges, but having this data might just make a difference.

Using a GIS, this kind of visual exploration is simple to do. One layer of information stacks on top of another. By changing the map symbols, altering the sequence of layers, or zooming in to specific parts of the map, patterns and relationships become easy to see.

Analyze geographic information

After creatively exploring the relationships between this and that, or here and there, focus on the information and maps that most seem to answer your questions. Using carefully constructed queries, you can highlight key comparisons, or expose patterns that had lain hidden during initial explorations. Focus on relationships between layers of information; make inferences about the distribution of things; calculate the degree to which the presence of something affects the presence or character of something else. Key in on the deeper questions—"Why is it there?" and "So what?" See if some predictions can be made. For instance, if you discover that most traffic accidents in your community occur at intersections along major streets running due east–west, what would you expect to find in other communities, and why?

The power of the computer becomes especially helpful in this analysis step. Since GIS data is made up of map representations and tables of characteristics, a GIS can handily solve queries and identify things. "Please computer find for me all cities of one million or more people where rainfall is less than 10 inches per year." Quick to find answers, the GIS is still dependent on you to shape the questions.

At this point in your inquiry, your aim should be to draw conclusions from what you have seen in the maps, charts, and queries and to answer your question. Sometimes you do not have the information you need to answer the question— that's OK. The important thing is that you now understand the issue better than before, and you have drawn some conclusions from your research, turning pieces of data into geographic knowledge.

Act upon geographic knowledge

You have used GIS to integrate data from multiple sources and to weave it into knowledge that enables you to act. Being geographically wise means acting on the geographic knowledge that you have gained.

Good citizens of the community and decision makers for the planet need to act according to an integrated understanding of the relationships between diverse forces. It is not enough simply to understand why things are where they are, and not even enough just to comprehend the impact. Decision makers engaged in geographic inquiry are equipped to make better choices for themselves and others based on the data and their study. Ideally, they will avoid "seat-of-their-pants" impulses because they have the appropriate knowledge to make a wise decision. Good citizens will share their geographic knowledge with a broader community, and help others act according to it. This may mean giving a presentation to the school about the health of nearby trees, or to the neighborhood about recent migrants. It may mean encouraging local businesses to provide resources for a community far away, or helping the state change its energy policies because of their effect beyond its borders. Understanding the widespread linkages and helping others see how their lives are affected means "thinking globally, acting locally." Acting on geographic knowledge means being willing to answer the question, "Now what?"

Throughout this book, these are the steps and ideas you will see. In some lessons, the five steps may not be obvious—clues such as "Now you are doing X" will not necessarily be present. Regardless, the components of thinking geographically are present. The process of geographic inquiry is not difficult, but it may be different from what you are accustomed to. Its method does not require a computer or sophisticated devices, but there is a tool that can simplify and hasten geographic investigations, and that tool is GIS. Like any tool, the GIS has no answers packed inside it. Instead, for those who engage the tool and the process of geographic inquiry, it provides a means to discover pathways through this most remarkable world of unending questions.

Correlation of National Geography Standards to *Mapping Our World* Lessons

STANDARD	MODULE 1	MODULE 2			MODULE 3			MODULE 4			MODULE 5			MODULE 6			MODULE 7		
	G	G	R	A	G	R	A	G	R	A	G	R	A	G	R	A	G	R	A
1 How to use maps and other geographic representations, tools, and technologies to acquire, process, and report information from a spatial perspective	✓	✓	✓	✓	✓	✓	✓	✓	✓	✓	✓	✓	✓	✓	✓	✓	✓	✓	✓
2 How to use mental maps to organize information about people, places, and environments in a spatial context																			
3 How to analyze the spatial organization of people, places, and environments on Earth's surface	✓										✓								
4 The physical and human characteristics of places	✓	✓	✓			✓						✓					✓		
5 That people create regions to interpret Earth's complexity		✓			✓														✓
6 How culture and experience influence people's perceptions of places and regions																			
7 The physical processes that shape the patterns of Earth's surface		✓	✓	✓			✓										✓		
8 The characteristics and spatial distribution of ecosystems on Earth's surface																			✓
9 The characteristics, distribution, and migration of human populations on Earth's surface									✓	✓									
10 The characteristics, distribution, and complexity of Earth's cultural mosaics													✓						
11 The patterns and networks of economic interdependence on Earth's surface														✓	✓				
12 The processes, patterns, and functions of human settlement								✓								✓			
13 How the forces of cooperation and conflict among people influence the division and control of Earth's surface											✓	✓	✓		✓				
14 How human actions modify the physical environment										✓									
15 How physical systems affect human systems																		✓	
16 The changes that occur in the meaning, use, distribution, and importance of resources																			
17 How to apply geography to interpret the past										✓									
18 How to apply geography to interpret the present and plan for the future		✓			✓			✓	✓	✓	✓	✓	✓	✓	✓	✓	✓	✓	✓

G = Global investigation   R = Regional case study   A = Advanced investigation

Mapping Our World

GIS Lessons for Educators

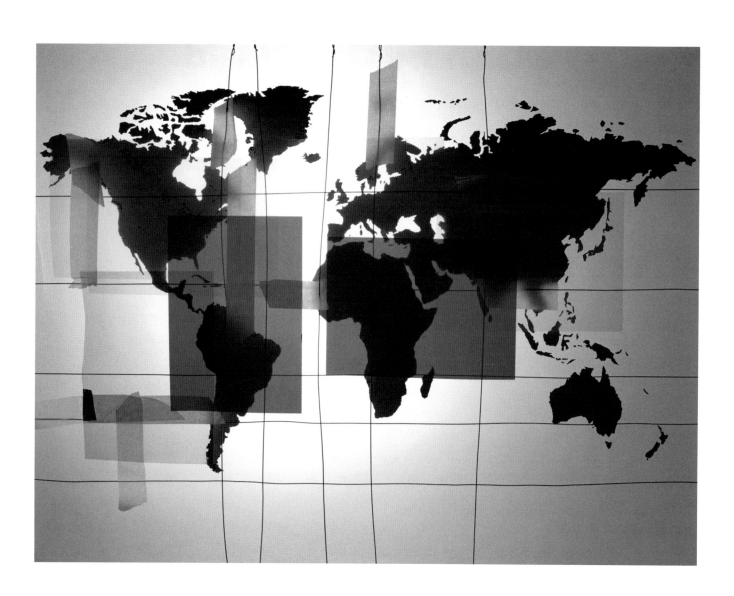

ArcView: The Basics

This module covers the concepts and skills that will be discussed and used in the rest of the book. As your students step through the activities, they will be introduced to the geographic inquiry method and acquire a working understanding of GIS theory and practice—the "how to" of ArcView as well as a sense of its conceptual framework, foundation, and purpose. By the end of the lesson, they will find themselves competent users of a basic GIS.

This module serves as a comprehensive introduction to the rest of the book. We suggest that you and your students complete it before moving on. Please note that the format of module 1 contains one lesson in two parts. Modules 2 through 7 each contain a global, a regional, and an advanced investigation lesson.

ArcView: The Basics

Lesson overview

This module introduces the basic concepts and skills of GIS that will prepare your students for the lessons ahead. The activities laid out here will instruct them in starting ArcView, navigating the computer to find the prepared projects and data, and compiling a group of functional skills to assist them in using ArcView. Students will also be learning the steps involved in the process of geographic inquiry at the same time.

The format of this module differs from that of the rest of the book: it has one lesson in two parts, not the global, regional, and advanced investigation lessons found in modules 2 through 7. We suggest that you have the students spend two class periods on these activities to gain a functional knowledge of GIS, ArcView skills, and the geographic inquiry method.

Estimated time   Two to three 45-minute class periods

Materials   ✔ Calculator

✔ Student handouts from this lesson to be copied:
- GIS Investigation sheets (pages 9 to 34)
- Student answer sheet (pages 35 to 42)
- Assessment(s) (pages 44 to 47)

Standards and objectives   *National geography standards*

GEOGRAPHY STANDARD	MIDDLE SCHOOL	HIGH SCHOOL
1 How to use maps and other geographic representations, tools, and technologies to acquire, process, and report information from a spatial perspective	The student knows how to make and use maps, globes, charts, models, and databases to analyze spatial distributions and patterns.	The student knows how to use technologies to represent and interpret Earth's physical and human systems.
3 How to analyze the spatial organization of people, places, and environments on Earth's surface	The student understands that places and features are distributed spatially across Earth's surface.	The student understands how spatial features influence human behavior.
4 The physical and human characteristics of places	The student understands how technology can shape the characteristics of places.	The student understands how the physical and human characteristics of place can change.

Standards and
objectives
(continued)

Objectives

The student is able to:

- Understand the basic concept of a geographic information system (GIS).
- Use a basic ArcView skill set to build a map.
- Use the five-step geographic inquiry model.
- Print maps.

GIS skills and tools

 Identify a feature on the map

  Zoom in to a desired section of the map or to the center of the map

  Zoom out to a desired section of the map or to the center of the map

 Zoom to the full extent of all themes

 Zoom to the previous map extent

 Pan to a different section of the map

 Add a theme to the view

 Find a feature in a theme

 Get help about ArcView

 Open a theme's table

 Clear all selected features so that none are selected

 Open a theme's properties and change the theme name

For more on geographic inquiry and these steps, see Geographic Inquiry: Thinking Geographically (pages xxi–xxiii).

Teacher notes

Lesson introduction

Begin this lesson by showing the multimedia clip entitled "What is GIS" in the GIS Resources section of the Teacher Resource CD. This clip will help students understand that each map in a GIS has spreadsheet or database information attached to it. There are a number of Adobe® PDF and Microsoft PowerPoint® presentations on the CD dealing with this topic as well.

Refer to the Geographic Inquiry section in this book to familiarize yourself with the geographic inquiry model. Have a brief discussion with your students about "thinking geographically."

Student activity

 Before completing this lesson with students, we recommend that you complete it as well. Doing so will allow you to modify the activity to accommodate the specific needs of your students.

Ideally each student should be at an individual computer, but the lesson can be modified to accommodate a variety of instructional settings.

On the first day, distribute the GIS Investigation sheets entitled "ArcView: The Basics, Part 1: Introducing the software." Explain that in this activity the students will begin to learn the basic ArcView skills they will need to create GIS maps. The worksheets will provide them with detailed instructions for their investigations. As they navigate through the lesson, they will be asked questions that will help keep them focused on key concepts. Some questions will have specific answers while others will require creative thought.

On the second day, distribute the GIS Investigation sheets entitled "ArcView: The Basics, Part 2: The geographic inquiry model." Explain to your students that in this lesson they will practice the ArcView skills they acquired in the first lesson, and be introduced to more advanced skills while exploring the geographic inquiry method.

 Teacher Tip: In part 2 of this activity, students are asked to use a calculator to divide two country statistics and come up with a third. Some of the countries have populations with nine or 10 digits. If your students have calculators that only allow them to enter eight-digit numbers, you may need to help students come up with a way around this problem. For example, they could divide both numbers by 1,000 before performing the calculation.

Steps 8 through 10 of part 2 do not involve the computer. You may wish to assign these steps as homework to be completed outside of class.

Things to look for while the students are working on this activity:

- Are students thinking spatially as they work through the procedure?
- Are students answering the questions as they work through the procedure?
- Are students using a variety of menus, buttons, and tools to answer the questions on the handout?
- Are students able to use the legends to interpret the data in the table of contents?
- Are students able to print out a map on the printer?

Conclusion

Before your students complete the assessment, conduct a brief discussion in which you ask them to brainstorm ideas about how GIS can be used in everyday life or how they could use GIS in their daily school assignments or classes. Ask them to describe the new geographic inquiry skills they learned as well as share how comfortable they are with using the ArcView technology. This discussion should also include an overview of which buttons they have used to build maps on screen, any ArcView operations that were confusing, and printer operations.

Assessment

Middle school: Highlights skills appropriate to grades 5 through 8

The middle school "ArcView: The Basics" assessment asks students to create and print a map. They will be expected to turn on three themes and zoom in to a location of their choosing, use the Identify tool to obtain three pieces of data about that area, and write a brief paragraph explaining how they created their maps.

High school: Highlights skills appropriate to grades 9 through 12

The high school "ArcView: The Basics" assessment asks students to create and print a map. They will be expected to turn on four themes and zoom in to a location of their choosing, use the Identify tool to obtain three pieces of data about that area, write a brief paragraph explaining how they created their maps, and describe what they learned geographically about the area on their map.

Extensions

- Create several views with data from other folders on the data CD to answer a question. Make a connection between the data to make sure that the maps they create are meaningful.

- Students can suggest other themes that might help explain the connection between the technological advances of telephone lines and cellular technology, and such political and social issues as GDP, education, health care, and so on.

- Students can develop a plan or outline for how to use GIS to fulfill a current class assignment.

- Ask students to choose a country in the news and use their GIS skills to find the country and to study the data associated with it.

- Check out the Resources by Module section of the Teacher Resource CD for print and media resources that educate the public on the uses of GIS, or visit *www.esri.com/mappingourworld* for a list of Internet links.

NAME _____   DATE _____

ArcView: The Basics

Answer all questions on the student answer sheet handout

Part 1: Introducing the software

This computer activity will introduce you to the basics of starting the ArcView program. You will be guided through the ABCs of using ArcView. If you read and do each step in this activity, you will find it much easier to complete the other activities in this book.

Step 1   Start ArcView

a   Double-click the ArcView icon on your computer's desktop. If you do not have the icon on your desktop, click Start, Programs, ESRI, and ArcView GIS.

b   If the ArcView GIS Education Edition welcome box appears, click the Try button.

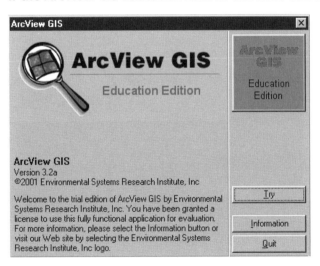

Step 2   **Open the module1.apr file**

a   In this exercise, a project file has been created for you. To open the file, click the **File** menu and click **Open Project**.

b   Navigate to the exercise data directory (**C:\esri\mapworld\mod1**). (Hint: First make sure the correct drive is selected in the Drives drop-down list. Then double-click folders to open them.)

c   Choose **module1.apr** from the list.

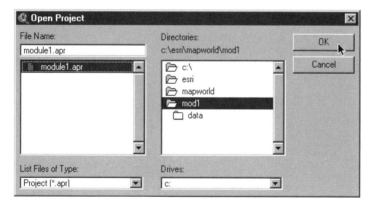

ArcView project files end with the three-letter extension **.apr**, which stands for ArcView project file.

d   Click OK.

When the project opens, you see a large window, called the ArcView window. Inside it is a smaller window, called the Project window. Its title is the name of the project file, module1.apr.

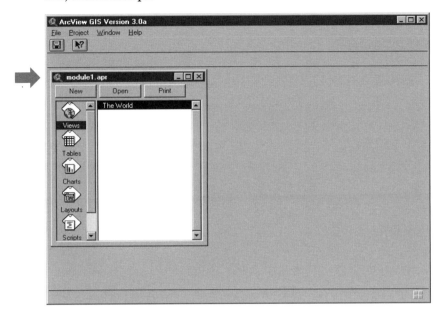

Step 3   Enlarge the ArcView window

When the ArcView project first opens, the ArcView window may be smaller than your full computer screen. If it is, you will want to enlarge it.

a   **In the upper right corner of your ArcView window there are three buttons. Click once on the middle button. It looks like a box.**

b   **The button you clicked is called the Maximize button. Now the ArcView window fills your whole screen. Notice that the size of the Project window did not change.**

c   Take a closer look at the Project window. You will notice five icons, or pictures, down the left side of the window. The icon names are Views, Tables, Charts, Layouts, and Scripts. These are the different kinds of documents an ArcView project can contain. When you want to see a list of these documents, click once on an icon.

Step 4   Open a View window

Now you will look at the map that has been constructed for you.

a   In ArcView, maps are displayed in views. Notice that The World is the only view document in the list, and it is already selected (highlighted).

b   Click the Open button.

A View window opens.

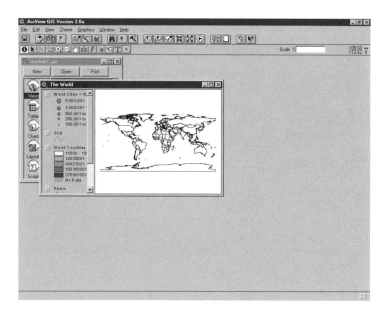

> *Note: If at any time you accidentally close the View window, you can always go back to the Project window and open the view from there.*

Step 5   **Stretch your View window**

You will enlarge your View window so it will be easier to see the map.

a   Place the cursor on the bottom right corner of the View window. The cursor changes to a diagonal double-headed arrow.

b   Click and drag the window down and to the right until the view almost fills the screen. Let go of the mouse button.

Make sure you can still see the Project window behind the top left corner of the View window.

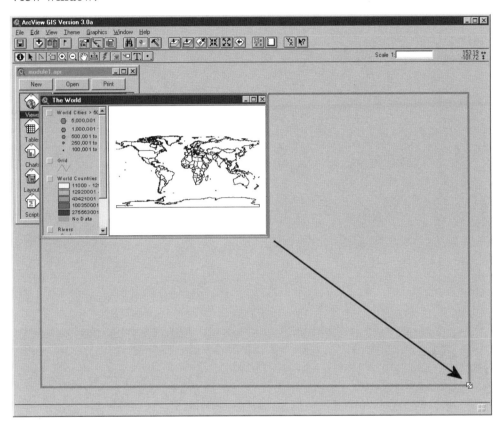

> *Important note: Enlarging view windows by stretching allows you to move easily between the Project and View windows. It is best **not** to enlarge views by clicking on the maximize button. You may maximize the entire ArcView window using the maximize button, however, as you did in step 2.*

c   Study the different parts of the View window. On the right side you see a map. On the left side you see a gray column that contains a list. This column is the table of contents. The items in the list represent the six different layers of information that can be shown on the map.

Step 6   **Work with themes**

Each of the layers in the table of contents is called a theme. The six themes in The World view are *World Cities > 50,000, Grid, World Countries Population, Rivers, Lakes,* and *Country Outlines.* In this step you will learn how to turn themes off and on and change their order.

a   Notice that each theme in the table of contents has a small box in front of it. Only the box for Country Outlines has a check mark in it.

b   Click the check mark next to Country Outlines. The check mark goes away. The display of Country Outlines in the map area also disappears.

c   Click the box next to the World Countries Population theme. (Don't worry that part of the theme name is cut off. You will learn how to widen the table of contents later in this activity.) The World Countries Population map is displayed. This is called turning on a theme.

d   Click the box next to the Rivers theme to turn it on.

e   Turn on the Lakes theme.

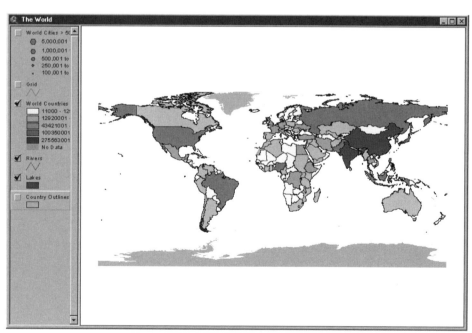

f   Examine the table of contents to answer the following question:

?   *Which themes are not visible on the map but are turned on in the table of contents?*

Next you will change the order of the themes.

g   Place your cursor on the Rivers theme in the table of contents.

h   Click and hold the mouse button. Drag the Rivers theme up above the World Countries Population theme. Let go of the mouse button.

?   *What happened on your map?*

i   Drag the Lakes theme above the World Countries Population theme.

?   *(1)   What happened on your map?*

?   *(2)   What would happen if you dragged the Lakes under World Countries Population again?*

> *Note: Whenever your maps don't appear as you think they should, check the following things:*
>
> 1. *Check to see if the theme you want is turned on.*
>
> 2. *Check the order of the themes in the table of contents. Themes that are represented by lines and points (streets, rivers, cities, etc.) will be covered up by themes that are represented by polygons (countries, states, etc.). You may need to drag line or point themes above the polygon themes in order to see them.*

Step 7   Identify a country and record country data

a   Look at the top of the ArcView window and find the four rows, or bars, pictured below. You will use these menus and buttons to perform the various GIS functions. Many of the things you do with the buttons and tools can also be accomplished with menu options.

You might think this is a lot of buttons to learn, but you will not be learning them all at once.

The value of GIS lies in the data (information) that is attached to each map. You will see how this works by using one of the tools to access data about countries.

b First you must tell ArcView which theme you want to know about. This is called making a theme active.

c Click anywhere on the World Countries Population theme in the table of contents except on its check box. Notice that it becomes raised like a button.

> **Note: Whenever you wish to get information about a particular theme, you must make it active. An active theme appears raised like a button.**

d The Identify tool lets you see data about your map just by clicking on places you are interested in. Find the Identify tool in the ArcView tool bar.

e When a tool is selected, it looks like it is pushed in. If your Identify tool already looks pushed in, proceed to the next step. Otherwise, click it to select it.

f Move your cursor over the map without clicking. Notice it changes to a plus with an "i" next to it.

g Click on the United States on your map.

The Identify Results window appears. It displays information about the United States. The information you see is all the data that is available about the United States in the World Countries Population theme.

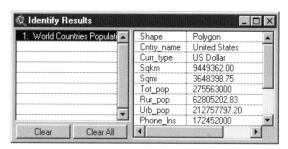

h In the Identify Results window, scan down the column that begins with the word Shape. Answer the following questions:

(1) What is the second listing in this column?

(2) What is the third listing in this column?

(3) What is the final listing in this column? (Hint: You will need to scroll down.)

i These words are names describing a characteristic, or attribute, of the United States. Attribute names are often abbreviated.

(1) What do you guess the field entitled "Sqmi" stands for?

(2) What is the number to the right of the field "Sqmi"?

Step 8   Compare the Identify Results data with the table data

  a   The picture below shows part of the table that contains the data attached to the World Countries Population theme. (You will open this table on the computer later in this lesson.)

  ❓   *Which row in this table has the attributes for the United States?*

Shape	Cntry_name	Curr_type	Sqkm	Sqmi	Tot_pop	Rur_pop	Urb_pop	Phone_Ins	Ce
Polygon	China	Renminbi Yuan	9367281.00	3616707.25	1261832000	823987681.10	437844318.90	70310000	
Polygon	India	Rupee	3159685.50	1219954.75	1014004000	725623736.00	288380264.00	17802000	
Polygon	United States	US Dollar	9449362.00	3648398.75	275563000	62805202.83	212757797.20	172452000	
Polygon	Indonesia	Rupiah	1887178.25	728639.44	224784000	134413428.50	90370571.53	4982000	
Polygon	Brazil	Cruzeiro Real	8504535.00	3283601.00	172860000	32359871.61	140500128.40	17039000	
Polygon	Russia	Ruble	16911282.00	6529445.00	146001000	32585764.99	113415235.00	26875000	
Polygon	Pakistan	Rupee	877524.69	338812.28	141554000	89116340.49	52437659.50	2557000	
Polygon	Bangladesh	Taka	138194.09	53356.74	129194000	101834741.00	27359258.98	-99	
Polygon	Japan	Yen	371030.50	143254.88	126550000	26729360.60	99820639.39	60381000	
Polygon	Nigeria	Naira	912245.00	352217.78	123338000	69083261.24	54254738.75	-99	
Polygon	Mexico	New Peso	1961254.13	757240.19	100350000	25704278.88	74645721.11	9254000	

  b   Compare the information in the Identify Results window with the information in the table and answer the following questions:

  ❓   *(1)   Another word for attributes is* fields. *The field names in the Identify Results window display in the column starting with the word Shape. Where are these field names displayed on the table?*

  ❓   *(2)   Find the field on the table that represents square miles of land. How many square miles of land are in the United States?*

  ❓   *(3)   Give a brief explanation of the relationship between the Identify Results window and the table.*

  c   Click the Close button at the top right corner of the Identify Results window.

  :⚡: *Caution: Be careful **not** to click the Close button on the ArcView window at the very top right corner of your screen because ArcView will close without saving your work.*

Step 9   Widen the view table of contents

       Now you will widen the table of contents in the view so that none of the theme legends are cut off.

  a   Move your cursor to the edge of the table of contents just to the right of the scroll bar. When it is in the right place, it should look like this: ◄|►

b Click and hold the mouse button. Drag the cursor to the right until you can see the full legend descriptions. Release the mouse button.

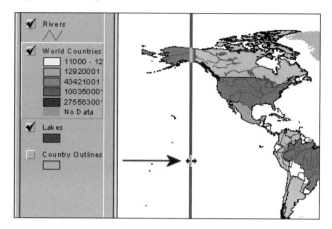

The table of contents becomes wider and the map becomes smaller.

Step 10 Explore city data on the world map

a Turn off the Lakes and Rivers themes by clicking the check mark in front of the theme name.

b Turn on the World Cities > 50,000 theme.

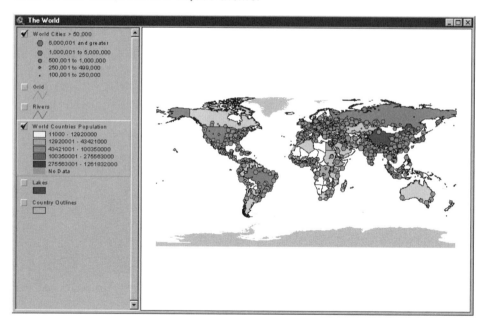

Your map displays all the world cities with populations greater than 50,000. There are so many cities that they are all jumbled together on this small map. You need to zoom in to a smaller portion of the world to see distinctions between the cities.

c Click the Zoom In tool to select it. Remember, it will look pushed in when it is selected. The Zoom In tool can be used two different ways.

One way to zoom in is to drag a box around the section you want to display. You will zoom in on Europe and Africa.

d   Place your cursor on Greenland.

e   Click and hold down the mouse button. Drag the cursor down and to the right. When your cursor is near Australia, release the mouse button.

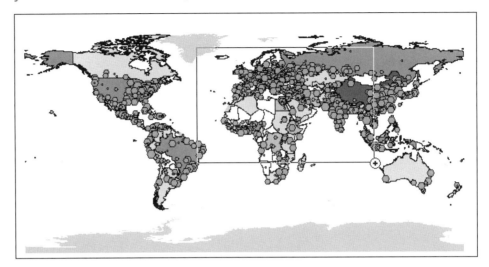

Another way to zoom in is to click on the place you want to be in the center of your map. Now you will zoom in closer to Europe.

f   Click slowly three times on the brown dots clustered around Europe.

> *Caution: Make sure to keep the mouse very still as you click. Otherwise, you may accidentally drag a tiny box and the map will zoom in too much. If this happens, go to the toolbar and click once on the Zoom to Previous Extent button.*

g   Make the World Cities > 50,000 theme active.

h   Use the Identify tool to find the name and country of any two cities you choose.

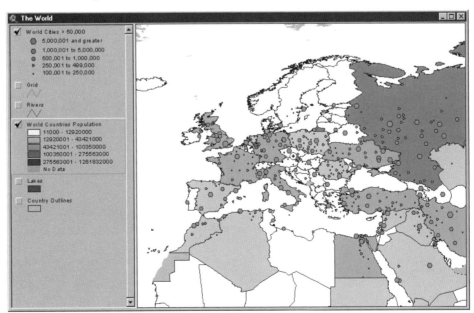

Step 11   Explore Europe with an attribute table

When you used the Identify tool in step 7, you saw that a country on the map (the United States) was connected to attribute information about that country. The attributes were displayed in the Identify Results window.

In GIS, each object on your map is called a *feature*. For example, the United States is a feature in the World Countries Population theme, and Paris is a feature in the World Cities > 50,000 theme. In this step, you will find out more about the connection between features and their attributes.

a   Make sure the **World Countries Population** theme is active.

b   Click the Open Theme Table button.

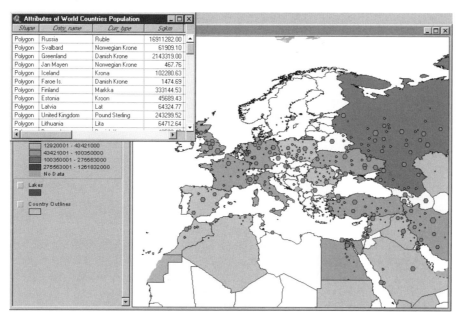

?   *What is the name of the table you opened?*

c   Click on **Russia** in the first row of the table.

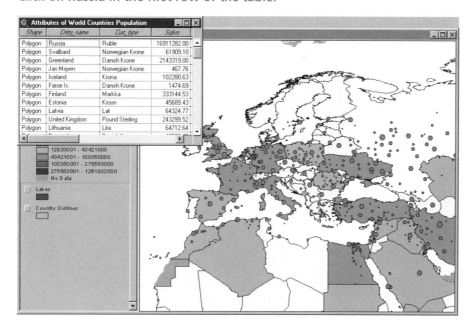

Notice that the row in the table turns yellow and so does the part of Russia you can see on the map.

d   Hold down the Shift key on your keyboard. In the table, click on Finland, Estonia, Latvia, and the United Kingdom.

?   *What happens to the map when you click on these rows in the table?*

e   Click on Iceland in the table.

?   *(1)   What happens to Russia and the other countries that were highlighted?*

?   *(2)   Did you see Iceland turn yellow on the map? If not, why not?*

f   Click the Zoom to Full Extent button. The map displays the whole world. Remember: Sometimes you may zoom in so close or zoom out so far that you lose your map entirely or can't tell where you are. You can always click the Zoom to Full Extent button to get back to a map you can recognize.

?   *Why can you see Iceland now when you couldn't see it in the previous step?*

g   Click the Close button on the table.

In later lessons, you will learn how to do more with tables. For example, you will learn how to sort a table and how to find countries in the table and on the map.

h   Click the Clear Selected Features button.

Step 12   Practice using the Identify tool

a   Click the Zoom In tool.

b   Click and drag a box around the continent of South America.

?   *What do you see on your map?*

c   Make sure the World Countries Population theme is active.

d   Use the Identify tool to identify the large South American country that is dark red.

?   *(1)   What country is it?*

?   *(2)   What is this country's total population?*

e   Close the Identify Results window.

f   Make the World Cities > 50,000 theme active.

g   Use the Identify tool to identify the large city in the northwest corner of the country you identified in step 12d.

?   *(1)   What city is it?*

?   *(2)   What population range is this city in?*

h   Look on the map for two cities in Brazil that are in a higher population range. (Hint: the brown symbol for those two cities will be larger than the one for the city you just identified.)

i   Zoom in closer to the two cities to separate them from surrounding cities for identification purposes.

?   *j*   Name these two large cities.

k   Close the Identify Results window.

Step 13   Practice using the Zoom Out tool

    a   Click the Zoom Out tool.

b   Drag a two-inch box anywhere on your map.

c   Drag another two-inch box anywhere on your map.

(1)   What does your map look like?

(2)   Which zoom button could you use to return your map to full size?

d   Zoom your map back to its original full size.

Step 14   Continue to explore the world map

a   Make the World Countries Population theme active.

    b   Click the Find button.

You will find the country of Sudan.

c   Click in the white box in the Find Text in Attributes dialog. Type **Sudan**.

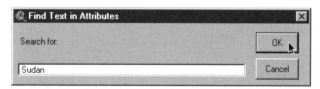

d   Click OK.

> *Note: If you get a message telling you "No more matches found," click OK and go back to step 14b. Make sure you spell the country name correctly.*

e   Notice that the country of Sudan turns yellow and moves to the middle of the map.

f   Turn off the World Cities > 50,000 theme.

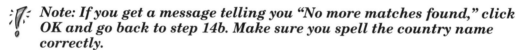   *g*   Use your Identify tool and click on Sudan.

h   Scroll down in the Identify Results window and answer the following questions:

(1)   How many tourists arrive in Sudan each year?

(2)   How many people live in Sudan?

(3)   Does this seem like a low or high number of tourists for this population?

i   Use the Find button to find **Qatar**.

Do you see Qatar on your map?

j   Click seven times on the Zoom In button. (Its arrows point inward.)

The Zoom In button is different from the Zoom In tool. Clicking the Zoom In button always zooms in to the middle of the map.

> *Remember, the Find tool placed Qatar in the middle of the map area. When you click the Zoom In button, you are zooming in on Qatar.*

k Use the Identify tool to answer the following questions:

 (1) *How many people live in Qatar?*

(2) *How many cell phones do they have?*

(3) *Divide the population of Qatar by the number of cell phones in Qatar. How many people are there for every cell phone in Qatar?*

(4) *What large country is directly west of Qatar?*

l Close the Identify Results window.

Whenever you wish to center the map differently, you can use the Pan tool to move the map around.

m Click the Pan tool.

n Click and hold on the country of Qatar. Drag the hand diagonally toward the bottom right corner of the View window. Let the mouse button go. Answer the questions below:

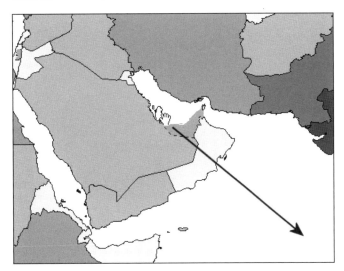

(1) *What boot-shaped country do you see on the map?*

(2) *What is the population of that country?*

(3) *How many cell phones does that country have?*

(4) *Divide the answer to question 2 by the answer to question 3. How many people are there for every cell phone in this country?*

o Pan east to Japan and answer the following questions:

(1) *What is the population of Japan?*

(2) *How many cell phones does Japan have?*

(3) *Divide the population of Japan by the number of cell phones in Japan. How many people are there for every cell phone in Japan?*

p Close the Identify Results window.

 q Click the Zoom Out button several times. (Its arrows point outward.)

What happened to your map?

The Zoom Out button always zooms out from the middle of the map.

r   Pan back to Qatar from Japan.

s   Using a zoom tool of your choice, zoom in until you can see that Qatar is still highlighted yellow.

t   Click the Clear Selected Features button.

What happened to Qatar?

Whenever you find or select a feature on your map, it will turn yellow to indicate that feature has been selected. If you want to turn the yellow off, click the Clear Selected Features button. (Make sure that whichever theme is yellow is the theme that is active before you click the button.)

Step 15   **Get help from the Help tool**

If you forget what a particular button is for or how to use a tool, you can ask ArcView to help you.

a   Click the Help button.

Notice your cursor turns into a diagonal arrow with a question mark.

b   Click the Find button.

A help page is displayed that tells you about the Find button.

c   Click the green link, "active theme(s)."

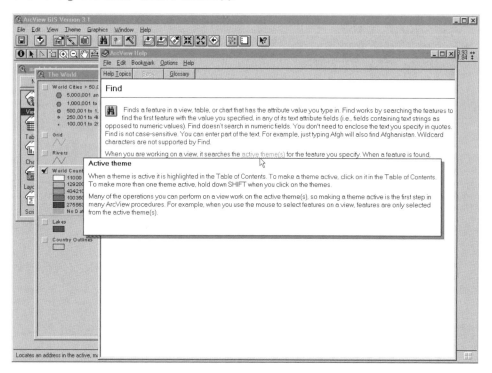

? *The pop-up window reminds you how to make a theme active. What does it say you must do to make more than one theme active at a time?*

d   Click anywhere in the Help window to make the pop-up window disappear.

e   Click the Close button on the top right corner of the ArcView Help window.

You can use the Help button to help you use any ArcView button, tool, or window.

f   Click the Help button.

g   Click any other button you want to know about.

h   When you are finished looking at the help, close the ArcView Help window.

Step 16   Print a map

In this step you will learn an easy way to print the map on your screen.

a   Use the Pan and Zoom tools to focus the map on a location of your choice.

b   Click the File menu, then click Print.

c   In the Print dialog, click the Setup button.

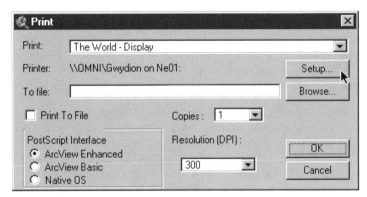

d   Ask your teacher which printer you should use. Select its name from the drop-down list in the Print Setup dialog.

e   Make sure the paper size is set to "Letter (8.5 x 11 in.)." You may need to select it from the drop-down list.

f Decide how you want the map to be placed on the paper. Click Portrait if you want the top to be a short side of the paper. Click Landscape if you want the top to be a long side.

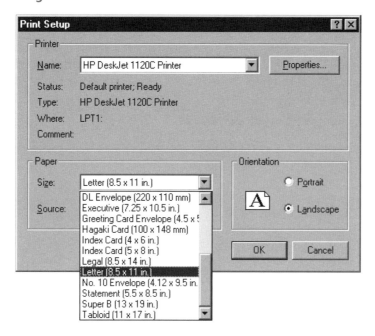

g Click OK on all the open Print windows.

h Your map should print after a few moments.

Step 17   Close the project and exit ArcView

a Click the File menu.

b Click Exit.

c Click No in the box that asks whether you want to save changes.

NAME _____   DATE _____   .

Part 2: The geographic inquiry model

Much like scientific analysis, geographic inquiry involves a process of asking questions and looking for answers. The geographic inquiry model is made up of the following five steps:

In this activity you will learn how maps and GIS can help you in the geographic inquiry process. At the same time, you will be practicing the ArcView skills you learned in part 1 and learning some new ones.

Step 1   Start ArcView

a   Double-click the ArcView icon on your computer's desktop.

b   If the Welcome to ArcView dialog appears (pictured below), click **Open an Existing Project** and click OK. If it doesn't appear, proceed to step 2.

Step 2   **Open the module1.apr file**

You will use the same ArcView project that you used in part 1 for this exercise.

a   To open the file, go to the File menu and choose **Open Project**.

b   Navigate to the exercise data directory (**C:\esri\mapworld\mod1**) and choose **module1.apr** from the list.

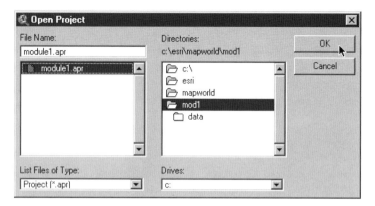

c   Click OK.

Step 3   **Adjust your windows and open the view**

a   Enlarge the ArcView window using the Maximize button.

b   Go to the Project window and open the view "The World."

c   Drag the corner of the View window and stretch it until it almost fills the ArcView window.

d   Widen the table of contents until the theme legends are not cut off. (Hint: Click and drag the right edge of the table of contents to the right.)

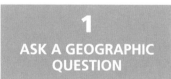

ASK A GEOGRAPHIC QUESTION

Step 4   **Ask a geographic question and develop a hypothesis**

It's important to think about the information on your maps both as you create them and when they've been completed. You might look at a map and then think of a question that it might help you answer. Or you might think of the question first, and then look for maps, or GIS themes, that might help you answer the question.

In this activity you will try to answer the following geographic question:

Geographic Question: Do the number of phone lines in a country increase proportionately with the number of people in the country?

?   *a*   What makes this a geographic question?

?   *b*   Just as with scientific inquiry, you'll begin by constructing a hypothesis. Write a hypothetical answer to the geographic question stated above.

Remember: A hypothesis is an educated guess. Your hypothesis may be right or wrong in the end. The goal is not to know the answer before you start your research, but to have a kind of "ballpark idea" or hunch about what the answer might be. Then you will set about proving or disproving it.

2
ACQUIRE GEOGRAPHIC RESOURCES

Step 5 **Add a theme to your view**

Next, you need to identify the kind of information that will help you explore your question. You want to display the data you identify as themes in your ArcView map.

 a Your view already has a theme with world countries and their population. What other attribute of countries do you need in order to investigate your hypothesis?

 b Click the Add Theme button.

An Add Theme dialog appears. It works the same way as the Open Project dialog you used at the beginning of this activity.

c Navigate in the right column to the module 1 data directory (**C:\esri\mapworld\mod1 \data**). (Hint: First make sure the correct drive is selected in the Drives drop-down list. Then double-click folders in the right column to open them.)

d Click **phone_line.shp**.

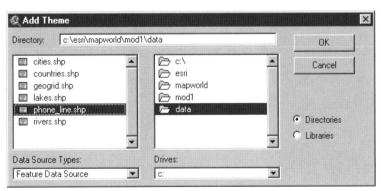

e Click OK.

? *What is the name of the theme that has been added to your table of contents?*

f Make the new theme active.

g Click the Theme Properties button.

h Locate the Theme Name box at the top of the Theme Properties dialog.

i   Change the theme name to **World Phone Lines**.

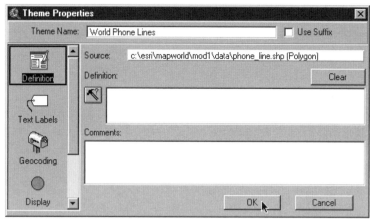

j   Click OK.

3
EXPLORE GEOGRAPHIC DATA

Step 6   Explore the World Phone Lines map

Now it's time to look at the World Phone Lines theme and think about what the map tells you.

a   Turn off the Country Outlines theme and turn on the World Phone Lines theme.

b   Look at the legend for the World Phone Lines theme and answer the following questions:

(1)   What color in the legend indicates countries with the fewest phone lines?

(2)   What color indicates countries with the most phone lines?

(3)   What color indicates countries with no data available for this theme?

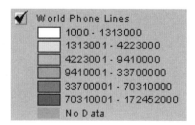

c   Notice that the colors in the legend change gradually from the lightest color to the darkest. The name for this type of legend is a graduated color legend.

What other theme in your view has a graduated color legend?

d   Look at the map to answer the following questions. You may need to turn themes on or off. You may need to use the Identify tool to find out a country's name. Answer the following questions:

   (1)   *Which country has the most phone lines?*

   (2)   *On which continent are most of the countries with the fewest phone lines?*

   (3)   *Which two countries have the largest populations?*

   (4)   *Name three countries that are in the same population class (color) as the United States.*

   (5)   *Name which of the three countries, if any, are in the same phone line class (color) as the United States.*

e   Turn on the World Phone Lines theme and make it active.

f   With the Identify tool, click on the United States.

g   Read the geographic question again.

   Geographic Question: Do the number of phone lines in a country increase proportionately with the number of people in the country?

h   Scroll down slowly in the Identify Results window and look at the different fields.

   What two fields might help in answering the geographic question?

i   Close the Identify Results window.

Step 7   **Research and record phone line and population data**

a   Click the Zoom to Full Extent button. This will return the map to its original size if you have zoomed or panned the map.

b   Click the Clear Selected Features button to deselect (turn the yellow off) any countries that might be selected.

c   Use your Identify tool and Find button to locate the countries in the table on the answer sheet. Record the population and phone lines for each country in the appropriate columns. (You will fill in the last column of the table later.)

   If you aren't sure what to do, steps 7d–7g show you how to find the information for the first country (China) as an example.

d Click the Find button and type **China** in the Search For box.

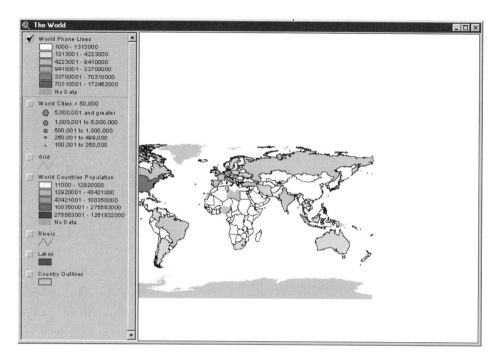

China is highlighted and centered in the view.

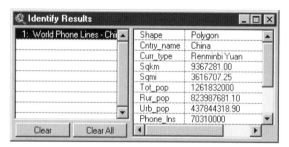

e Click the Identify tool and click China on the map.

f Find the total population attribute in the Identify Results window. Record the number in the Country Population column next to China.

g Scroll down in the Identify Results window and find the phone line attribute. Record the number in the Phone Lines column for China.

h Repeat steps 7d–7g for the other countries in the table on the answer sheet.

Step 8   Close the project and exit ArcView

You won't need to use ArcView to complete the rest of this lesson, so you'll close your project at this time.

a Click the File menu.

b Click Exit.

c Click No in the box that asks whether you want to save changes.

Ask your teacher whether you should continue doing steps 9–11 in class, or whether to do them as homework.

Step 9   **Calculate the number of people per phone line**

Now you will use your calculator to fill out the last column of the table.

a   Enter into your calculator the population figure for China.

> *Note: Your calculator may not let you type in a number as large as the one for China's population. If it doesn't, ask your teacher to help you calculate the statistic for countries with large populations.*

b   Divide it by the number of phone lines for China.

c   Round off the result to the nearest two decimal places and record it in the last column on the line for China. (Hint: The answer for China should be 17.95.)

d   Repeat the calculation for the other countries in the table.

4
ANALYZE GEOGRAPHIC INFORMATION

Step 10   **Analyze results of your research**

a   In the table on the answer sheet, the list on the left ranks the countries by population from highest to lowest. Use the data in your table to rank the countries on the right from the lowest number of people per phone line to the highest number of people per phone line. Then draw lines connecting the same country in each list.

b   Use the data in your table to answer the following questions:

(1)   *Which country has the fewest people per phone line? How many people have to share a phone line in this country?*

(2)   *How does the country in question 1 rank in population size with the other seven countries in your table?*

(3)   *Which country has the most people per phone line? How many people have to share a phone line in this country?*

(4)   *How does the country in question 3 rank in population size with the other seven countries in your table?*

(5)   *What is the population of Japan? How many people have to share a phone line in Japan?*

(6)   *What country has the second-most phone lines after the United States? How does its number of people per phone line compare with the seven other countries in your table?*

(7)   *Russia and Pakistan have about the same number of people. Why do you suppose these two countries have such a different number of people who have to share a phone line? What factors do you think contribute to this disparity?*

(8)   *Read the geographic question again. What do you think the answer to the geographic question is?*

Geographic Question: Do the number of phone lines in a country increase proportionately with the number of people in the country?

(9)   *Compare your initial hypothesis with your answer in question 8. How does your hypothesis compare with your answer to the geographic question?*

5
ACT UPON GEOGRAPHIC KNOWLEDGE

Step 11   Develop a plan of action

The last step of the geographic inquiry process is to act on what you have learned. Your action plan might be simply to repeat the process; thinking about what you've learned often leads to deeper, more complex, and interesting geographic questions.

For this step, choose one of the following four countries: China, Brazil, Indonesia, or the United States.

Imagine that you are an expert specializing in telecommunications. You need to devise a plan for your chosen country that deals effectively with the basic concern of your original geographic question.

To develop an effective plan, you may need to conduct further research on the phone system within your country. If you decided, for instance, that increasing the number of phone lines operating in your chosen country would improve the quality of life there, you could come up with a written plan of action for telecommunications officials, pointing out areas of strength and weakness, and explaining where and why expansion would be most beneficial.

?   *a*   **Use the information in your table to describe the current phone line situation in your chosen country. Record your chosen country on the answer sheet.**

?   *b*   **Do you think that increasing the number of phone lines operating in your chosen country would improve the quality of life there? Why or why not?**

?   *c*   **List three concerns you have about increasing the number of phone lines in your chosen country.**

?   *d*   **List two new geographic questions that you would like to investigate to help you develop a sound plan.**

NAME _____   DATE _____

Student answer sheet
Module 1
ArcView: The Basics

Part 1: Introducing the software

Step 6   Work with themes
f   Which themes are not visible on the map but are turned on in the table of contents?

_____

_____

h   What happened on your map?

_____

i-1   What happened on your map?

_____

i-2   What would happen if you dragged the Lakes under World Countries Population again?

_____

_____

Step 7   Identify a country and record country data
h-1   What is the second listing in this column?   _____

h-2   What is the third listing in this column?   _____

h-3   What is the final listing in this column? (Hint: You will need to scroll down.)

_____

i-1   What do you guess the field entitled "Sqmi" stands for?

_____

i-2   What is the number to the right of the field "Sqmi"?

_____

Step 8   Compare the Identify Results data with the table data
a   Which row in this table has the attributes for the United States?   _____

b-1   Where are these field names displayed in the table?

_____

b-2   How many square miles of land are in the United States?

_____

b-3   Give a brief explanation of the relationship between the Identify Results window and the table.

_____

_____

_____

Step 10   Explore city data on the world map

h   Use the Identify tool to find the name and country of any two cities you choose.

CITY NAME	COUNTRY WHERE THE CITY IS LOCATED

Step 11   Explore Europe with an attribute table

b   What is the name of the table you opened?

_____

d   What happens to the map when you click on these rows in the table?

_____

_____

e-1   What happens to Russia and the other countries that were highlighted?

_____

_____

e-2   Did you see Iceland turn yellow on the map? If not, why not?

_____

_____

_____

f   Why can you see Iceland now when you couldn't see it in the previous step?

_____

_____

_____

Step 12   Practice using the Identify tool

b   What do you see on your map?

_____

d-1   What country is it? _____

d-2   What is this country's total population? _____

g-1   What city is it? _____

g-2   What population range is this city in? _____

j   Name these two large cities.

_____

_____

Step 13   Practice using the Zoom Out tool

c-1   What does your map look like?

_____

c-2   Which zoom button could you use to return your map to full size?

_____

Step 14   Continue to explore the world map

h-1   How many tourists arrive in Sudan each year?   _____

h-2   How many people live in Sudan?   _____

h-3   Does this seem like a low or high number of tourists for this population?   _____

i   Do you see Qatar on your map?   _____

k-1   How many people live in Qatar?   _____

k-2   How many cell phones do they have?   _____

k-3   How many people are there for every cell phone in Qatar?   _____

k-4   What large country is directly west of Qatar?   _____

n-1   What boot-shaped country do you see on the map?   _____

n-2   What is the population of that country?   _____

n-3   How many cell phones does that country have?   _____

n-4   How many people are there for every cell phone in this country?   _____

o-1   What is the population of Japan?   _____

o-2   How many cell phones does Japan have?   _____

o-3   How many people are there for every cell phone in Japan?   _____

q   What happened to your map?

_____

t   What happened to Qatar?

_____

Step 15   Get help from the Help tool

c   What does it say you must do to make more than one theme active at a time?

_____

Part 2. The geographic inquiry model

Step 4   Ask a geographic question and develop a hypothesis

a   What makes this a geographic question?

_____

_____

b   Hypothesis:

_____

_____

_____

_____

_____

_____

Step 5   Add a theme to your view

a   What other attribute of countries do you need in order to investigate your hypothesis?

_____

e   What is the name of the theme that has been added to your table of contents?

_____

Step 6   Explore the World Phone Lines map

b-1   What color in the legend indicates countries with the fewest phone lines?

_____

b-2   What color indicates countries with the most phone lines?

_____

b-3   What color indicates countries with no data available for this theme?

_____

c   What other theme in your view has a graduated color legend?

_____

d-1   Which country has the most phone lines?

_____

d-2   On which continent are most of the countries with the fewest phone lines?

_____

d-3   Which two countries have the largest populations?

_____

_____

d-4   Name three countries that are in the same population class (color) as the United States.

_____

_____

_____

d-5   Name which of the three countries, if any, are in the same phone line class (color) as the United States.

_____

_____

_____

h   What two fields might help in answering the geographic question?

_____

Step 7   Research and record phone line and population data

c   Use your Identify tool and Find button to locate the countries in the table below. Record the population and phone lines for each country in the appropriate columns. (You will fill in the last column of the table later.)

COUNTRY NAME	COUNTRY POPULATION	PHONE LINES	NUMBER OF PEOPLE PER PHONE LINE
China	1,261,832,000	70,310,000	17.95
India			
United States			
Indonesia			
Brazil			
Russia			
Pakistan			
Japan			

Step 10   Analyze results of your research

a   In the table below, the list on the left ranks the countries by population from highest to lowest. Use the data in your table to rank the countries on the right from the lowest number of people per phone line to the highest number of people per phone line. Then draw lines connecting the same country in each list.

RANKED BY POPULATION (HIGHEST TO LOWEST)	RANKED BY NUMBER OF PEOPLE PER PHONE LINE (LOWEST NUMBER OF PEOPLE PER PHONE LINE TO HIGHEST)
China	
India	
United States	
Indonesia	
Brazil	
Russia	
Pakistan	
Japan	

b-1   Which country has the fewest people who have to share a phone line?

_____

How many people is that?   _____

b-2   How does the country in question 1 rank in population size with the other seven countries in your table?

_____

b-3   Which country has the most people who have to share a phone line?

_____

How many people is that?   _____

b-4   How does the country in question 3 rank in population size with the other seven countries in your table?

_____

b-5   What is the population of Japan?   _____

How many people have to share a phone line in Japan?   _____

b-6   What country has the second-most phone lines after the United States?

_____

How does its number of people per phone line compare with the seven other countries in your table?

_____

b-7   Russia and Pakistan have about the same number of people. Why do you suppose these two countries have such a different number of people who have to share a phone line?

_____

What factors do you think contribute to this disparity?

_____

_____

_____

b-8   What do you think the answer to the geographic question is?

_____

_____

_____

_____

_____

b-9   How does your hypothesis compare with your answer to the geographic question?

_____

_____

_____

_____

_____

Step 11   Develop a plan of action

a   Name your chosen country. _____

Use the information in your table to describe the current phone line situation in your chosen country.

_____

_____

_____

_____

_____

b   Do you think that increasing the number of phone lines operating in your chosen country would improve the quality of life there? Why or why not?

_____

_____

_____

_____

_____

c   List three concerns you have about increasing the number of phone lines in your chosen country.

_____

_____

_____

_____

_____

d   List two new geographic questions that you would like to investigate to help you develop a sound plan.

_____

_____

_____

_____

_____

NAME _____ DATE _____

ArcView: The Basics
Middle school assessment

Open the ArcView project **module1.apr**. Use the ArcView skills you have learned in this module to do the following things:

1 Create a map with at least three different themes.
2 Zoom in on the map to an area of the world of your choosing.
3 Find out three pieces of specific information about the area you chose. (Hint: Use the Identify button.)
4 Write a geographic question that involves one of the pieces of information that you listed in question 3.
5 Print the map and attach it to this page.

ArcView: The Basics

Assessment rubric

Middle school

STANDARD	EXEMPLARY	MASTERY	INTRODUCTORY	DOES NOT MEET REQUIREMENTS
The student knows how to make and use maps, globes, graphs, charts, models, and databases to analyze spatial distributions and patterns.	Creates and prints a map with more than three themes and focused on a portion of the world using a GIS.	Creates and prints a map with three different themes and focused on a portion of the world using a GIS.	Creates and prints a map with one or two different themes using a GIS.	Has difficulty creating the map without assistance and does not print it out.
The student knows and understands that places and features are distributed spatially across Earth's surface.	Identifies more than three pieces of information about the area of the world covered by his or her map and develops a geographic question based on that information.	Identifies three pieces of information about a particular area of the world and develops a geographic question based on that information.	Identifies one or two pieces of information about a particular area of the world and attempts to create a geographic question based on that information.	Identifies some information about a place, but does not create a geographic question based on the information gathered.
The student knows the role of technology in shaping the characteristics of places.	Successfully completes the assessment and develops a plan in step 11 of the lesson that illustrates an understanding of the importance of GIS technologies in analyzing the aspects of a region or place and solving geographic problems and questions.	Successfully completes the assessment and develops a clear and concise plan in step 11 of the lesson that illustrates an understanding of how GIS technologies contribute to geographic understanding of places and development of plans for changing them.	The student has a beginning understanding of the importance of GIS and related technologies in solving geographic questions.	The student does not see the relationship between GIS and related technologies in solving geographic questions.

This is a four-point rubric based on the National Standards for Geographic Education. The "Mastery" level meets the target objective for grades 5–8.

NAME _____   DATE _____

ArcView: The Basics
High school assessment

Open the ArcView project **module1.apr**. Use the ArcView skills you have learned in this module to do the following things:

1 Create a map with at least three different themes.
2 Zoom in on the map to an area of the world of your choosing.
3 Find out three pieces of specific information about the area you chose. (Hint: Use the Identify button.)
4 Write a geographic question that involves one of the pieces of information that you listed in question 3.
5 Write a brief paragraph explaining what you learned geographically about the area you zoomed to on your map.
6 Print the map and attach it to this page.

ArcView: The Basics

Assessment rubric

High school

STANDARD	EXEMPLARY	MASTERY	INTRODUCTORY	DOES NOT MEET REQUIREMENTS
The student knows how to use technologies to represent and interpret Earth's physical and human systems.	Creates and prints a map with more than three themes and focused on a portion of the world using a GIS.	Creates and prints a map with three different themes and focused on a portion of the world using a GIS.	Creates and prints a map with one or two different themes using a GIS.	Has difficulty creating the map without assistance and does not print it out.
The student knows and understands the spatial behavior of people.	Identifies more than three pieces of information about the area of the world covered by his or her map and develops a geographic question about the human impact to that place, based on the information.	Identifies three pieces of information about a particular area of the world and develops a geographic question about the human impact to that place based on the information.	Identifies one or two pieces of information about a particular area of the world and attempts to create a geographic question based on that information.	Identifies some information about a place, but does not create a geographic question based on the information gathered.
The student knows and understands the changing physical and human characteristics of place.	In step 11 of the lesson, writes a clear and concise plan of action that takes into account the physical and human characteristics of a place or places.	In step 11 of the lesson, writes a clear and concise paragraph on the geographic characteristics of a particular place.	In step 11 of the lesson, identifies some physical and human characteristics of a place and attempts to formalize understanding in paragraph form.	In step 11 of the lesson, lists one or two characteristics of a place but does not show geographic understanding of the place.

This is a four-point rubric based on the National Standards for Geographic Education. The "Mastery" level meets the target objective for grades 9–12.

Physical Geography I Landforms and Physical Processes

Explore the powerful forces that originate in the earth's interior and shape the landforms that characterize its surface.

The Earth Moves: A global perspective

Students will observe patterns of earthquake and volcanic activity on the earth's surface and the relationship of those patterns to the location of diverse landforms, plate boundaries, and the distribution of population. Based on their exploration of these relationships, students will form a hypothesis about the earth's distribution of earthquake and volcanic activity and identify world cities that face the greatest risk from those phenomena.

Life on the Edge: A regional case study of East Asia

Students will investigate the Pacific Ocean's "Ring of Fire," with particular focus on earthquake and volcanic activity in East Asia, where millions of people live with the daily threat of significant seismic or volcanic events. Through the analysis of volcanic location and earthquake depth, students will identify zones of subduction at tectonic plate boundaries and the location of populations in the greatest danger of experiencing a volcanic eruption or a major earthquake.

Mapping Tectonic Hot Spots: An advanced investigation

This lesson is designed to target the self-motivated, independent GIS students in your class. They will use the Internet to acquire the most recent data on earthquakes and volcanoes worldwide. By exploring this data through a GIS, students will construct a current World Tectonic Hot Spots map. This lesson could be an independent research project completed over several days.

The Earth Moves
A global perspective

Lesson overview

Students will observe patterns of seismic and volcanic activity around the world and the relationship of those patterns to the location of diverse landforms, plate boundaries, and the distribution of population. Based on their exploration of these relationships, students will form a hypothesis about the planet's distribution of earthquake and volcanic activity, and identify those cities that face the greatest risk from those phenomena.

Estimated time   Two to three 45-minute class periods

Materials   ✔ Large map of the world

✔ Ten adhesive dots or map pins (two colors) for every four students

✔ Colored pencils for each student

✔ Student handouts from this lesson to be copied:
  - World map (pages 55 to 57)
  - GIS Investigation sheets (pages 59 to 69)
  - Student answer sheets (pages 71 to 73)
  - Assessment(s) (pages 75 to 79)

Standards and objectives

National geography standards

	GEOGRAPHY STANDARD	MIDDLE SCHOOL	HIGH SCHOOL
1	How to use maps and other geographic representations, tools, and technologies to acquire, process, and report information from a spatial perspective	The student knows and understands how to use maps to analyze spatial distributions and patterns.	The student knows and understands how to use geographic representations and tools to analyze and explain geographic problems.
4	The physical and human characteristics of places	The student knows and understands how to analyze the physical characteristics of places.	The student knows and understands the changing physical characteristics of places.
7	The physical processes that shape the patterns of Earth's surface	The student knows and understands how physical processes shape patterns in the physical environment.	The student knows and understands the spatial variation in the consequences of physical processes across Earth's surface.
15	How physical systems affect human systems	The student knows and understands how natural hazards affect human activities.	The student knows and understands how humans perceive and react to natural disasters.

Standards and objectives (continued)

Objectives

The student is able to:

- Locate zones of significant seismic and volcanic activity on the earth's surface.
- Describe the relationship between zones of high seismic activity, volcanic activity, and the location of tectonic plate boundaries.
- Identify world regions and cities where human populations face the greatest threat from earthquakes.

GIS skills and tools

 Add themes to the view

 Open the attribute table for a theme

  Sort data in ascending and descending order

      Zoom in and out of the view

 Select records

 Use the Identify tool to learn more about a selected record

 Use the Query Builder to select data

 Clear all selections made

 Pan the view to see different areas of the map

 Use the Find tool to locate a specific feature

- Turn themes on and off
- Add and view image source data
- Change the order of the table of contents to change the map display

For more on geographic inquiry and these steps, see Geographic Inquiry: Thinking Geographically (pages xxi to xxiii).

Teacher notes

Lesson introduction

Begin the lesson by distributing the provided outline maps of the world. Have the students individually write a "V" on locations on the map where they believe volcanoes are located, and an "E" on locations they think earthquakes typically occur. They should select at least eight volcanic areas and eight areas for earthquake activity. Once students have done this individually, divide them into groups of three or four and have them discuss their ideas within their small group.

Each group should take five minutes to consider the following questions and record their answers in the space provided on the worksheet:

- Are there any similarities between your group members' maps? What are they?
- How did the members of your group choose where to put a "V" and "E" on their maps?
- Are there locations where earthquakes and volcanoes occur close together? List them.
- Are there major cities close to these locations? What are they?

As a class, the students will incorporate each group's map into one large class map. Tell the students that one color of map pins will represent earthquakes and the other color will represent volcanoes. At the end of five minutes, each group should take 10 adhesive dots or map pins (five of each color) and place them on a large world map in the classroom. One color should be placed at locations where group members agree that there are volcanoes and the other color should be placed where group members agree that there are earthquakes. Before beginning the next part of the lesson, briefly discuss the patterns reflected on the map.

Student activity

 Before completing this lesson with students, we recommend that you complete it as well. Doing so will allow you to modify the activity to accommodate the specific needs of your students.

After the initial discussion, have the students work on the computer component of the lesson. Ideally, each student should be at an individual computer, but the lesson can be modified to accommodate a variety of instructional settings.

Distribute this lesson's GIS Investigation sheets to your students. Explain that in this activity, they will use GIS to investigate where earthquakes and volcanoes occur on the earth's surface. The worksheets will provide them with detailed instructions for their investigations. As they investigate, they will create a hypothesis about where these major physical events occur, identify zones of volcanic and seismic activity, and identify those densely populated places that face the greatest risk from these seismic activities.

In addition to the instructions, the investigation sheets include questions to help students focus on key concepts. Some questions will have specific answers while others will require creative thought. Answers to these questions are located after the student answer sheets.

Things to look for while the students are working on this activity:
- Are the students using a variety of tools?
- Are the students answering the questions as they work through the procedure?
- Students should be referring to their original maps and notes from class discussion.

Conclusion Before beginning the assessment, engage students in a discussion of the observations and discoveries that they made during their exploration of the maps. Ask students to compare their initial ideas—as reflected by the pattern of colored dots on the classroom's world map—with the insights they acquired in the course of their GIS investigation.

- Has this investigation raised any questions that they would like to explore further?
- How can GIS help world cities better prepare for seismic events?
- How have their ideas about earthquakes and volcanoes changed since the start of the lesson?

Assessment *Middle school: Highlights skills appropriate to grades 5 through 8*

Part 1 of The Earth Moves assessment for middle school students asks them to create an informative paper map based on their GIS investigation. Part 2 is a set of four questions that students will answer on a separate sheet of paper. An assessment rubric to evaluate student performance is provided at the end of the lesson.

High school: Highlights skills appropriate to grades 9 through 12

Part 1 of The Earth Moves assessment for high school students asks them to create an informative paper map based on their GIS investigation. Part 2 is a set of three questions that students will answer on a separate sheet of paper. An assessment rubric to evaluate student performance is provided at the end of the lesson.

Extensions
- By using ArcView, students can create their own digital maps for the assessment. They can mark all of the required features and print the map using the layout document.
- Investigate your local area with ArcView, and ask students to create a new theme that identifies natural hazards in your region.
- Look at historical data for notable earthquakes and volcanic eruptions and incorporate this data as new themes. From this information, ask students to make predictions on the next significant eruption or seismic event.
- Analyze fault line data included with this lesson to see if it provides additional insight to the location and movement of plate boundaries.
- Check out the Resources by Module section of the Teacher Resource CD for print and media resources on the topics of earthquakes, volcanoes, and plate tectonics or visit *www.esri.com/mappingourworld* for Internet links.

NAME _____   DATE _____

The Earth Moves
A map investigation

Part 1: Individual map

Volcanoes and earthquakes happen all over the planet. On the map provided, identify places where you think these physical events typically occur. Mark eight locations with a "V" for potential volcano sites, and mark eight locations with an "E" for places where you think earthquakes typically occur.

Part 2: Small group discussion

In your small group, compare your maps and answer the following questions:

Step 1   Are there any similarities between your group members' maps? What are they?

_____

_____

_____

_____

_____

_____

_____

Step 2   How did the members of your group choose where to put a "V" and "E" on their maps?

_____

_____

_____

_____

_____

_____

_____

Step 3 Are there locations on your maps where earthquakes and volcanoes occur close
together? List them.

_____

_____

_____

_____

_____

_____

_____

_____

Step 4 Are there major cities close to these locations? What are they?

_____

_____

_____

_____

_____

_____

_____

_____

NAME _____

DATE _____

The Earth Moves

Outline map of the world

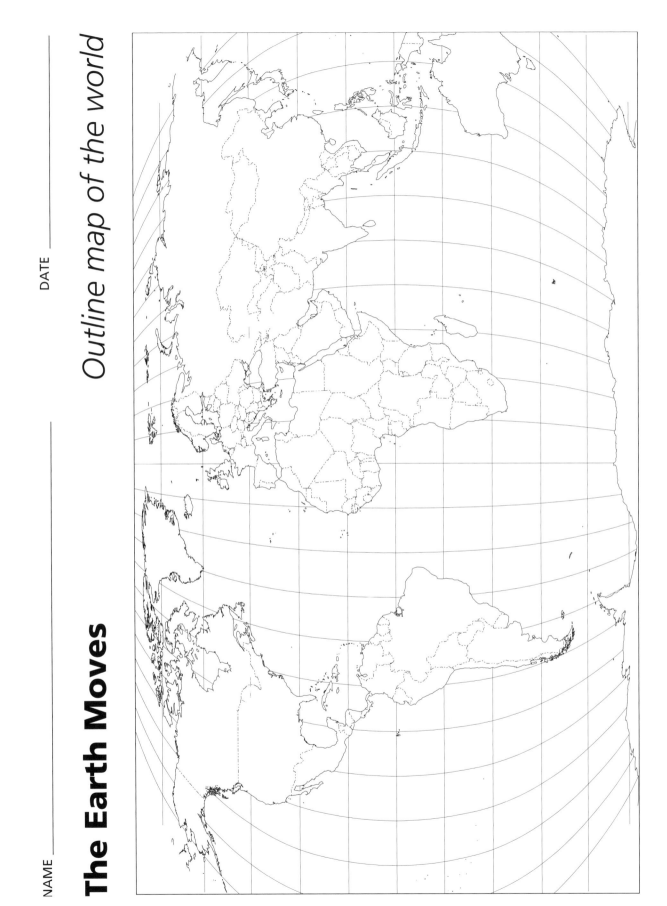

MODULE 2 • PHYSICAL GEOGRAPHY I: LANDFORMS AND PHYSICAL PROCESSES

NAME _____   DATE _____

The Earth Moves
A GIS investigation

ACQUIRE

ASK   EXPLORE

ACT   ANALYZE

Answer all questions on the student answer sheet handout

Step 1   Start ArcView

a   Double-click the ArcView icon on your computer's desktop.

b   If the Welcome to ArcView dialog appears (pictured below), click **Open an Existing Project** and click OK. If it doesn't appear, proceed to step 2.

Step 2   Open the global2.apr file

a   In this exercise, a project file has been created for you. To open the file, go to the **File** menu and choose **Open Project**.

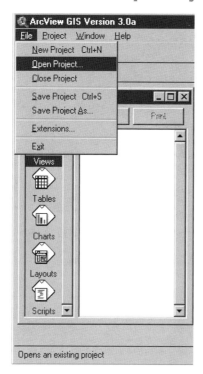

b   Navigate to the exercise data directory (**C:\esri\mapworld\mod2**) and choose **global2.apr** from the list.

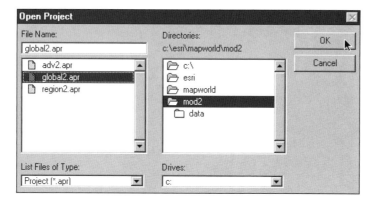

c   Click OK.

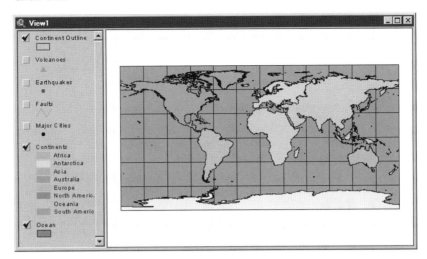

When the project opens, you see a map in the View window with three themes turned on (Continent Outline, Continents, and Ocean). The check mark next to the theme name tells you the theme is turned on and visible in the view.

Step 3   **Look at earthquake location data**

In this step, you will compare your original theories about earthquake and volcano locations to actual data using GIS.

    a   Turn on the Earthquakes theme by clicking the box to the left of the name in the table of contents.

This places a check mark in the box and adds a layer of points showing the locations of earthquakes on the map.

b Evaluate the map and write your answers to the following questions on the answer sheet:

? *(1) Do earthquakes occur in the places you predicted? List the regions you predicted correctly for earthquake locations.*

? *(2) What patterns do you see in the map?*

Step 4 Sort and analyze earthquake magnitudes

You can take a closer look at the data behind the dots by looking at the attribute table of the Earthquakes theme. The attribute table contains specific information about the features in a theme. In the Earthquakes theme, each point represents an earthquake with a magnitude of greater than 4.0 on the Richter scale. In this step, you will use the attribute table to focus on the 15 strongest earthquakes.

a Make the Earthquakes theme active by clicking once on the Earthquakes theme in the table of contents. (Be careful not to click the box with the check mark because the theme will no longer be displayed in the view.)

You know the theme is active because the box around it appears slightly raised above the other themes in the table of contents.

 b Click the Open Theme Table button.

This shows you all the attribute data associated with the purple earthquake points on the map.

⚡ ***Do not maximize this table. It will prevent you from easily returning to the view.***

c Look at the table. Scroll down to see more records.

Remember: Each record in this table represents one point on the map.

d Click the field (column heading) labeled Mag to select it.

Attributes of Earthquakes

Shape	Date	Lat	Lon	Depth	Mag
Point	20000101	41.9270	20.5430	10.00	4.80
Point	20000101	-11.3480	164.5680	33.00	4.30
Point	20000101	-26.0120	-68.8030	105.00	4.50
Point	20000101	36.8740	69.9470	54.30	5.10
Point	20000101	-60.7220	153.6700	10.00	6.00
Point	20000101	37.0270	69.9640	33.00	4.40
Point	20000101	-17.0730	167.0260	33.00	4.30
Point	20000101	8.3700	126.3360	100.70	4.50
Point	20000101	-23.3190	-179.9780	537.80	4.60
Point	20000101	-3.2070	148.8770	10.00	4.50
Point	20000101	57.4490	-154.3200	58.00	4.00

The heading appears darker when the field is selected. This field represents the magnitude of the earthquakes.

e Scroll up to the top of the table. Now you will put the magnitudes in order from the largest magnitude to the smallest.

f   Click the Sort Descending button.

The records have been rearranged from largest to smallest. Now you will select the 15 largest earthquakes.

g   Hold down the Shift key, click on the first record in the table (now it is highlighted in yellow), and drag your mouse until the first 15 records are highlighted in yellow.

To make sure that you have highlighted 15 earthquakes, look in the upper left hand corner of the ArcView window. It should display:

   If you select too many records, click the Clear Selection button to clear the selections and try again.

When you select a record in the attribute table, its point on the map will be highlighted also.

h   Click the title bar of the view so you can see where the 15 strongest earthquakes are located.

Use the Zoom tools and buttons to get a better view. Here is a brief explanation of those tools and buttons:

The Zoom In tool will allow you to drag a box around the area you want to zoom in on. You can also click the Zoom In tool and then click the map to get a closer look. This is a very useful tool when you want to see more detail in the map.

The Zoom Out tool will allow you to drag a box around the area you want to zoom out from. You click the Zoom Out tool and then click the map to see a bigger area.

The Zoom In to Extent button zooms in to the center of the view.

The Zoom Out to Extent button zooms out from the center of the view.

The Zoom to Full Extent button will zoom your view to the extent of all the themes. In this example, it will zoom out so you can see all of the continents. Click this button when you want to view the entire world map.

The Pan tool repositions the map. Use the Pan tool by clicking the spot you'd like to see and dragging it to the center of the display. The view will redraw itself around the area you panned.

Use these buttons and tools to explore the locations of the selected earthquakes.

?   *How do the 15 selected locations compare to your original paper map? List three ways.*

i   Click the Zoom to Full Extent button to see the world in the view.

Step 5   Look at volcano data

a   Turn off the Earthquakes theme and turn on the Volcanoes theme.

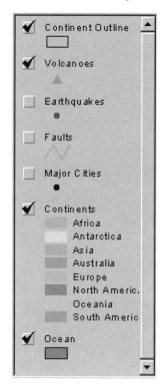

?   *(1)   How do the volcano locations compare with your original predictions? List the regions of volcanic activity you predicted correctly.*

?   *(2)   What patterns do you see in the volcano points and how do they compare with the earthquake patterns? (Hint: Turn the Earthquakes theme on and off.)*

The volcano data includes information on the status of each volcano (active, inactive, and potentially active). You will focus on the active volcanoes.

Step 6   Select all active volcanoes

    a   Make the Volcanoes theme active and click the Open Theme Table button.

The Type field of the table tells you if each volcano is Active, Potentially active, or Solfatara (emits gases, but is otherwise inactive).

b   Click on the Type field heading.

    c   Click the Sort Ascending button. Scroll down and you will notice that there are many active volcanoes.

It would not be fun to highlight all of these as you did with the Earthquakes theme. This is a smart database—we can ask it to select all of the active volcanoes by using the Query Builder.

    d   Click the Query Builder button.

e   **Double-click** Type in the left hand Fields box. **Single-click** the equals sign (=). Finally, **double-click** "Active" in the right-hand Values box.

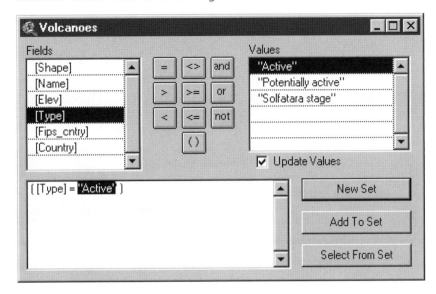

f   Click New Set. All of the active volcanoes are selected and highlighted yellow.

> *Hint: If you receive a syntax error, check that your equation is exactly like the one in the graphic above. If it isn't, delete the equation and try again.*

g   Close the Query dialog and click the view title bar to see the map. Use the Zoom and Pan tools to explore where the active volcanoes are located.

?   *(1)   Does this data provide any patterns that were not evident before? Identify those patterns.*

?   *(2)   Create a hypothesis as to why volcanoes and earthquakes happen where they do.*

Step 7   Identify the active volcanoes on different continents

In order to learn more about the active volcanoes, you can use the Identify tool.

a   Click the Identify tool and move your cursor over the view area. Notice how the cursor has a small "i" next to it.

b   Click on an active volcano in the view.

The Identify Results dialog appears, showing you the name of the volcano, its elevation, type, and country location. For example:

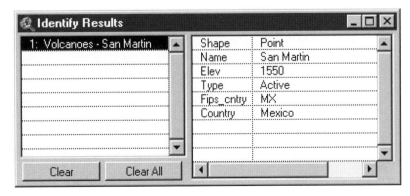

c   Close the Identify Results window and zoom in to the continent of your choice.

d   Use the Identify tool to find the name, elevation, activity level, and country location of three volcanoes.

?   *Write that information on the answer sheet.*

e   Close the Identify Results dialog.

f   Click the Clear Selection button to deselect all of the active volcanoes.

Step 8   **Add the plate boundaries theme**

The earth is always changing. The crust of the earth is composed of several tectonic plates that are always on the move. Movement occurs at the boundaries between the plates and on the surface of the plates themselves.

?   *Based on the location of the earthquakes and volcanoes, where do you think the plate boundaries are? Draw them on your paper map.*

There are four basic types of plate boundaries. In this part of the GIS activity, you will investigate where these boundaries exist and how they affect the landforms close to them. Here's a quick review of the different types of plate boundaries:

- *Divergent boundary* — One or two plates are splitting apart. New crust is being formed from the center of the earth, causing the plate to spread. Rift valleys are one example of this type of plate movement.

- *Convergent boundary* — Two plates are colliding, forcing one plate to dip down underneath another one. The plate that is folding under has old crust that is being destroyed, while the plate on top has mountains and volcanoes being formed. In the ocean, these appear as trenches.

- *Transform boundary* — Plates are sliding against each other causing large faultlines and mountains to form. Here, the crust is neither created nor destroyed.

- *Plate boundary zones* — Plate boundaries appear erratic (zigzagged). Scientists believe there are actually microplates in these areas, but it is unclear what effect these zones have on the physical environment.

a   Turn off the Volcanoes theme.

b   Click the Add Theme button.

c   Navigate to the exercise data directory (**C:\esri\mapworld\mod2\data**). Scroll down and select **plat_lin.shp**.

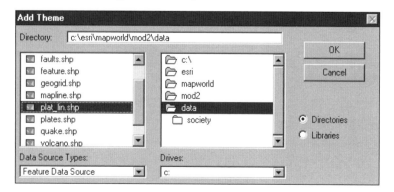

d   Click OK.

The plat_lin.shp theme is added to your table of contents.

e   Click the box next to plat_lin.shp to turn on the theme.

f   Double-click the plat_lin.shp theme in the table of contents. The Legend Editor opens.

g   Click the Load button. Click plat_lin.avl from the column and click OK. The legend changes color and thickness.

h   Click Apply and close the Legend Editor.

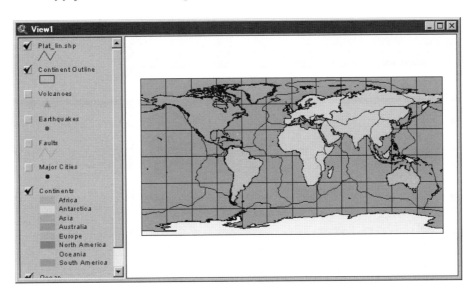

?   *Compare the actual plate boundaries to the ones you drew on your paper map. Record all similarities and differences.*

Step 9   Add a theme and an image file

In order to get a closer look at landforms and boundaries, you will add two more themes:

- *feature.shp:* a theme containing major physical features of the planet.
- *wsiearth.jpg:* a color-shaded relief map of the earth, made from a satellite image.

a   Click the Add Theme button and navigate to your data folder (**C:\esri\mapworld \mod2\data**). Double-click **feature.shp** to add it to your view.

b   Turn on feature.shp.

Adding an image file is a little different from adding a feature theme. You will have to change the file type before you can select an image file.

c   Click the Add Theme button and navigate to your data folder (**C:\esri\mapworld \mod2\data**).

You notice that the file you're looking for (wsiearth.jpg) is not listed.

d   Click the Data Source Type arrow and select Image Data Source.

Any file that is .tif or .jpg is an image file and will always be added as an Image Data Source.

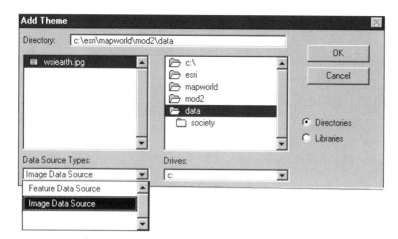

e   Select wsiearth.jpg and click OK to add it to your view. Turn the theme on.

f   Click the Feature.shp theme and drag it above the Wsiearth.jpg theme so you can see both themes in the view. Drag Plat_lin.shp so it's at the top of the table of contents.

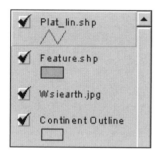

g   Look at the view. Are there any areas where major landforms, plate boundaries, and seismic activity overlap?

h   Use the Identify tool to find out the names of all major landforms formed at plate boundaries. Write them in the table on the answer sheet and label them on your paper map.

Hint: Make the Features.shp theme active when you want to find the names of the major landforms.

Step 10   **Identify major cities at high and low risk for seismic activity**

Next, you will find cities with a high and low risk of earthquake and volcanic activity.

a   Move the Major Cities theme to the top of the table of contents and turn it on. Turn off Wsiearth.jpg.

b   With the Major Cities theme active, use the Zoom, Pan, and Identify tools to find the names of specific cities that are high-risk or low-risk for a seismic event. Write those names in the table on the answer sheet.

> ***Hint: Remember to turn themes on and off as needed and to move the themes around in the table of contents.***

Step 11   **Exit ArcView**

In this GIS investigation, you used different themes to determine where earthquake and volcanic activity is located around the world. From this information, you were able to determine cities at high or low risk for these natural disasters.

a   Ask your teacher for instructions on where to save this ArcView project and on how to rename the project.

b   If you are not going to save the project, exit ArcView by choosing Exit from the File menu. Click No when you are asked if you want to save changes to global2.apr.

NAME _____ DATE _____

Student answer sheet
Module 2
Physical Geography I: Landforms and Physical Processes

Global perspective: The Earth Moves

Step 3 Look at earthquake location data

b-1 Do earthquakes occur in the places you predicted? List the regions you predicted correctly for earth-quake locations.

_____
_____
_____
_____
_____

b-2 What patterns do you see in the map?

_____
_____
_____
_____
_____

Step 4 Sort and analyze earthquake magnitudes

h How do the 15 selected locations compare to your original paper map? List three ways.

_____
_____
_____
_____

Step 5 Look at volcano data

a-1 How do the volcano locations compare with your original predictions? List the regions of volcanic activity you predicted correctly.

_____
_____
_____
_____

a-2   What patterns do you see in the volcano points and how do they compare with the earthquake patterns? (Hint: Turn the Earthquakes theme on and off).

_____

_____

_____

_____

_____

Step 6   Select all active volcanoes

g-1   Does this data provide any patterns that were not evident before? Identify those patterns.

_____

_____

_____

_____

_____

g-2   Create a hypothesis as to why volcanoes and earthquakes happen where they do.

_____

_____

_____

_____

_____

Step 7   Identify the active volcanoes on different continents

d   Use the Identify tool to find the name, elevation, activity level, and country location of three volcanoes. Write that information in the space below.

_____

_____

_____

_____

_____

Step 8   Add the plate boundaries theme

h   Compare the actual plate boundaries to the ones you drew on your paper map. Record all similarities and differences.

_____

_____

_____

_____

_____

Step 9   Add a theme and an image file

h   Use the Identify tool to find out the names of all major landforms formed at plate boundaries. Write them below and label them on your paper map.

NAME OF LANDFORM	FORMED BY
Mid-Atlantic Ridge	The separation of South American and African plates

Step 10   Identify major cities at high and low risk for seismic activity

b   Find the names of specific cities that are high-risk or low-risk for a seismic event. Write those names in the table.

HIGH RISK	LOW RISK
1	1
2	2
3	3
4	4
5	5

NAME ————————————— DATE —————————————

The Earth Moves

Outline map of the world

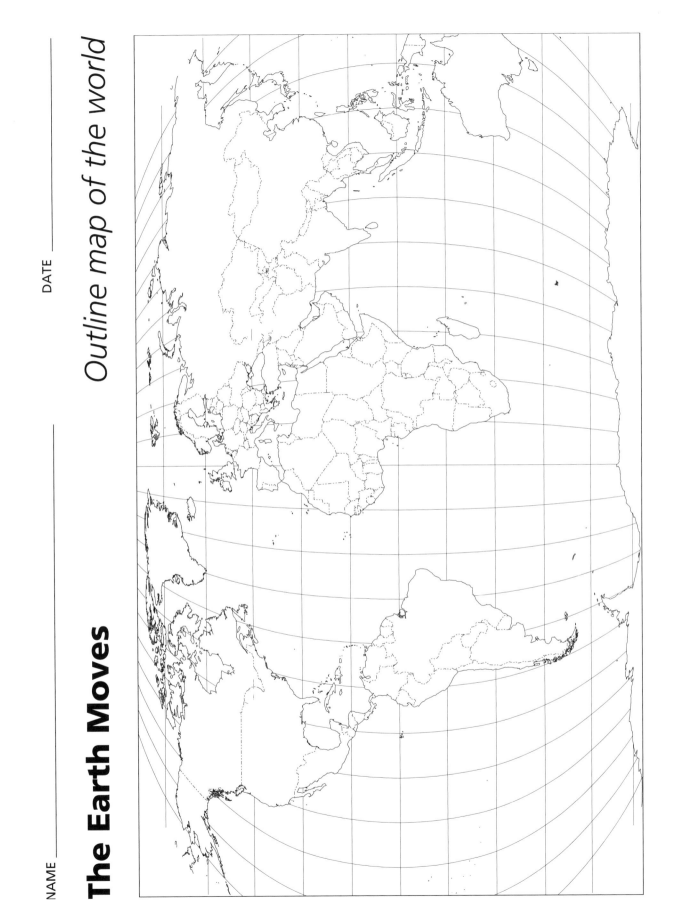

MODULE 2 • PHYSICAL GEOGRAPHY I: LANDFORMS AND PHYSICAL PROCESSES

NAME _____   DATE _____

The Earth Moves
Middle school assessment

Part 1: Create an informative map

On the paper outline map provided, do the following three things:

1 Mark and identify all of the plate boundaries.
2 Identify three to five major landforms that are associated with the physical forces of volcanism and earthquakes by writing their locations on your map.
3 From the following list, label five world cities with high risk for a volcanic or seismic disaster.

San Francisco, USA	Bombay, India	Jakarta, Indonesia
Manila, Philippines	Bogotá, Columbia	Addis Ababa, Ethiopia
Mexico City, Mexico	Rome, Italy	Seattle, USA
Houston, Texas	Madrid, Spain	Tokyo, Japan
Hong Kong, China	Reykjavik, Iceland	
Sydney, Australia	Cairo, Egypt	

You may use the ArcView project and class notes to help you create this map. In the next section, you will be asked to explain why you chose the locations you did. Be sure to think carefully about your cartographic choices.

 Hint: You can use the Find tool in ArcView to locate the world cities.

Part 2: Map analysis

On a separate piece of paper, answer each question below in paragraph form.

1 Describe the relationships you see between tectonic plate boundaries and areas at high risk for volcanic and seismic activity.
2 Explain why you selected *each* city on your map as a place at high risk for an earthquake or volcanic eruption.
3 Rank your five selected cities in the order of highest risk. The number one city will be the most at risk. Explain why you ranked them that way.
4 Compare and contrast your outline map from before the GIS investigation to the map you created in this assessment. How have your ideas changed?

The Earth Moves

Assessment rubric

Middle school

STANDARD	EXEMPLARY	MASTERY	INTRODUCTORY	DOES NOT MEET REQUIREMENTS
The student knows and understands how to use maps to analyze spatial distributions and patterns.	Uses GIS to analyze volcanic and earthquake data to identify five world cities most at risk for volcanic or seismic disasters, provides ample evidence for his or her choices, and draws conclusions on commonalities between the places.	Uses GIS to analyze volcanic and earthquake data to identify five world cities most susceptible to volcanic or earthquake disasters and provides ample evidence to support his or her decisions.	Uses GIS to identify four to five world cities most susceptible to volcanic or earthquake disasters and provides limited support for his or her decisions.	Uses GIS to identify some world cities susceptible to volcanic or earthquake disasters, but does not provide evidence to support his or her choices.
The student knows and understands how to analyze the physical characteristics of places.	Identifies at least five major landforms and the volcanic or seismic process that created them. Correctly identifies all plate boundaries.	Identifies three to five major landforms that were created because of volcanic or seismic activity.	Identifies two to three major landforms that were created by volcanic or seismic activity.	Identifies one major landform created by volcanic or earthquake activity.
The student knows and understands how physical processes shape patterns in the physical environment.	Clearly describes the relationship between zones of high earthquake and volcanic activity and the location of tectonic plate boundaries through the use of a variety of media.	Describes the relationship between zones of high earthquake and volcanic activity and the location of tectonic plate boundaries.	Provides limited evidence of the relationship between zones of high earthquake and volcanic activity and the location of tectonic plate boundaries.	Does not show evidence of understanding the relationship between zones of high earthquake and volcanic activity and the location of tectonic plates.
The student knows and understands how natural hazards affect human activities.	Ranks top five high-risk cities and provides ample evidence, supported with outside sources, for his or her decisions.	Ranks top five high-risk cities and provides a clear explanation for his or her choices.	Ranks the top five cities according to degree of risk, but does not provide an explanation for his or her decision.	Does not rank the cities in any particular order. Does not provide any explanation.

This is a four-point rubric based on the National Standards for Geographic Education. The "Mastery" level meets the target objective for grades 5–8.

NAME _____   DATE _____

The Earth Moves
High school assessment

Part 1: Create an informative map

On the paper outline map provided, do the following three things:

1 Mark and identify all of the plate boundaries.
2 Identify three to five major landforms that are associated with the physical forces of volcanism and earthquakes by writing their locations on your map.
3 From the following list, identify five world cities with high risk for a volcanic or seismic disaster:

San Francisco, USA   Bombay, India   Jakarta, Indonesia
Manila, Philippines   Bogotá, Columbia   Addis Ababa, Ethiopia
Mexico City, Mexico   Rome, Italy   Seattle, USA
Houston, USA   Madrid, Spain   Tokyo, Japan
Hong Kong, China   Reykjavik, Iceland
Sydney, Australia   Cairo, Egypt

You may use the ArcView project and class notes to help you create this map. In the next section, you will be asked to explain why you chose the locations you did. Be sure to think carefully about your cartographic choices.

 Hint: You can use the Find tool in ArcView to locate the world cities.

Part 2: Map analysis

On a separate piece of paper, answer each question below in paragraph form.

1 Provide evidence for *each* of the cities you selected as high risk for a major earthquake or volcanic eruption.
2 Choose one of the cities on your map and research how this city is prepared (or not prepared) for a seismic disaster. Write a complete paragraph outlining the city's preparedness.
3 For each of the major landforms you selected, hypothesize how you think they formed. Use your knowledge of the four major types of plate boundaries to answer this question.

The Earth Moves

Assessment rubric

High school

STANDARD	EXEMPLARY	MASTERY	INTRODUCTORY	DOES NOT MEET REQUIREMENTS
The student knows and understands how to use geographic representations and tools to analyze and explain geographic problems.	Uses GIS to analyze volcanic and earthquake data to identify five or more world cities at high risk for volcanic or earthquake activity, and provides a detailed explanation of how he or she came to these conclusions.	Uses GIS to analyze volcanic and earthquake data to identify five world cities at high risk for volcanic or earthquake activity, and provides a brief narrative stating the logic for their conclusions.	Uses GIS to analyze volcanic and earthquake data to identify three to five world cities at high risk for volcanic or earthquake activity, and provides little explanation for their conclusions.	Uses GIS to analyze volcanic and earthquake data to identify one to five cities, but does not provide any explanation for their conclusions.
The student knows and understands the spatial variation in the consequences of physical processes across Earth's surface.	Clearly explains the correlation between the spatial distribution of volcanoes and earthquakes in relation to plate boundaries, and ties this in to their understanding of physical landforms.	Explains the correlation between the spatial distribution of volcanoes and earthquakes in relation to plate boundaries.	Identifies the correlation between volcanoes, earthquakes, and plate boundaries, but has difficulty explaining the relationship.	Identifies the relationship between the spatial distribution of volcanoes and earthquakes, but does not correlate them to the location of plate boundaries.
The student knows and understands the changing physical characteristics of places.	Clearly describes through various media how the physical process of plate tectonics shape five major landforms.	Describes with ample evidence how the physical process of plate tectonics shaped three to five major landforms.	Describes in some detail how the physical process of plate tectonics shaped two to three major landforms.	Identifies a few major landforms, but provides little description of how plate tectonics shaped them.
The student knows and understands how humans perceive and react to natural disasters.	Researches and explains how several of their high-risk cities have prepared for a major volcanic or seismic event.	Researches and explains how two of their high-risk cities have prepared for a major volcanic or seismic event.	Researches and provides limited explanation for how one of their high-risk cities has prepared for a major seismic or volcanic event.	Provides some explanation of how a high-risk city has prepared for a major seismic or volcanic event, but does not provide research to support it.

This is a four-point rubric based on the National Standards for Geographic Education. The "Mastery" level meets the target objective for grades 9–12.

Life on the Edge
A regional case study of East Asia

Lesson overview

Students will investigate the Pacific Ocean's "Ring of Fire," with particular focus on earthquake and volcanic activity in East Asia, where millions of people live with the daily threat of significant seismic or volcanic events. Through the analysis of volcanic location and earthquake depth, students will identify zones of subduction at tectonic plate boundaries and the location of populations in the greatest danger of experiencing a volcanic eruption or a major earthquake.

Estimated time   Two to three 45-minute class periods

Materials   ✔ Colored pencils

✔ Student handouts from this lesson to be copied:
- Map of East Asia (page 85)
- GIS Investigation sheets (pages 87 to 94)
- Student answer sheets (pages 95 to 96)
- Assessment(s) (pages 97 to 101)

Standards and objectives

National geography standards

GEOGRAPHY STANDARD	MIDDLE SCHOOL	HIGH SCHOOL
1 How to use maps and other geographic representations, tools, and technologies to acquire, process, and report information from a spatial perspective	The student knows and understands how to use maps and databases to analyze spatial distributions and patterns.	The student knows and understands how to use geographic representations and tools to analyze and explain geographic problems.
4 The physical and human characteristics of places	The student knows and understands how to analyze the physical characteristics of places.	The student knows and understands the changing physical characteristics of places.
7 The physical processes that shape the patterns of Earth's surface	The student knows and understands how to predict the consequences of physical processes on Earth's surface.	The student knows and understands the spatial variation in the consequences of physical processes across Earth's surface.
15 How physical systems affect human systems	The student knows and understands how natural hazards affect human activities.	The student knows and understands how humans perceive and react to natural disasters.

<table>
<tr><td>*Standards and objectives (continued)*</td><td>

Objectives

The student is able to:

- Locate zones of significant earthquake and volcanic activity in East Asia.
- Describe the relationship between zones of high earthquake activity, volcanic activity, and the location of tectonic plate boundaries.
- Identify subduction zones along plate boundaries.
- Identify densely populated areas that are most at risk for volcanic and/or seismic disasters.

</td></tr>
</table>

GIS skills and tools

   Add themes to the view

   Measure distances between points in the view

   Label objects in the view

   Select records

   Use the Identify tool to learn more about a selected record

    Sort data in ascending and descending order

        Zoom in and out of the view

   Pan the view to see different areas of the map

- Turn themes on and off
- Add and view image source data
- Change the order of the table of contents to change the map display
- Interpret the table of contents and legend

For more on geographic inquiry and these steps, see Geographic Inquiry: Thinking Geographically (pages xxi to xxiii).

Teacher notes

Lesson introduction

First, provide a brief overview of the region of East Asia and the Ring of Fire. Emphasize East Asia's dense population and its place in the Ring of Fire. Distribute the map handout to each student. The map contains outlines of countries, plate boundaries, and major cities in East Asia.

Ask your students to draw outlines of the areas where they believe there is the greatest risk for a major geophysical disaster. Give the students about three to five minutes to make their initial predictions. If time permits, ask some students to share their predictions with the class, or break out into small groups for discussion. Later in the lesson, students will be asked to make a hazards zone map of the Ring of Fire.

Student activity

 Before completing this lesson with students, we recommend that you work through it yourself. Doing so will allow you to modify the activity to accommodate the specific needs of your students.

After the initial discussion, students will work on the computer component of the lesson. Distribute the student GIS Investigation sheets, which provide step-by-step instructions for the activity. In this investigation, students will analyze several bits of data including information on population density, plate boundaries, volcanoes, and earthquake activity for the year 2000. They will identify areas of subduction along plate boundaries based on characteristics of these zones.

In addition to detailed instructions, the worksheets include questions to help students focus on key concepts. Some questions will have specific answers, while others will require creative thought. Students will also add information to their outline map of East Asia.

Are the students following each step?
- Are the students answering the questions as they work through the procedure?
- Are the students referring to their original maps and notes from class discussion?
- Do some students complete the exercise quickly? If yes, refer to the Extensions section for ideas on how to make the GIS Investigation more challenging.

Conclusion

After the GIS Investigation, lead a class discussion that compares student predictions and student findings after the exercise.

How closely did their predictions match what they learned?
- Compare student observations and questions generated by their investigations of the maps.
- Have students identify what data was most important in determining hazard zones for this region. One way to do this is to list types of data on the board and have the students rank them in order of importance.
- Discuss why some data was more helpful than others.
- Ask the students if there is other data that they think would be helpful to them and have them explain why.

Assessment   *Middle School: Highlights skills appropriate to grades 5 through 8*

Part 1 of the Life on the Edge middle school assessment asks students to create a Hazards Map of East Asia, identifying zones of high and low risk for volcanic and seismic events. In addition, students are required to create a legend for their unique map. Part 2 of the assessment requires students to analyze their maps when answering the four questions. An assessment rubric to evaluate student performance is provided at the end of the lesson.

High School: Highlights skills appropriate to grades 9 through 12

Part 1 of the Life on the Edge assessment for high school students asks them to create a Hazards Map of East Asia, identifying zones of low, medium, and high risk for volcanic and seismic events. In addition, students are required to create a detailed legend for their unique map. Part 2 of the assessment requires students to analyze their maps when answering a set of three questions. One of the questions asks students to develop an emergency action plan for a major city in a high-risk zone for a volcanic or seismic event. An assessment rubric to evaluate student performance is provided at the end of the lesson.

Extensions

- Students can select a city within East Asia and create a buffer zone around a volcano to determine the area of potential damage if there is a major eruption. They can then create a disaster plan for that city in the event that such an eruption occurs.

- Students can create a new theme that points out locations of natural hazards in their local area.

- Analyze historical data for the most notable earthquakes and volcanic eruptions and create new themes from this data. Students can then make predictions on the next significant eruption or seismic event for East Asia.

- Students can create their own digital maps for the assessment. By using ArcView, they can mark all of the required features and print the map using the Layout function.

- Students can present their maps to the class, comparing their initial hazards maps to their final theories.

- Check out the Resources by Module section of the Teacher Resource CD for print and media resources on the topics of earthquakes, volcanoes, plate tectonics, and East Asia or visit *www.esri.com/mappingourworld* for Internet links.

NAME _____   DATE _____

Life on the Edge
A map investigation

East Asia sits on the western edge of the Ring of Fire. This region experiences an enormous amount of geophysical activity both seismic and volcanic in nature. On the outline map of East Asia, use your colored pencils to outline the areas where you think the greatest risk of volcanic or earthquake disasters is. The map includes country borders, some major cities, and thicker outlines that indicate plate boundaries.

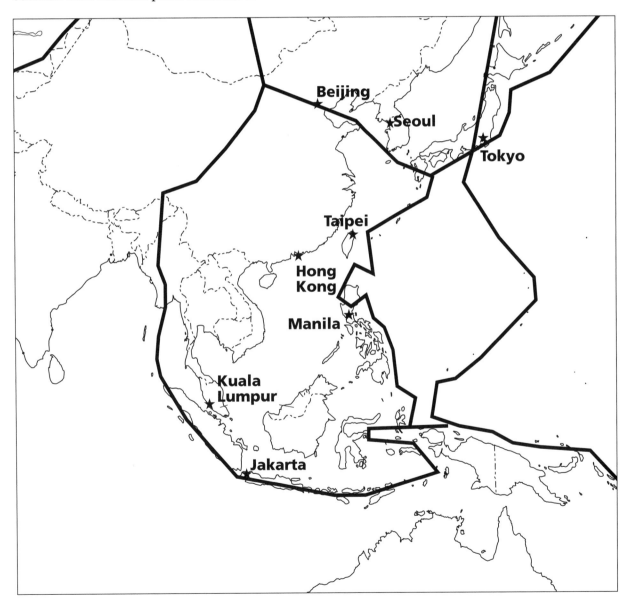

MODULE 2 • PHYSICAL GEOGRAPHY I: LANDFORMS AND PHYSICAL PROCESSES

NAME _____   DATE _____

Life on the Edge
A GIS investigation

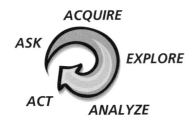

Answer all questions on the student answer sheet handout

In this investigation, you will investigate the Pacific Ocean's "Ring of Fire," with particular focus on earthquake and volcanic activity in East Asia, where millions of people live with the daily threat of significant seismic or volcanic events.

Step 1   Start ArcView

   a   Double-click the ArcView icon on your computer's desktop.

   b   If the Welcome to ArcView dialog appears (pictured below), click **Open an Existing Project** and click OK. If it doesn't appear, proceed to step 2.

Step 2   Open the region2.apr file and look at cities data

   a   In this exercise, a project file has been created for you. To open the file, click on the **File** menu and choose **Open Project**.

   b   Navigate to the exercise directory for this module (**C:\esri\mapworld\mod2 \region2.apr**) and choose **region2.apr** from the list.

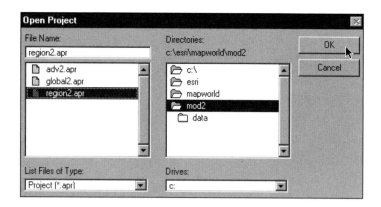

c   Click OK.

When the project opens, you see a map in the View window with three themes turned on (Major Cities, Countries, and Ocean). The check mark next to the theme name tells you the theme is turned on and visible in the view.

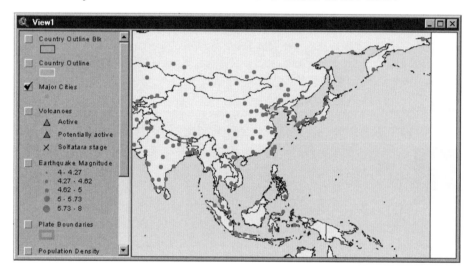

In order to identify the name of each city marked on the map, you can use the Identify tool.

d   Click Major Cities in the table of contents. A raised box will appear around the theme name. This indicates that the theme is active.

 e   Click the Identify tool. Notice how the cursor has a small "i" next to it when it's over the view area. Click on any green dot in the view.

An Identify Results dialog appears that is similar to the graphic below:

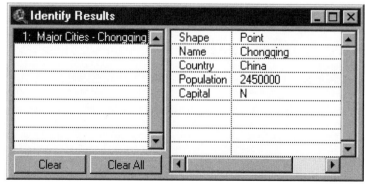

> *The right side of the Identify Results window tells you the name of the*
> *city, the country it's in, its population, and whether or not it's the*
> *capital city. In the example above, N = No (Chongqing is not the*
> *capital of China).*

? **f** *Use the Identify tool to locate one city within India and Japan. Record each city's*
name and population in the table on your answer sheet.

g When you are finished, close the Identify Results window by clicking the ✕ in the
upper right corner of the window.

Step 3 Look at population density

a Turn on the Population Density theme by clicking in the box next to its name.

A check mark appears and the theme is drawn in the view. The darker areas on the
map represent areas of high population density. In this theme, density is measured
by the number of people per square kilometer of space.

b Scroll down the table of contents so you can see the legend for the Population
Density theme.

c Make sure the Major Cities theme is still active.

d Use the Identify tool to locate two world cities in East Asia in areas where the population density is greater than 200 people per square kilometer.

? *Record the names of these two cities.*

Step 4 Look at earthquake magnitudes

a Turn on the Earthquake Magnitude theme.

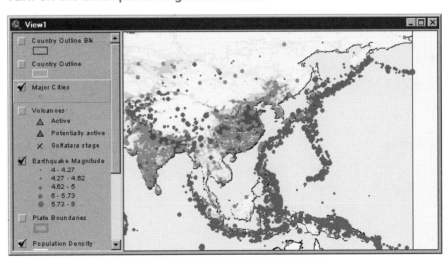

This theme adds a layer of graduated points to the map and displays earthquake
locations according to magnitude. The larger dots represent greater earthquake
magnitudes.

> *Note: Only earthquakes with a magnitude greater than 4.0 on the*
> *Richter scale are included in this theme.*

?   *(1)   Where did the largest earthquakes occur?*

?   *(2)   Did large earthquakes occur near densely populated areas? Where?*

Step 5   Measure the distance between active volcanoes and nearby major cities

a   Turn off the Earthquake Magnitude theme.

b   Turn on the Volcanoes theme.

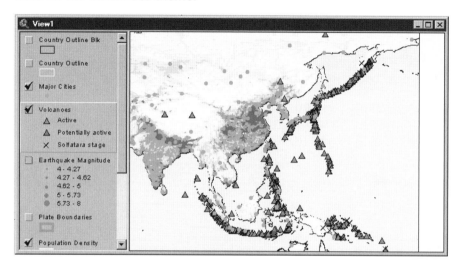

There are three types of volcanoes identified by this theme: active, potentially active, and solfatara stage (venting primarily hot gases). Note the different symbols used for each type of volcano.

c   Choose a major city that's located near an active volcano. Click the Zoom In tool and then click your chosen city's dot in the view. Your view is now centered on that city.

You will use the Measure tool to measure the distance between the active volcano and the major city you selected.

d   Click the Measure tool.

e   Click the volcano that is close to your chosen major city. Now drag your mouse pointer over to the city. A line is attached from the point where you first clicked as you move your mouse.

f   Once you have reached the city, **double-click** to end the line.

In the lower left corner of your view, you see the distance you measured.

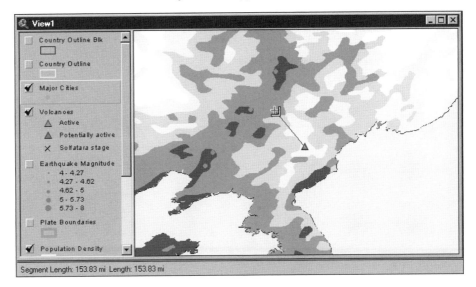

 g Use the Measure tool to measure the distance from other cities to nearby active volcanoes.

? *(1)* *Are there many active volcanoes located close to highly populated areas? What is the closest distance you found? Record the name of the volcano and the city and their distance apart.*

? *(2)* *What patterns do you see in the volcano points and how do they compare with the earthquake patterns? (Hint: You should turn the Earthquake Magnitude theme on and off.)*

h Make Study Area the active theme.

 i Click the Zoom to Active Theme button. This takes you back to the view of East Asia.

Step 6 Look at plate boundaries

a Turn off all themes except Countries and Oceans. Turn on the Plate Boundaries theme and make it the active theme by clicking on its name in the table of contents.

Now you will label the plate boundaries in the East Asia region.

 b Click the Label tool.

c In the view, click the inside of one of the plates. A label with the plate name appears on the map.

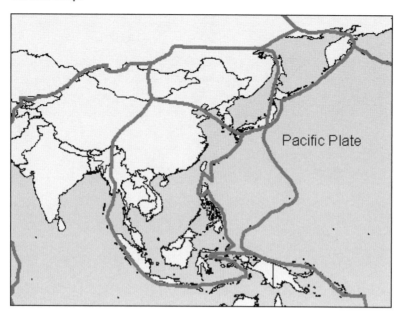

Pacific Plate

d Label all of the plates in East Asia.

e Record the labels on your paper outline map.

f To turn off the Label tool, click the Pointer tool.

Many of the plate boundaries in the Pacific Rim have areas called subduction zones. These are places along plate boundaries where one plate is diving underneath another one. Subduction zones can be identified by underwater trenches and island arcs that are formed at these boundaries.

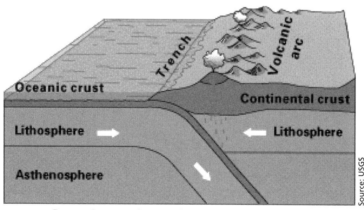

Oceanic-continental convergence

In order to get a closer look at landforms and boundaries, you will add the wsiearth.jpg theme. This is a composite satellite image of the earth. It will allow you to see the detailed physical features of the earth.

Step 7   Add an image file

a   Click the Add Theme button and navigate to your data folder (**C:\esri\mapworld \mod2\data**). You notice that the wsiearth.jpg is not listed.

b   Click the arrow to change the Data Source Type to Image Data Source.

c   Select **wsiearth.jpg** and click OK to add it to your view. Turn on wsiearth.jpg.

d   It displays on top of the other themes because it is at the top of the table of contents. Drag the Plate Boundaries theme up to the top of the table of contents so you can see it too.

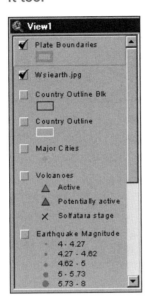

e   Look in the view to determine where subduction is occurring. Remember: Subduction zones are characterized by deep trenches and volcanic island arcs, and occur along plate boundaries.

> *Hint: You may need to turn the Plate Boundaries theme on and off.*

f   On your paper map, draw the zones of subduction.

Step 8   Investigate your map

a   Use the Zoom and Pan tools to focus on different areas of the map to get a closer look at all the physical features.

Here is a brief explanation of the Zoom and Pan tools:

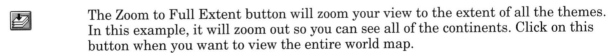

The Zoom to Full Extent button will zoom your view to the extent of all the themes. In this example, it will zoom out so you can see all of the continents. Click on this button when you want to view the entire world map.

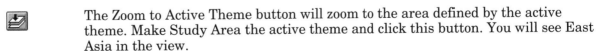

The Zoom to Active Theme button will zoom to the area defined by the active theme. Make Study Area the active theme and click this button. You will see East Asia in the view.

The Zoom In tool will allow you to drag a box around the area you want to zoom in on. You can also click the Zoom In tool and then click on the map to get a closer look. This is a very useful tool when you want to see more detail in the map.

The Zoom Out tool will allow you to drag a box around the area you want to zoom out from. You click the Zoom Out tool and then click on the map to see a bigger area.

The Zoom In to Extent button zooms in to the center of the view.

The Zoom Out to Extent button zooms out from the center of the view.

The Pan tool repositions the map. Use the Pan tool by clicking the spot you'd like to see and dragging it to the center of the display. The view will redraw itself around the area you panned.

b Use the Identify tool to find out the names of volcanoes or cities near them. To use the tool, make sure the theme you want to identify is active, click the Identify tool, and then click an object in the view.

Remember: The map is drawn beginning with the theme listed at the bottom of the table of contents first and ending with the theme listed at the top. Move the theme layers around in the table of contents by clicking and dragging to see all of the themes display.

c Check with your teacher if there is an area of East Asia that you need to explore. When you're finished exploring the map, proceed to step 9.

Step 9   Exit ArcView

In this GIS investigation, you analyzed earthquake and volcano data specific to East Asia. You identified major cities with high population densities and measured their distance to active volcanoes. You worked with an image file and plate boundary data to determine where zones of subduction occur in East Asia.

a Ask your teacher for instructions on where to save this ArcView project and on how to rename the project.

b If you are not going to save the project, exit ArcView by choosing Exit from the File menu. Click No when you are asked if you want to save changes to region2.apr.

NAME _____   DATE _____

Student answer sheet

Module 2
Physical Geography I: Landforms and Physical Processes

Regional case study: Life on the Edge

Step 2   Open the region2.apr file and look at cities data

f   Use the Identify tool to locate one city within each country listed in the table below and record that city's population.

CITY NAME	COUNTRY NAME	CITY POPULATION
Kunming	China	1,280,000
	India	
	Japan	

Step 3   Look at population density

d   Use the Identify tool to locate two world cities in East Asia in areas where the population density is greater than 200 people per square kilometer. Record below.

WORLD CITIES THAT HAVE A POPULATION DENSITY GREATER THAN 200 PEOPLE PER SQUARE KILOMETER

Step 4   Look at earthquake magnitudes

a-1   Where did the largest earthquakes occur?

_____

_____

_____

_____

_____

a-2   Did large earthquakes occur near densely populated areas? Where?

_____

_____

_____

_____

_____

Step 5   Measure the distance between active volcanoes and nearby major cities

g-1   Are there many active volcanoes located close to highly populated areas? What is the closest distance you found? Record the name of the volcano and the city, and their distance apart.

_____

_____

_____

_____

g-2   What patterns do you see in the volcano points and how do they compare with the earthquake patterns?

_____

_____

_____

_____

_____

NAME _____   DATE _____

Life on the Edge
Outline map of East Asia

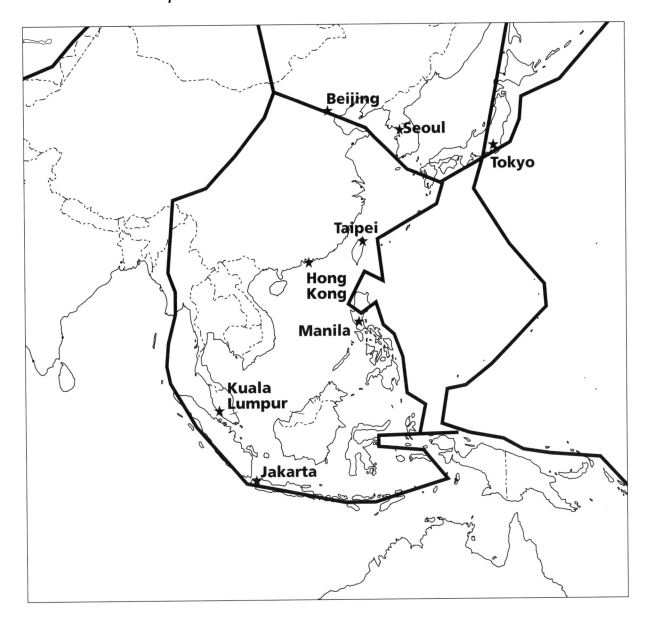

NAME _____   DATE _____

Life on the Edge
Middle school assessment

Part 1: Create a hazards map

On the paper outline map provided, do the following three things:
1   Mark and identify zones at high and low risk for volcanic activity.
2   Mark and identify zones at high and low risk for seismic activity.
3   Create a map legend that identifies the four zones described above.

You may use the ArcView project and class notes to help you create this map and determine zones of high and low risk. In the next section, you will be asked to explain why you chose the locations you did. Be sure to think carefully about your cartographic choices.

   Hint: You can use the Find tool in ArcView to quickly locate different world cities.

Part 2: Map analysis

On a separate piece of paper, answer each question below in paragraph form:
1   What criteria did you use to define your hazard zones for volcanic and earthquake activities?
2   Describe the relationships you see between tectonic plate boundaries and areas at high risk for volcanic and seismic activity.
3   Describe the landforms that are located in and around the Pacific Rim. Identify their common characteristics.
4   Choose a city in one of the high-risk zones. Develop an emergency action plan for that city. Be sure to take into consideration the population of the city and possible evacuation plans.

Life on the Edge

Assessment rubric

Middle school

STANDARD	EXEMPLARY	MASTERY	INTRODUCTORY	DOES NOT MEET REQUIREMENTS
The student knows and understands how to use maps and databases to analyze spatial distributions and patterns.	Uses GIS to analyze volcanic and earthquake data to create an original hazards map, identifying zones of high and low volcanic and seismic activity.	Uses GIS to analyze volcanic and earthquake data to identify zones of significant seismic and volcanic activity and compares this with his or her original findings.	Uses GIS to identify the locations of major volcanoes and earthquakes.	Has difficulty correctly identifying major volcanoes and earthquakes.
The student knows and understands how to analyze the physical characteristics of places.	Describes in detail the characteristics of the Pacific Rim landforms and identifies the physical processes that formed them.	Clearly describes the characteristics of landforms of the Pacific Rim.	Describes the characteristics of the Pacific Rim landforms using little detail.	Identifies landforms of the Pacific Rim, but does not provide any specific characteristics.
The student knows and understands how physical processes shape patterns in the physical environment.	Clearly describes the relationship between zones of high earthquake and volcanic activity and the location of tectonic plate boundaries through the use of a variety of media.	Describes the relationship between zones of high earthquake and volcanic activity and the location of tectonic plates.	Provides limited evidence of the relationship between zones of high earthquake and volcanic activity and the location of tectonic plate boundaries.	Does not show evidence of understanding the relationship between zones of high earthquake and volcanic activity and the location of tectonic plates.
The student knows and understands how natural hazards affect human activities.	Using a variety of media, the student illustrates use of higher-order thinking to determine areas at high risk of human loss due to geographic hazards in East Asia.	Uses higher-order thinking to determine areas at high risk of human loss due to geographic hazards in East Asia, and provides a clear explanation.	Identifies areas of risk for possible human loss, but does not clearly explain the relationship between densely populated areas and geographic hazards of East Asia.	Identifies areas at risk for loss of human life in the region of East Asia, but does not provide explanation for his or her decision.

This is a four-point rubric based on the National Standards for Geographic Education. The "Mastery" level meets the target objective for grades 5–8.

NAME _____   DATE _____

Life on the Edge
High school assessment

Part 1: Create a hazards map

On the paper outline map provided, do the following four things:

1 Mark and identify zones of high, medium, and low risk for volcanic activity.
2 Mark and identify zones of high, medium, and low risk for seismic activity.
3 Create a complete map legend.
4 Include all major cities, plate boundaries, and subduction zones.

Before you draw your map, you need to determine the criteria for each level of risk (low, medium, and high). Use the following items to determine the different levels of risk:

1 Population density
2 Magnitude of earthquakes from year 2000 data
3 Location of active volcanoes and their proximity to major population centers
4 Location of plate boundaries and subduction zones

You may use the ArcView project and class notes to help you create this map and determine zones of high and low risk. In the next section, you will be asked to explain why you chose the locations you did. Be sure to think carefully about your cartographic choices.

 Hint: You can use the Find tool in ArcView to quickly locate different world cities.

Part 2: Map analysis

On a separate piece of paper, answer each question below in paragraph form:

1 Define each level of hazard risk (high, medium, and low) you used in your map. Be sure to include each factor used to determine the different levels of risk.
2 Describe the relationships you observe between tectonic plate boundaries and areas at high risk for volcanic and seismic activity.
3 Explain how subduction zones affect the physical features of at least three places in the Pacific Rim.
4 Choose a major city in a high-risk zone. Develop an emergency action plan for the city. Be sure to take into consideration the population of the city and possible evacuation plans.

Life on the Edge

Assessment rubric

High school

STANDARD	EXEMPLARY	MASTERY	INTRODUCTORY	DOES NOT MEET REQUIREMENTS
The student knows and understands how to use geographic representations and tools to analyze and explain geographic problems.	Uses GIS to analyze volcanic and earthquake data to identify five or more world cities at high risk for volcanic or earthquake activity, and provides a detailed explanation of how he/she came to these conclusions.	Uses GIS to analyze volcanic and earthquake data to create an original hazards map of zones of high, medium, and low risk for volcanic and seismic activity and provides a brief narrative stating the logic of their choice.	Uses GIS to analyze volcanic and earthquake data to create an original hazards map of zones of high, medium, and low risk for volcanic and seismic activity and provides little explanation of how the map was created.	Uses GIS to analyze volcanic and earthquake data, but does not correctly identify zones of various levels of activity.
The student knows and understands the spatial variation in the consequences of physical processes across Earth's surface.	Clearly explains the correlation between the spatial distribution of volcanoes and earthquakes in relation to plate boundaries, and incorporates it into their understanding of physical landforms.	Explains the correlation between the spatial distribution of volcanoes and earthquakes in relation to plate boundaries.	Identifies the correlation between volcanoes, earthquakes, and plate boundaries, but has difficulty explaining the relationship.	Identifies the relationship between the spatial distribution of volcanoes and earthquakes, but does not correlate them to the location of plate boundaries.
The student knows and understands the changing physical and human characteristics of places.	Correctly identifies all areas of subduction in East Asia and explains in detail through words and images how this affects the physical aspects of at least three places in the region.	Correctly identifies all areas of subduction in East Asia, and explains how this affects the physical aspects of three places in the region.	Correctly identifies all areas of subduction in East Asia and explains how this affects the physical aspects of two to three places in the region.	Correctly identifies a few areas of subduction in East Asia and does not explain their effects on the physical environment.
The student knows and understands how humans perceive and react to natural disasters.	Using a variety of media, illustrates use of analytical thinking to formulate a thorough emergency action plan for a major city at high risk for a seismic event.	Uses analytical thinking to formulate an emergency action plan for a major city at high risk for a seismic event.	Identifies a major city at high risk for a seismic event and formulates an emergency action plan that lacks detail and evidence of higher-order thinking.	Identifies a major city at high risk for a seismic event, but does not include an emergency action plan.

This is a four-point rubric based on the National Standards for Geographic Education. The "Mastery" level meets the target objective for grades 9–12.

Mapping Tectonic Hot Spots
An advanced investigation

Lesson overview

This lesson is designed to target the self-motivated, independent GIS students in your class. They will use the Internet to acquire the most recent data on earthquakes and volcanoes worldwide. By exploring this data through a GIS, students will construct a current World Tectonic Hot Spots map. This lesson could be an independent research project completed over several days.

Estimated time   Three to four 45-minute class periods

Materials   ✔ Student handouts from this lesson to be copied:
- Volcano data sheet (page 106)
- GIS Investigation sheets (pages 107 to 111)

Standards and objectives   *National geography standards*

GEOGRAPHY STANDARD	MIDDLE SCHOOL	HIGH SCHOOL
1 How to use maps and other geographic representations, tools, and technologies to acquire, process, and report information from a spatial perspective	The student knows and understands how to use maps and databases to analyze spatial distributions and patterns.	The student knows and understands how to use technologies to represent and interpret Earth's physical systems.
7 The physical processes that shape the patterns of Earth's surface	The student knows and understands how physical processes shape patterns in the physical environment.	The student knows and understands the spatial variation in the consequences of physical processes across Earth's surface.
15 How physical systems affect human systems	The student knows and understands how natural hazards affect human activities.	The student knows and understands how humans perceive and react to natural disasters.
18 How to apply geography to interpret the present and plan for the future	The student knows and understands how the interaction of physical and human systems may shape present and future conditions on Earth.	The student knows and understands how to use geographic perspectives to analyze problems and make decisions.

Objectives

The student is able to:

- Use the Internet to locate data for recent earthquake and volcanic activity.
- Use a GIS to map data acquired from the Internet.
- Compare the pattern of recent earthquake and volcanic events with the patterns of seismic events in the past.
- Identify world regions and cities where human populations face the greatest threat from current seismic events.

GIS skills and tools

- Locate data on the Internet and prepare it in a comma-delimited text file
- Add tables to a project
- Add event themes
- Convert event themes to shapefiles
- Classify data according to different specifications
- Create a layout
- Print a presentation-style map

For more on geographic inquiry and these steps, see Geographic Inquiry: Thinking Geographically (pages xxi to xxiii).

Teacher notes

Lesson introduction

Lead a class discussion that addresses the following questions:

- Are you aware of any earthquake or volcanic activity on the earth within the last month?
- Write down the number of earthquakes you believe occurred throughout the world or in a specific area within the last month.

Explain to your students that they will use the Internet to find data on the most recent seismic events and they will use a GIS to map that data. This exercise will require the student to work independently, with little guidance.

Student activity

 Before completing this lesson with students, we recommend that you complete it as well. Doing so will allow you to modify the activity to accommodate the specific needs of your students.

Distribute this lesson's student GIS Investigation sheets, along with the volcano data sheet handout. Students should follow the guidelines to retrieve earthquake and volcano data from the Internet. They will create a text file of that data, introduce that data into ArcView, and create point themes locating recent earthquake and volcano events. Once the data is in ArcView, your students will be able to perform different analyses on it.

Conclusion

Before beginning the assessment, engage students in a discussion of patterns of recent seismic activity revealed by their maps. Do the mapped events occur along known plate boundaries? Are some boundary zones more active at the present time than others? Are any patterns revealed by the current data? Where is future research and analysis needed?

Assessment

Students will create a layout titled "Tectonic Hot Spots, (Month, Year)." For example, Tectonic Hot Spots, June 2001 would be an appropriate title for data collected during June 2001. In addition to the map, you may wish to ask students to submit a written report that summarizes their observations and explains their analysis of the data. A written report could include any or all of the following points:

- Briefly summarize the earthquake and volcano activity for the thirty-day period.
- Were you surprised at the frequency or the location of tectonic activity in the past month? Why or why not?
- Compare the earthquake and volcano activity in the thirty-day period with the earthquakes 2000 data from the Earthquakes theme. What relationships can be observed between these two data sets?

 Note: Due to the independent nature of this lesson, there is no supplied assessment rubric. You are free to design assessment rubrics that meet the needs of your specific adaptation.

Extensions

- Have the students investigate their local area to find out if there were ever any volcanic or seismic activities in their hometown. Identify how these events (even if they were thousands of years ago) helped to shape the landscape of their area.
- Find the closest active volcano and/or closest fault line to your town. How does this affect your community and the communities that are closest to the volcano/fault line?
- Conduct research into current films and literature to find books and movies that deal with volcanoes and/or earthquakes. Do these films paint a realistic picture based on the research the students have conducted?
- Check out the Resources by Module section of the Teacher Resource CD for print and media resources on the topics of earthquakes, volcanoes, and plate tectonics or visit *www.esri.com/mappingourworld* for Internet links.

NAME _____

DATE _____

Mapping Tectonic Hot Spots

Volcano data sheet

VOLCANO	COUNTRY	DATE OF ACTIVITY	LATITUDE (DD)	LONGITUDE (DD)

NAME _____ DATE _____

Mapping Tectonic Hot Spots
An advanced investigation

 Note: Due to the dynamic nature of the Internet, the URLs listed in this lesson may have changed, and the graphics shown below may be out of date. If the URLs do not work, refer to this book's Web site for an updated link: www.esri.com/mappingourworld.

ACQUIRE
ASK
EXPLORE
ACT
ANALYZE

In this investigation, you will use the Internet to acquire the most recent data on earthquakes and volcanoes worldwide. By exploring this data through a GIS, you will construct a current World Tectonic Hot Spots map.

Step 1 **Locate earthquake data on the World Wide Web**

The World Wide Web is a great source for geographic data. In this step, you will go to the U.S. Geological Survey (USGS) Web site to find earthquake data.

a Open your Web browser and go to the USGS National Earthquake Information Center at gldss7.cr.usgs.gov/neis/epic/epic.html.

b For Search area, select **Global (Worldwide)**.

c For Output File Type, select **Spreadsheet format (comma delimited)**.

d For Search Parameters, select **USGS/NEIC (PDE) 1973 - Present**.

e For Optional Search Parameters, type the range of dates for the preceding 30 days. (For example, if today is July 11, request data from 2001/6/11 to 2001/7/11). Set the Minimum Magnitude to 5 and the Maximum Magnitude to 9.

Optional Search Parameters:

Date
2001 Starting Year 6 Starting Month 11 Starting Day
2001 Ending Year 7 Ending Month 11 Ending Day

Magnitude
5 Minimum Magnitude 9 Maximum Magnitude

Depth
___ Minimum Depth ___ Maximum Depth

Intensity
___ Minimum Intensity ___ Maximum Intensity

Submit Search   Clear form

f Do not type any values for Depth or Intensity. Click Submit Search.

Step 2 Prepare data for ArcView

a   Use your mouse to highlight and select all the data shown. Be sure to include the field names, but do not include other text on the Web page.

b   From the Edit menu, select Copy.

c   Open a text editor like WordPad or Notepad. From the Edit menu, choose Paste.

For example:

```
Quake01.txt - Notepad                               _ □ ×
File  Edit  Search  Help
Year,Month,Day,Time(hhmmss.mm)UTC,Latitude,Longitude,Magnitude,Depth
 2001,06,11,004209.73, -6.55, 154.99,5.2, 33
 2001,06,11,121847.21,  3.30, 127.13,5.0, 33
 2001,06,12,142215.92, 14.54, -45.03,5.2, 10
 2001,06,12,224127.88, 44.19, 148.58,5.4, 33
 2001,06,13,034929.02,-18.83,-173.35,5.3, 33
 2001,06,13,041046.33,-27.21, -66.50,5.2, 16
 2001,06,13,131755.15, 24.46, 122.40,5.6, 77
 2001,06,13,222703.71, 23.89, 123.42,5.0, 33
 2001,06,14,023525.81, 24.51, 122.03,5.9, 32
 2001,06,14,035532.20,-19.04,-173.32,5.3, 33
 2001,06,14,122704.43,-22.05,-179.46,5.7,604
 2001,06,14,133926.88, 51.19,-179.85,5.4, 33
```

d   Delete the "(hhmmss.mm)UTC" section of the Time column heading. Shorten Latitude to **Lat**, Longitude to **Long**, and Magnitude to **Mag**.

e   Ask your teacher where to save this document. Save it as a text file and name it **Quake01.txt**. Exit the text editor program.

Step 3 Start ArcView and add your data

a   Start ArcView.

b   Open the **adv2.apr** project from your exercise directory (**C:\esri\mapworld\mod2 \adv2.apr**).

When the project opens, you see the map in the view. Two themes are turned on (Continent Outline and Ocean).

 c   From the Project window, click the Tables icon.

d   Click the Add button to open the Add Table dialog box. In the lower left, change List Files of Type to **Delimited Text (*.txt)**. Navigate to the location of **quake01.txt** and click OK.

You will use the Latitude and Longitude coordinates in this table to create a new theme of point features that indicate the location of the earthquakes.

e   Close the quake01 table and make the view active.

Step 4   Add an Event Theme

a   From the View menu, choose Add Event Theme. The Add Event Theme dialog box appears.

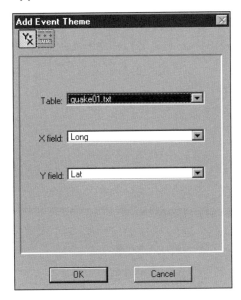

b   Confirm that the Table is quake01.txt, the X Field is Long, and the Y Field is Lat.

c   Click OK to create a new theme from the data in your quake01.txt table. Quake01.txt appears in the table of contents.

d   Turn the theme on to see where these recent earthquakes have occurred. Because ArcView randomly selects a symbol color, the theme may be difficult to see.

e   Change the symbol color to see the points better.

f   From the Theme menu, click Convert to Shapefile.

g   Save it to a temporary location (e.g., C:\Temp directory or a personal folder) and name it according to the dates of the earthquake data. For example, if data covered June 11–July 11, 2001, name the shapefile **Qk61101**. Delete Quake01.txt from the view.

Step 5   Save your work

a   Save your project. Ask your teacher where you should save your work and how you should name the project. It is recommended that you save your project as **ABadv2.apr** where your initials are AB.

Step 6   Locate and format volcano data

You will now follow a different procedure to collect and map volcano data.

a   Open your Web browser and go to the Weekly Volcanic Activity Report produced by the Smithsonian and USGS at www.nmnh.si.edu/gvp/usgs.

b   Open the links for all New Activity/Unrest and Ongoing Activity. Example:

New Activity/Unrest: | <u>Loihi Seamount</u>, USA |
Ongoing Activity: | <u>Etna</u>, Italy | <u>Kilauea</u>, USA | <u>Pinatubo</u>, Philippines | <u>Popocatépetl</u>, México | <u>Shiveluch</u>, Russia | <u>Soufrière Hills</u>, Montserrat | <u>Tungurahua</u>, Ecuador |

c Record the data on your volcano data sheet.

d Open Notepad or another text editor program. Use the information on your volcano data sheet to create a comma-delimited file.

> *Note: When creating this file, be sure to eliminate all spaces. Also, convert all south and west latitudes and longitudes to negative numbers. See example for guidelines:*

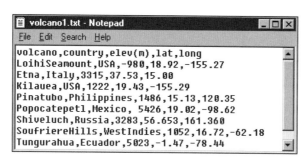

```
volcano,country,elev(m),lat,long
LoihiSeamount,USA,-980,18.92,-155.27
Etna,Italy,3315,37.53,15.00
Kilauea,USA,1222,19.43,-155.29
Pinatubo,Philippines,1486,15.13,120.35
Popocatepetl,Mexico, 5426,19.02,-98.62
Shiveluch,Russia,3283,56.653,161.360
SoufriereHills,WestIndies,1052,16.72,-62.18
Tungurahua,Ecuador,5023,-1.47,-78.44
```

e Ask your teacher where to save the file. Save it as a delimited file and name it **volcano01.txt**.

f Follow the procedures outlined for the earthquake data to create a volcano01.txt point theme on your map. You will need to select volcano01.txt in the Add Event Theme dialog box.

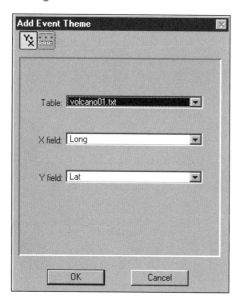

g Convert the volcano01.txt event theme to a shapefile and delete the event theme from the project.

h Save your project according to your teacher's instructions. Be sure to keep the name you assigned earlier in the investigation.

Step 7   **Explore the Data**

a   Explore the data you have mapped by classifying the data points in a variety of ways: earthquake magnitude, earthquake depth, date of most recent volcanic eruption, and so on. Your teacher may provide additional patterns to analyze.

As you explore the data, consider the following questions:

- Do the mapped events occur along known plate boundaries? Where?
- Are some boundary zones more active at the present time than others? Which ones?
- Where did the largest earthquakes occur?
- Are any of the volcanoes currently showing activity near cities or located in areas with high population density? Identify them.

Step 8   **Map your observations and save the project**

a   Create a layout and title it **Tectonic Hot Spots, (Month, Year)**. For example, Tectonic Hot Spots, June 2001 would be an appropriate title for a map with data from June 2001.

Remember to include the following items in your layout:
- Title
- Legend
- Map scale
- North arrow
- Author of map
- Date of map creation
- Data sources

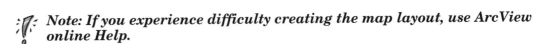 *Note: If you experience difficulty creating the map layout, use ArcView online Help.*

b   Print your map.

c   Save your project according to your teacher's instructions.

Physical Geography II

Ecosystems, Climate, and Vegetation

Physical Geography II explores the variety of characteristics that influence climate. These include latitude, elevation, landforms, proximity to ocean, and the El Niño and La Niña events.

Running Hot and Cold: A global perspective

Students will explore characteristics of the earth's tropical, temperate, and polar zones by analyzing monthly and annual temperature patterns in cities around the world. In the course of their investigation, students will observe temperature patterns associated with changes in latitude as well as differences caused by factors such as elevation and proximity to the ocean.

Seasonal Differences: A regional case study of South Asia

Students will observe patterns of monsoon rainfall in South Asia and analyze the relationship of those patterns to the region's physical features. The consequences of monsoon season on human life will be explored by studying South Asian agricultural practices and patterns of population distribution.

Sibling Rivalry: An advanced investigation

Students will study climatic phenomena El Niño and La Niña by downloading map images from the GLOBE (Global Learning and Observations to Benefit the Environment) Web site. They will incorporate these images into an ArcView project and identify patterns and characteristics of these phenomena, then assess the impact these anomalies have on the global and local environment.

Running Hot and Cold
A global perspective

Lesson overview

Students will explore characteristics of the earth's tropical, temperate, and polar zones by analyzing monthly and annual temperature patterns in cities around the world. In the course of their investigation, students will observe temperature patterns associated with changes in latitude as well as differences caused by factors such as elevation and proximity to the ocean.

Estimated time　　Two to three 45-minute class periods

Materials　　✔ Colored pencils

✔ Student handouts from this lesson to be copied:
  - Student worksheet (page 119)
  - Running Hot and Cold outline map of the world (page 120)
  - GIS Investigation sheets (pages 121 to 130)
  - Answer sheet (pages 131 to 137)
  - Assessment(s) (pages 138 to 141)

Standards and objectives

National geography standards

GEOGRAPHY STANDARD	MIDDLE SCHOOL	HIGH SCHOOL
1　How to use maps and other geographic representations, tools, and technologies to acquire, process, and report information from a spatial perspective	The student understands how to use maps, charts, and databases to analyze spatial distributions and patterns.	The student understands how to use technologies to represent and interpret Earth's physical and human systems.
5　That people create regions to interpret Earth's complexity	The student understands the elements and types of regions.	The student understands the structure of regional systems.
7　The physical processes that shape the patterns of Earth's surface	The student understands how Earth–Sun relationships affect physical processes and patterns on Earth.	The student understands spatial variation in the consequences of physical processes across Earth's surface.

Standards and objectives (continued)

Objectives

The student is able to:

- Locate tropical, temperate, and polar zones.
- Describe the characteristic yearly and monthly temperature patterns in those zones.
- Describe the influence of latitude, elevation, and proximity to the ocean on yearly temperature patterns.
- Compare and explain monthly temperature patterns in the northern and southern hemispheres.

GIS skills and tools

 Label a feature in the view

 Add text to a view

 Identify attributes of a feature

 Add a theme to a view

 Select a feature

 Clear the selected features

 Pan the view

 Find a feature

 Open the attribute table

 Sort records in ascending order

 Sort records in descending order

 Promote selected records to the top of the table

 Select a text label in the view and move it

 Zoom to the full extent of the view

For more on geographic inquiry and these steps, see Geographic Inquiry: Thinking Geographically (pages xxi to xxiii).

Teacher notes

Lesson introduction

Begin the lesson by asking students to name places that they believe to be the coldest and hottest on the planet. Briefly compare their choices and the reasoning behind them. Distribute the Running Hot and Cold student worksheet. Working in pairs or small groups, students should identify the planet's three hottest cities in July and the three coldest cities in January. At the end of five minutes, each group should share their lists with the rest of the class. Use the blackboard or an overhead projector to tally the cities mentioned as each group reports. Based on the tally, circle the cities that were listed most often. Explain that they are going to do an activity that will explore temperature patterns in cities around the world. As they complete the GIS Investigation, they will have an opportunity to check their answers on this worksheet and reconsider them in view of what they learn.

Before beginning the computer activity, engage students in a discussion about the cities that are circled on the list.

* Why do you think this city is one of the coldest or hottest?
* What countries are these cities located in?
* Has anyone ever visited one of these cities?

Student activity

 Before completing this lesson with students, we recommend that you complete it as well. Doing so will allow you to modify the activity to accommodate the specific needs of your students.

After completing and discussing the Running Hot and Cold student worksheet, have your students work on the computer component of the lesson. Ideally, each student should be at an individual computer, but the lesson can be modified to accommodate a variety of instructional settings.

Distribute the GIS Investigation sheets to the students. Explain that in this activity they will use GIS to observe and analyze yearly and monthly temperature patterns in cities around the world. The worksheet will provide them with detailed instructions for their investigations. As they investigate, they will identify global and regional temperature variations and speculate on possible reasons for the patterns that they observe.

Teacher Tip: Step 8 asks students to rename their project and save it. Make sure you direct your students on how to most effectively rename their project and where to save it.

In addition to instructions, the handout includes questions to help students focus on key concepts.

Things to look for while the students are working on this activity:
* Are the students using a variety of tools?
* Are the students answering the questions as they work through the procedure?
* Are the students experiencing any difficulty navigating between windows in the project?

Conclusion   When the class has finished the GIS Investigation, lead a discussion that summarizes the conclusions they reached. Be sure to address latitude in the Northern Hemisphere, latitude in the Southern Hemisphere, proximity to ocean, and elevation as factors that influence temperature. After students have had an opportunity to share their conclusions, discuss the similarities and differences among the ideas presented. Allow students to question each other and clarify confusing or contradictory statements. Develop a consensus about how each factor influences temperature.

Assessment   In the middle and high school assessments, students will use the GIS Investigation to draw conclusions about the factors that influence temperature patterns. They will write an essay offering data and examples from the GIS Investigation that support these statements.

Middle school students will be required to make a paper or ArcView-generated map that illustrates the points of their essay. High school students will be required to use ArcView only to make their maps.

Extensions
* Collect additional temperature data for cities in one specific world region. Use that data to create a regional version of this project and analyze that data to create a regional temperature profile.
* Investigate the phenomena of global warming. Use the Internet to collect monthly temperature data for recent years in one or more of the cities included in the project. Compare actual recorded temperatures to average monthly temperatures to see if current temperatures are warmer than average. Compare changes in one region with global changes to see if there are differences.
* Collect rainfall data for the cities included in this project. Use this data in combination with the temperature data to create an ArcView layout that illustrates typical climate patterns for each of the climate types.
* Check out the Resources by Module section of the Teacher Resource CD for print and media resources on the topics of climate and global temperatures or visit *www.esri.com/mappingourworld* for Internet links.

NAME _____   DATE _____

Running Hot and Cold
A map investigation

Directions: In the spaces below, list the three cities on the map on the next page that you believe are the hottest in July and the three cities that you believe are the coldest in January.

Hottest in July:

_____

_____

_____

Coldest in January:

_____

_____

_____

Running Hot and Cold

Outline map of the world

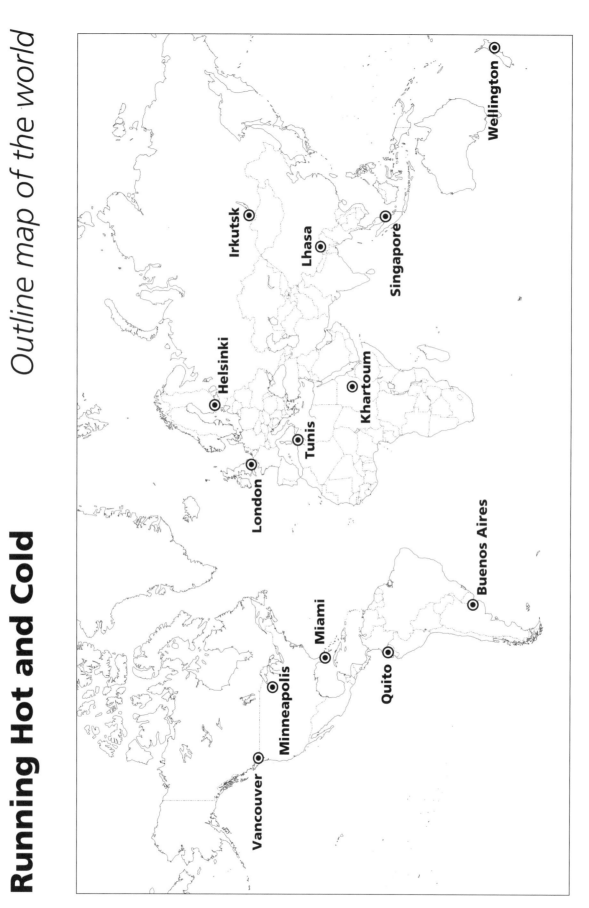

NAME _____   DATE _____

Running Hot and Cold
A GIS investigation

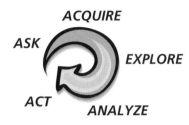

Answer all questions on the student answer sheet handout

In this activity, you will analyze monthly and annual temperature patterns in cities around the world. You will explore how latitude, elevation, and proximity to the ocean influence temperature patterns in the world's tropical, temperate, and polar zones.

Step 1   Start ArcView

a   Double-click the ArcView icon on your computer's desktop.

b   If the Welcome to ArcView dialog (pictured below) appears, click **Open an Existing Project** and click OK. If it doesn't appear, proceed to step 2.

Step 2   Open the global3.apr file

a   In this exercise, a project file has been created for you. To open the file, go to the **File** menu and choose **Open Project**.

b   Navigate to the exercise data directory (**C:\esri\mapworld\mod3\data**) and choose **global3.apr** from the list.

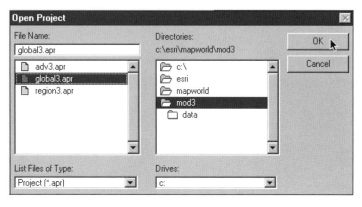

c   Click OK.

d When the project opens, you will see a world map displaying the average yearly temperature for 96 world cities.

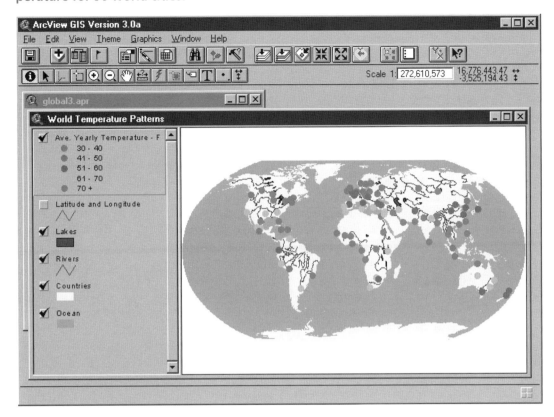

Step 3 Observe annual world temperature patterns

The symbols on the map represent cities around the world. The color of each symbol reflects an average of temperatures recorded throughout the year in that city.

a Look at the global temperature patterns displayed in the map.

b Are the temperatures listed as degrees Fahrenheit or degrees Celsius?

c Write three observations about the pattern of temperatures displayed on the map. Your observations should be global in scope, not focused on a specific country or city.

d Click the check mark next to the Ave. Yearly Temperature theme to turn it off.

Step 4 Label the latitude zones

a Turn on the Latitude and Longitude theme.

b From the Window menu, select Show Symbol Window.

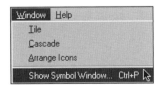

c Click the Font button and select Arial Black with a Size of 9.

d Close the Font Palette by clicking the ✕ in the upper right corner. Make Latitude and Longitude the active theme.

e From the Theme menu, click Auto-Label. The Auto-Label dialog displays.

f Click the box next to Allow Overlapping Labels.

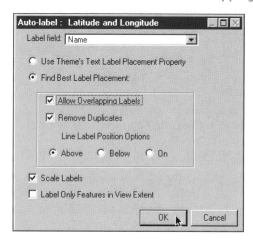

g Click OK.

All of the major latitude and longitude lines are labeled in the view. For the purposes of this exercise, the areas between the major latitude lines represent five zones of latitude. The table below names each latitude zone and the area it covers.

NAME OF LATITUDE ZONE	AREA IT COVERS IS BETWEEN THESE LATITUDES
North Polar Zone	Arctic Circle and North Pole
North Temperate Zone	Tropic of Cancer and Arctic Circle
Tropical Zone	Tropic of Cancer and Tropic of Capricorn
South Temperate Zone	Tropic of Capricorn and Antarctic Circle
South Polar Zone	Antarctic Circle and South Pole

Now you will label each of these zones on your map.

h   Click the Text tool.

i   Click the map somewhere between the Arctic Circle and the North Pole. The Text Properties dialog appears.

j   Type **North Polar Zone** in the space and click OK. The North Polar Zone is now labeled in the view.

k   Label each of the remaining zones using the same procedure you used to label the North Polar Zone.

l   From the Theme menu, select Remove Labels to remove all the latitude and longitude labels. Now you are left with the new latitude zones labels.

m   Click the Pointer tool. Click and drag any label in the view that you want to move. Move the labels so they cover ocean area and very little of the continents.

n   Turn on the Ave. Yearly Temperature theme and make it active. Observe the temperature patterns as they correspond to the latitude zones.

o   Click the Identify tool.

p   Use the Identify tool to find information on cities and complete the table on your answer sheet. Click on a specific city to obtain the necessary information.

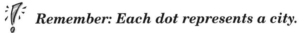   ***Remember: Each dot represents a city.***

(1)   Why do you think there aren't any cities in the North or South polar zones?

(2)   How is the North Temperate Zone different from the South Temperate Zone?

q   Turn off Ave. Yearly Temperature.

Step 5   Observe climate distribution

a   Click the Add Theme button.

b   Navigate to the module 3 data directory (**C:\esri\mapworld\mod3\data**). Scroll down, select **climate.shp**, and click OK.

The Climate.shp theme displays the regions of the world characterized by different types of climate.

c   Turn on Climate.shp.

d   Click and drag the Latitude and Longitude theme to the top of the table of contents. Move the Ave. Yearly Temperature theme to the top as well.

(1)   Complete the table on your answer sheet.

(2)   Which zone has the greatest number of climates?

e   Turn on the Avg. Yearly Temperature theme.

Give an example of a city in each of the climate zones listed in the answer sheet.

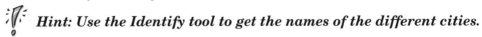

f   Ask your teacher for instructions on how to rename and save this project. Record the new name of the project and its new location on your answer sheet.

g Ask your teacher if you should exit ArcView now. Proceed to step 6b if you are continuing work.

h From the File menu, click Exit. When asked if you want to save changes, click No.

Step 6   Observe monthly temperature patterns in the Northern Hemisphere

a Start ArcView and navigate to the directory where you saved this project. Refer to your answer sheet for its name and location. Open the project.

b Close the World Temperatures Patterns view by clicking on the ✕ in the upper right corner of the view.

Now you see a chart titled Monthly Temperatures and a view titled World Cities. The chart shows the monthly temperatures for Boston. Notice that the symbol for Boston is highlighted in the view.

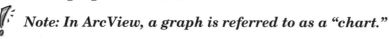 *Note: In ArcView, a graph is referred to as a "chart."*

 c Click the Select feature tool.

d Click on Miami at the southern tip of Florida.

Notice that both the map and the chart have changed.

? (1) *What does the chart show now?*

? (2) *What city is highlighted in the view?*

e Hold down the Shift key and select Boston again in the view.

? (1) *What does the chart show now?*

? (2) *What city or cities are highlighted in the view?*

f Use the Monthly Temperatures chart to compare the pattern of monthly temperatures in Miami to the pattern of monthly temperatures in Boston.

? *Complete the table on your answer sheet.*

g Hold down the Shift key and click on the city directly north of Boston.

? (1) *What is the name of the city?*

? (2) *How does its monthly temperature pattern differ from Boston's?*

h Hold down the Shift key and click on the closest city south of Miami.

? (1) *What is the name of the city?*

? (2) *How does its monthly temperature pattern differ from Miami's?*

? *i* List the name of each of the cities displayed in the chart and complete the information in the table on the answer sheet.

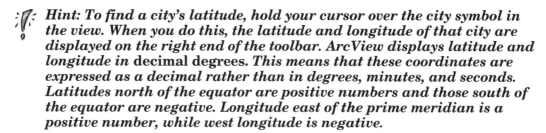 *Hint: To find a city's latitude, hold your cursor over the city symbol in the view. When you do this, the latitude and longitude of that city are displayed on the right end of the toolbar. ArcView displays latitude and longitude in decimal degrees. This means that these coordinates are expressed as a decimal rather than in degrees, minutes, and seconds. Latitudes north of the equator are positive numbers and those south of the equator are negative. Longitude east of the prime meridian is a positive number, while west longitude is negative.*

?   *j*   Based on the information displayed in the chart, the map, and the table on your answer sheet, state a hypothesis about how the monthly temperature patterns change as latitude increases.

Step 7   **Test your hypothesis**

   a   Click the Pan tool. Click the map and pan over to Western Europe.

   b   Click the Find tool. The Find Text in Attributes dialog displays.

    c   Type **Stockholm** in the text field and click OK. The view recenters around Stockholm and the city is highlighted yellow.

    d   Click the Select Feature tool. Hold down the Shift key and select three more European cities that are increasingly south of Stockholm. The city names appear in the chart and the cities are highlighted in the view.

?   *(1)*   *Complete the table on the answer sheet.*

?   *(2)*   *Do the cities you selected confirm or dispute your hypothesis? Explain.*

Step 8   **Analyze temperature patterns in the Southern Hemisphere**

You've already made a hypothesis about how latitude affects monthly temperature patterns in the Northern Hemisphere. Now you will explore the effect of latitude on the monthly temperature patterns within the Southern Hemisphere.

    a   Click the Pan tool and reposition your view so it is centered on Australia.

    b   Click the Find tool to locate the city of **Darwin**.

   c   Click the Select Feature button. Hold down the Shift key and click the three cities on Australia's eastern and southern coasts.

?   *Complete the table on your answer sheet.*

?   *d*   Based on the information displayed on the chart, the map, and the table you just completed, compare the monthly temperature patterns in the Southern Hemisphere to those in the Northern Hemisphere.

?   *Formulate a hypothesis about the relationship between monthly temperature patterns and increases in latitude.*

Step 9   **Test your hypothesis on how latitude affects monthly temperature patterns in the Southern Hemisphere**

    a   Use the Find tool to locate **Cape Town**, a city in Africa.

    b   Select two or three more African cities that are located between Cape Town and the equator.

?   *(1)*   *Complete the table on your answer sheet.*

?   *(2)*   *Based on your observations, do the cities you selected confirm or dispute your hypothesis about how latitude affects monthly temperature patterns in the Southern Hemisphere? Explain.*

c Click the Clear All Selected Features button and reposition your map so it's centered on North America.

d Ask your teacher if you should stop and exit ArcView. If you should stop, click Exit from the File menu. You should not save your changes. If you do not need to exit ArcView, proceed to the next step.

Step 10   Investigate the ocean's influence on temperature

In addition to latitude and hemisphere, a city's proximity to the ocean also influences its temperature. Now you will investigate how the ocean influences the air temperature of coastal cities.

a One at a time, select all the cities in Canada.

?   *(1) In which Canadian city would you experience the coldest winter temperatures?*

?   *(2) In which Canadian city would you experience the warmest winter temperatures?*

?   *(3) Looking at the map, why do you think the warmest city has temperatures that are so much warmer than the others in the winter? (Hint: How is this city different from all the others in terms of its location?)*

b Reposition your map so that it is centered on Western Europe.

c Make sure that Cities is the active theme and click the Open Theme Table button.

d Click the title bar for the Attributes of Cities table and move it to the left so you can still see the view.

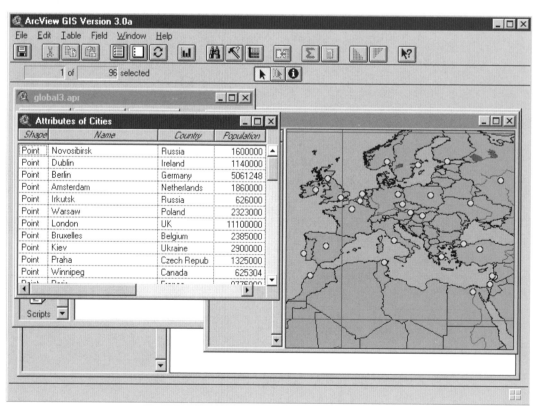

e Click the column heading called **Name**. It turns a darker gray and looks like a button that's been pushed in.

f   Click the Sort Ascending button to make the list alphabetical.

g   Scroll down the list, hold the Shift key down, and select the following cities: London, Amsterdam, Berlin, Warsaw, and Kiev. Take note of where each one is on the map as you select it.

All of the cities are highlighted in the table, in the view, and in the chart.

h   Click the chart and view title bars so the attribute table appears behind them. Analyze the chart and view.

(1)   Complete the table on your answer sheet.

(2)   What do these cities have in common as to their location on the earth?

(3)   Which cities have the mildest temperatures?

(4)   What happens to the winter temperatures as you move from London to Kiev?

(5)   Why do you think some cities have milder temperatures than the others?

i   Take a look at how you answered the questions about the Canadian city temperatures (step 10b).

Based on your observations of Canada and Western Europe, state a hypothesis about the influence of proximity to the ocean (or distance from it) on patterns of temperature.

Step 11   Investigate the impact of elevation on temperature patterns

Elevation of a city significantly affects the temperature of that city. In this step, you will investigate the relationship between elevation and temperature.

  a   Click the Zoom to Full Extent button.

b   Click the title bar for the Attributes of Cities table to bring it forward.

c   Hold the Shift key down and click in the table on the following cities to select them, taking note of where each one is on the map as you select it: Kisangani, Libreville, Quito, Singapore.

d   Make the chart and the view visible by clicking their title bars.

(1)   Complete the table on your answer sheet.

(2)   What do these cities have in common as to their location on the earth?

(3)   What temperature pattern do these four cities have in common?

(4)   How is Quito different from the other three?

(5)   Since all these cities are located on or very near the equator, what other factor could explain the difference in their temperature patterns?

e   Click the Attributes of Cities title bar.

  f   Click the Promote button. All the selected records are brought to the top of the table.

Analyze the attribute table and complete the table on your answer sheet.

g   Close the Attributes of Cities table.

h   Compare the elevation table to the chart on your computer.

?   *Based on your observation of temperatures along the equator and the information in the table, state a hypothesis about the influence of elevation on patterns of temperature.*

i   Clear all selections.

Step 12   Revisit your initial ideas

Before you began this GIS Investigation, you were asked to identify the three coldest cities in January and the three hottest cities in July from a map on which those cities were labeled.

a   Take out the paper map you used at the beginning of the GIS Investigation.

b   Select the 13 cities listed on your paper map in the Attributes of Cities table.

c   Click the first column heading named **J**.

d   Click the Sort Ascending button to sort the table from lowest to highest January temperatures.

e   Click the Promote button.

?   *f*   On your answer sheet, rank the 13 cities from lowest to highest according to their average January temperatures.

g   Click the third column heading named **J**. (This is for July.)

h   Click Sort Descending to sort July temperatures from highest to lowest.

i   Click the Promote button.

?   *j*   On your answer sheet, rank the 13 cities from highest to lowest according to their average July temperatures.

k   Compare your original predictions from your paper map with the correct answers on your answer sheet.

?   *l*   On your answer sheet, put a check mark (✔) next to those answers that you predicted correctly.

m   Clear the selections in the table. Close the theme table and return to the view.

       There are no cities selected in the view at this time.

Step 13   Exit ArcView

Throughout this GIS Investigation, you explored temperature data from 96 world cities. You explored the different latitude zones and have identified the variety of climates in each zone. You made different hypotheses to explain temperature patterns, and tested each hypothesis. Now you know how latitude, hemisphere, proximity to the ocean, and elevation affect temperature patterns around the world.

a   Exit ArcView by choosing Exit from the File menu.

b   Do not save this project.

NAME _____ DATE _____

Student answer sheet

Module 3
Physical Geography II:
Ecosystems, Climate, and Vegetation

Global perspective: Running Hot and Cold

Step 3 Observe annual world temperatures

b Are the temperatures listed as degrees Fahrenheit or degrees Celsius? _____

c Write three observations about the pattern of temperatures displayed on the map.

_____

_____

_____

Step 4 Label the latitude zones

p Use the Identify tool to find information on cities and complete the table below.

ZONE	TYPICAL TEMPERATURE RANGE	EXAMPLE CITY (IT REFLECTS TYPICAL TEMPERATURES OF THAT ZONE)	ANOMALIES (CITIES THAT DO NOT FIT THE PATTERN OF THEIR ZONE)
Tropical			
North Temperate Zone			
South Temperate Zone			

p-1 Why do you think there aren't any cities in the North or South polar zones?

_____

_____

_____

p-2   How is the North Temperate Zone different from the South Temperate Zone?

_____

_____

_____

Step 5   Observe climate distribution

d-1   Complete the table.

LATITUDE ZONES	CHARACTERISTIC CLIMATE(S)
Tropical zones	
Temperate zones	
Polar zones	

d-2   Which zone has the greatest number of climates?

_____

e   Give an example of a city in each of the following climate zones:

Arid	_____
Tropical Wet	_____
Tropical Wet and Dry	_____
Humid Subtropical	_____
Mediterranean	_____
Marine	_____
Humid Continental	_____
Subarctic	_____
Highland	_____

f   Write the new name you gave the project and where you saved it.

_____   _____

(Name of project.   **(Navigation path to where project is saved.**
For example: kzeregion5.apr)   **For example: C:\student\kze\mapworld)**

Step 6   Observe monthly temperature patterns in the Northern Hemisphere

d-1   What does the chart show now?

_____

d-2   What city is highlighted in the view?   _____

e-1   What does the chart show now?

_____

e-2   What city or cities are highlighted in the view?   _____

f   Complete the table below.

CITIES	COLDEST MONTH	LOWEST TEMPERATURE	HOTTEST MONTH	HIGHEST TEMPERATURE	TEMPERATURE RANGE OVER 12 MONTHS
Boston					
Miami					

g-1   What is the name of the city? _____

g-2   How does its monthly temperature pattern differ from Boston's?

_____

_____

h-1   What is the name of the city? _____

h-2   How does its monthly temperature pattern differ from Miami's?

_____

_____

i   List the name of each of the cities displayed in the chart and complete the information in the table below.

CITY	LATITUDE	COLDEST MONTH	LOWEST TEMPERATURE (°F)	HOTTEST MONTH	HIGHEST TEMPERATURE (°F)	TEMPERATURE RANGE OVER 12 MONTHS
Boston						
Miami						

j   Based on the information displayed in the chart, the map, and the table on your answer sheet, state a hypothesis about how the monthly temperature patterns change as latitude increases.

_____

_____

Step 7   Test your hypothesis

d-1   Complete the table below.

CITY	LATITUDE
Stockholm	

d-2 Do the cities you selected confirm or dispute your hypothesis? Explain.

_____

_____

_____

Step 8 Analyze temperature patterns in the Southern Hemisphere

c Complete the table below.

CITY	LATITUDE	COLDEST MONTH	LOWEST TEMPERATURE (°F)	HOTTEST MONTH	HIGHEST TEMPERATURE (°F)	TEMPERATURE RANGE OVER 12 MONTHS
Darwin						

d Compare the monthly temperature patterns in the Southern Hemisphere to those in the Northern Hemisphere.

_____

_____

_____

Formulate a hypothesis about the relationship between monthly temperature patterns and increases in latitude.

_____

_____

_____

Step 9 Test your hypothesis on how latitude affects monthly temperature patterns in the Southern Hemisphere

b-1 Complete the table below.

CITY	LATITUDE
Cape Town	

b-2 Based on your observations, do the cities you selected confirm or dispute your hypothesis about how latitude affects monthly temperature patterns in the Southern Hemisphere? Explain.

_____

_____

_____

Step 10   Investigate the ocean's influence on temperature

a-1   In which Canadian city would you experience the coldest winter temperatures?

_____

a-2   In which Canadian city would you experience the warmest winter temperatures?

_____

a-3   Looking at the map, why do you think the warmest city has temperatures that are so much warmer than the others in the winter?

_____

_____

_____

_____

h-1   Complete the table below.

CITY	LATITUDE
London	
Amsterdam	
Berlin	
Warsaw	
Kiev	

h-2   What do these cities have in common as to their location on the earth?

_____

h-3   Which cities have the mildest temperatures?

_____

h-4   What happens to the winter temperatures as you move from London to Kiev?

_____

h-5   Why do you think some cities have milder temperatures than the others?

_____

i   Based on your observations of Canada and Western Europe, state a hypothesis about the influence of proximity to the ocean (or distance from it) on patterns of temperature.

_____

_____

_____

_____

Step 11   Investigate the impact of elevation on temperature patterns

d-1   Complete the table below.

CITY	LATITUDE
Kisangani	
Libreville	
Quito	
Singapore	

d-2   What do these cities have in common as to their location on the earth?

_____

d-3   What temperature pattern do these four cities have in common?

_____

d-4   How is Quito different from the other three?

_____

d-5   Since all these cities are located on or very near the equator, what other factor could explain the difference in their temperature patterns?

_____

f   Analyze the attribute table and complete the table below.

CITY	ELEVATION (FEET)
Kisangani	
Libreville	
Quito	
Singapore	

h   Based on your observation of temperatures along the equator and the information in the table above, state a hypothesis about the influence of elevation on patterns of temperature.

_____

_____

_____

_____

Step 12   Revisit your initial ideas

f   Rank the 13 cities from lowest to highest according to their average January temperatures.

1. _____   8. _____

2. _____   9. _____

3. _____   10. _____

4. _____   11. _____

5. _____   12. _____

6. _____   13. _____

7. _____

j   Rank the 13 cities from highest to lowest according to their average July temperatures.

1. _____   8. _____

2. _____   9. _____

3. _____   10. _____

4. _____   11. _____

5. _____   12. _____

6. _____   13. _____

7. _____

l   Put a check mark (✔) next to those answers that you predicted correctly.

NAME _____   DATE _____

Running Hot and Cold
Middle school assessment

Part 1

Use the ArcView project and your answer sheet to complete the tables below. For each city, circle each factor that influences its temperature pattern.

THREE HOTTEST CITIES IN JULY	CIRCLE THE FACTORS THAT INFLUENCE TEMPERATURE PATTERNS		
	Latitude	Proximity to the ocean	Elevation
	Latitude	Proximity to the ocean	Elevation
	Latitude	Proximity to the ocean	Elevation

THREE COLDEST CITIES IN JANUARY	CIRCLE THE FACTORS THAT INFLUENCE TEMPERATURE PATTERNS		
	Latitude	Proximity to the ocean	Elevation
	Latitude	Proximity to the ocean	Elevation
	Latitude	Proximity to the ocean	Elevation

Part 2

Use your GIS Investigation, global3.apr, and other resources such as an atlas to write an essay that compares monthly and annual temperature patterns typical of the Tropical Zone and the North and South temperate zones. Your essay should provide example cities and data to support your conclusions.

Create a paper map or an ArcView-generated map that illustrates the conclusions you make in your essay.

Running Hot and Cold

Assessment rubric

Middle school

STANDARD	EXEMPLARY	MASTERY	INTRODUCTORY	DOES NOT MEET REQUIREMENTS
The student understands how to use maps, graphs, and databases to analyze spatial distributions and patterns.	Uses GIS to analyze various aspects of climate such as monthly and annual temperature and precipitation and identifies cities with specific climate characteristics. Creates a detailed map that illustrates the points highlighted in the essay.	Uses GIS to analyze various aspects of climate such as monthly and annual temperature and precipitation and identifies cities with specific climate characteristics. Creates a map that illustrates points highlighted in the essay.	With some assistance, can use GIS to analyze various aspects of climate such as monthly and annual temperature and precipitation. Correctly identifies some cities with specific climate characteristics. Creates a map that illustrates some of the points highlighted in the essay.	Has difficulty using GIS to analyze various aspects of climate and identifying cities with specific climate characteristics. Creates a map, but has difficulty illustrating the points highlighted in the essay.
The student understands the elements and types of regions.	Writes an essay and creates a map that shows clear understanding of the climate patterns for various zones of Earth, including the Tropics and the North and South temperate zones.	Writes an essay and creates a map that shows an understanding of the climate patterns for various zones of Earth, including the Tropics and the North and South temperate zones.	Writes an essay and creates a map that shows some understanding of climate patterns, but does not clearly define differences between different zones, or only identifies characteristics of some zones.	Writes an essay and creates a map that shows limited understanding of climate patterns and cannot identify differences between various zones.
The student understands how Earth–Sun relationships affect physical processes and patterns on Earth.	Identifies key reasons and provides clear examples of why cities experience variations in climate patterns at different latitudes, in different hemispheres, at different elevations, and at different distances from the ocean.	Identifies key reasons why cities experience variations in climate patterns at different latitudes, in different hemispheres, at different elevations, and at different distances from the ocean.	Identifies some key reasons why cities experience variations in climate patterns due to some of the following: different latitudes, in different hemispheres, at different elevations, or at different distances from the ocean.	Identifies one or two reasons why cities experience variations in climate patterns at different latitudes, in different hemispheres, at different elevations, or at different distances to the ocean.

This is a four-point rubric based on the National Standards for Geographic Education. The "Mastery" level meets the target objective for grades 5–8.

NAME _____   DATE _____

Running Hot and Cold
High school assessment

Part 1

Use the ArcView project and your answer sheet to complete the tables below. For each city, circle each factor that influences its temperature pattern.

THREE HOTTEST CITIES IN JULY	CIRCLE THE FACTORS THAT INFLUENCE TEMPERATURE PATTERNS		
	Latitude	Proximity to the ocean	Elevation
	Latitude	Proximity to the ocean	Elevation
	Latitude	Proximity to the ocean	Elevation

THREE COLDEST CITIES IN JANUARY	CIRCLE THE FACTORS THAT INFLUENCE TEMPERATURE PATTERNS		
	Latitude	Proximity to the ocean	Elevation
	Latitude	Proximity to the ocean	Elevation
	Latitude	Proximity to the ocean	Elevation

Part 2

Use your GIS Investigation, global3.apr, and other resources such as an atlas to write an essay that compares monthly and annual temperature patterns typical of the Tropical Zone and the North and South temperate zones. Your essay should provide example cities and data to support your conclusions.

Create an ArcView-generated map that illustrates the conclusions you make in your essay.

Running Hot and Cold

Assessment rubric

High school

STANDARD	EXEMPLARY	MASTERY	INTRODUCTORY	DOES NOT MEET REQUIREMENTS
The student understands how to use technologies to represent and interpret Earth's physical and human systems.	Uses GIS to analyze various aspects of climate such as monthly and annual temperature, precipitation, elevation, and proximity to the ocean and identifies cities with specific climate characteristics. Uses GIS to create a detailed map that illustrates the points highlighted in the essay.	Uses GIS to analyze various aspects of climate such as monthly and annual temperature, precipitation, elevation, and proximity to the ocean and identifies cities with specific climate characteristics. Uses GIS to create a map that illustrates points highlighted in the essay.	With some assistance, can use GIS to analyze various aspects of climate such as monthly and annual temperature and precipitation. Correctly identifies some cities with specific climate characteristics. Uses GIS to create a map that illustrates some of the points highlighted in the essay.	Has difficulty using GIS to analyze various aspects of climate and identifying cities with specific climate characteristics. Uses GIS to create a map, but has difficulty illustrating the points highlighted in the essay.
The student understands the structure of regional systems.	Writes an essay and creates a GIS-generated map that shows clear understanding of the climate patterns for various zones of Earth, including the Tropics and the North and South temperate zones.	Writes an essay and creates a GIS-generated map that shows an understanding of the climate patterns for various zones of Earth, including the Tropics and the North and South temperate zones.	Writes an essay and creates a GIS-generated map that shows some understanding of climate patterns, but does not clearly define differences between different zones, or only identifies characteristics of some zones.	Writes an essay and creates a GIS-generated map that shows limited understanding of climate patterns and cannot identify differences between various zones.
The student understands spatial variation in the consequences of physical processes across Earth's surface.	Identifies key reasons and provides clear examples of why cities experience variations in climate patterns at different latitudes, in different hemispheres, at different elevations, and at different distances from the ocean.	Identifies key reasons why cities experience variations in climate patterns at different latitudes, in different hemispheres, at different elevations, and at different distances from the ocean.	Identifies some key reasons why cities experience variations in climate patterns due to some of the following: different latitudes, in different hemispheres, at different elevations, or at different distances from the ocean.	Identifies one or two reasons why cities experience variations in climate patterns at different latitudes, in different hemispheres, at different elevations, or at different distances from the ocean.

This is a four-point rubric based on the National Standards for Geographic Education. The "Mastery" level meets the target objective for grades 9–12.

Seasonal Differences

A regional case study of South Asia

Lesson overview

Students will observe patterns of monsoon rainfall in South Asia and analyze the relationship of those patterns to the region's physical features. The consequences of monsoon season on human life will be explored by studying South Asian agricultural practices and patterns of population distribution.

Estimated time	Two to three 45-minute class periods
Materials	✔ Four large pieces of butcher paper
	✔ Four or more markers
	✔ Student handouts from this lesson to be copied:
	• GIS Investigation sheets (pages 147 to 154)
	• Student answer sheets (pages 155 to 159)
	• Assessment(s) (pages 160 to 163)

Standards and objectives

National geography standards

	GEOGRAPHY STANDARD	MIDDLE SCHOOL	HIGH SCHOOL
1	How to use maps and other geographic representations, tools, and technologies to acquire, process, and report information from a spatial perspective	The student understands how to use maps, graphs, and databases to analyze spatial distributions and patterns.	The student understands how to use technologies to represent and interpret Earth's physical and human systems.
4	The physical and human characteristics of places	The student understands how physical processes shape places.	The student understands the changing human and physical characteristics of places.
15	How physical systems affect human systems	The student understands how variations within the physical environment produce spatial patterns that affect human adaptation.	The student understands how the characteristics of different physical environments provide opportunities for or place constraints on human activities.

Objectives

The student is able to:
• Describe the patterns of monsoon rainfall in South Asia.
• Explain the influence of landforms on patterns of precipitation in South Asia.
• Describe the impact of South Asia's climate and physical features on agriculture and population density in the region.

GIS skills and tools

 Add themes to the view

 Select features from the active theme

 Clear selected features

 Measure distance in the view

 Add a theme to the view

 Change the theme name

- Understand the relationship between a map and a chart in an ArcView project

For more on geographic inquiry and these steps, see Geographic Inquiry: Thinking Geographically (pages xxi to xxiii).

Teacher notes

Lesson introduction

Tell your students that they are going to explore seasonal differences in South Asia. They may be surprised to learn that students in South Asia would probably describe their year in terms of three seasons rather than four. Engage them in a discussion of local and personal perceptions and assumptions about seasons. Tack up four large pieces of butcher paper and have them list images, descriptions, and memories relating to each season.

- How does the physical environment change from season to season?
- How are those changes reflected in the activities, foods, and clothing they may have listed on the sheets?
- To what extent do seasonal changes in their environment affect their day-to-day lives?

Student activity

 Before completing this lesson with students, we recommend that you complete it yourself. Doing so will allow you to modify the activity to suit the specific needs of your students.

After the initial discussion, have the students work on the computer component of the lesson. Ideally, each student should be at an individual computer, but the lesson can be modified to accommodate a variety of instructional settings.

Distribute the GIS Investigation sheets to your students. Explain that in this activity, they will use GIS to observe and analyze the variable patterns of rainfall in South Asia that result from the region's seasonal monsoon winds. In South Asia it is rainfall, rather than temperature, that defines the seasons. The activity sheets will provide them with detailed instructions for their investigations. As they investigate, they will explore the relationship between South Asia's monsoon rains and its physical features and analyze the climate's impact on agriculture and population.

In addition to instructions, the handout includes questions to help students focus on key concepts. Some questions will have specific answers while others require creative thought.

Things to look for while the students are working on this activity:

- Are the students using a variety of GIS tools?
- Are the students answering the questions as they work through the procedure?
- Are the students experiencing any difficulty working with multiple windows and toggling between windows in the project?

Conclusion

Use a projection device to display the region3.apr in the classroom. As a group, compare student observations and conclusions from the lesson. Students can take turns being the "driver" on the computer to highlight patterns and relationships that are identified by members of the class. Focus on the following concepts about South Asia's monsoon climate in your discussion.

- Rainfall is limited to one season of the year in South Asia except in the desert west, where little rain falls at all. (This would be an excellent point at which to elaborate on the seasonal shift in monsoon winds that produces the patterns of rainfall students observed in the project.)

Conclusion
(continued)

- Typically, the rainy season lasts from June through September, although the actual length of the season and amounts of rainfall vary across the subcontinent. (Be sure to note the orographic patterns of precipitation along India's southwest coast and in northeast India on the southern slopes of the Himalayas.)
- Agricultural activities are directly related to patterns of rainfall.
- In general, population density varies with patterns of rainfall. However, the importance of South Asia's rivers as an additional source of water for agriculture is apparent from the high density of population along their paths.

Close the lesson by challenging the students to identify the three seasons in South Asia. In general these seasons are the following:

- The rainy season (approximately June–September)
- The dry lush season after the rains when everything is growing and green (approximately October–January)
- The dry dusty season before the rains come (approximately February–May)

Assessment

Middle school: Highlights skills appropriate to grades 5 through 8

Students will assume the role of an American student living for a year in South Asia as an exchange student. They can choose to live in or around Bombay, New Delhi, or Calcutta. They will write a letter to friends back home on October 1, January 1, April 1, and July 1. Their letter will describe seasonal changes in their location and ways that their daily lives and the lives of people around them reflect those changes.

High school: Highlights skills appropriate to grades 9 through 12

Students will assume the role of an American student traveling for a year in South Asia. They will write a letter to friends back home on October 1, January 1, April 1, and July 1. Each letter will be from a different South Asian city. Their letter will describe seasonal characteristics in each city and ways that their daily lives and the lives of people around them reflect those characteristics.

Extensions

- Use the Internet to find rainfall data for South Asian cities in specific years such as 1990, 1995, and 2000 to see if the average patterns observed in this project are relatively consistent or if they vary significantly from year to year.
- Import downloaded data into ArcView.
- Research South Asian farming methods to find out how activities such as planting and harvesting are coordinated with patterns of rainfall.
- Research the monthly and yearly rainfall patterns in your own location and compare these to the patterns observed in South Asia.
- Check out the Resources by Module section of the Teacher Resource CD for print and media resources on the topics of South Asia and monsoons or visit *www.esri.com/mappingourworld* for Internet links.

NAME _____   DATE _____

Seasonal Differences
A GIS investigation

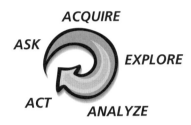

ACQUIRE

ASK

EXPLORE

ACT

ANALYZE

Answer all questions on the student answer sheet handout

In this activity, you will analyze the variable patterns of precipitation in South Asia that result from the region's seasonal monsoon winds. As you investigate those patterns, you will explore relationships between rainfall and physical features and analyze the climate's impact on agriculture and population.

Step 1   Start ArcView

a   Double-click the ArcView icon on your computer's desktop.

ArcView

b   If the Welcome to ArcView dialog appears (pictured below), click **Open an Existing Project** and click OK. If it doesn't appear, proceed to step 2.

Step 2   Open the region3.apr file

   a   In this exercise, a project file has been created for you. To open the file, go to the **File** menu and choose **Open Project**.

   b   Navigate to the exercise data directory (**C:\esri\mapworld\mod3**) and choose **region3.apr** from the list.

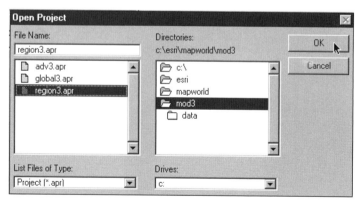

c   Click OK.

When the project opens, you see three open windows: a view called South Asia, a chart called Monthly Rainfall, and a chart called Annual Rainfall.

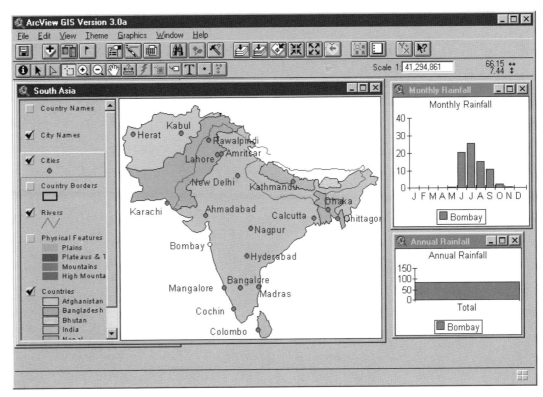

Note: In ArcView, graphs are referred to as charts.

d   Click the Maximize button in the upper right corner of your ArcView window.

e   Move the charts over to the right and stretch the View window to make the map bigger.

Step 3   Observe patterns of rainfall

The map in this project allows you to explore and compare variations in the patterns of rainfall throughout the South Asian region. Look at the map and notice that the city of Bombay is selected—it is highlighted yellow. The charts to the right display rainfall information for the selected city—in this case, Bombay.

a   Analyze the charts and answer the following questions on your answer sheet.

(1)   *Which month gets the most rainfall in Bombay?*

(2)   *Which months appear to get little or no rainfall in Bombay?*

(3)   *Approximately how much rainfall does Bombay get each year (in inches)?*

(4)   *Write a sentence summarizing the overall pattern of rainfall in Bombay in an average year.*

b   Click the Select Feature tool. Click a dot for another city.

(1)   *How did this change the map?*

(2)   *How did this change the charts?*

c   Click the city of Mangalore to select it in the map.

?   *Analyze the charts and fill in the Mangalore section of the table on your answer sheet.*

d   Hold the Shift key down and click the cities of Bombay and Ahmadabad.

?   *(1)   Complete the table on the answer sheet.*

?   *(2)   As you move northward along the subcontinent's west coast, how does the pattern of rainfall change?*

?   *(3)   Although the monthly rainfall amounts differ, what similarities do you see among the overall rainfall patterns of these three cities?*

Step 4   Compare coastal and inland cities

   a   Make sure the Select Feature tool is still active and select Bangalore.

?   *Use the Monthly Rainfall and Yearly Rainfall charts to complete the table on your answer sheet.*

b   Hold down the Shift key and select Mangalore.

?   *How does the rainfall pattern of Bangalore compare with that of Mangalore?*

   c   Click the Measure tool. Your cursor turns into a right-angle ruler with crosshairs.

d   Click the dot that represents Bangalore once, then move it to the dot that represents Mangalore and double-click.

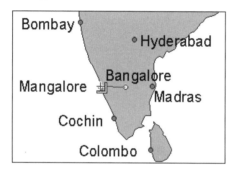

Note: If you accidentally clicked the wrong spot, you can double-click to end the line and start over.

A segment length appears on the bottom left of the View window.

?   *What is the distance between the two cities?*

Although Bangalore is located only a short distance inland from Mangalore, it receives far less rainfall than the coastal city.

e   Turn on the Physical Features theme.

?   *How can this data help you explain the differences between patterns of rainfall in inland Bangalore and coastal Mangalore?*

f   Turn off Physical Features and make sure Cities is the active theme.

Step 5   Compare eastern and western Asian cities

   a   Click the Select Feature tool. One at a time, select the Afghan cities of Kabul and Herat.

       (1)   *Analyze the charts and complete the table on your answer sheet.*

       (2)   *Describe the pattern of rainfall in these two cities.*

       (3)   *How do you think Afghanistan's rainfall pattern will affect the way of life in that country?*

   b   Select the eastern cities of Calcutta and Dhaka.

       (1)   *Analyze the charts and complete the table on your answer sheet.*

       (2)   *Describe the pattern of rainfall in these two cities.*

   c   Hold down the Shift key and select four cities (Calcutta, Dhaka, Herat, and New Delhi).

      What is happening to the patterns of rainfall as you move from west to east across South Asia?

   d   Clear all of your selections and stretch the View window so it's much larger and partially covering the charts.

Step 6   Observe yearly precipitation

You've already looked at the monthly precipitation patterns for individual cities across South Asia. In this step, you will add data and look at the total yearly rainfall for regions of South Asia.

   a   Click the Add Theme button.

   b   Navigate to the data directory (**C:\esri\mapworld\mod3\data**).

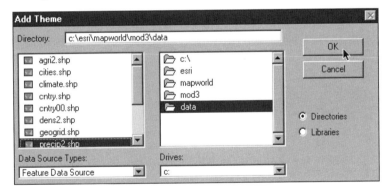

   c   Scroll down and select **precip2.shp**. Click OK.

d Drag the new theme, Precip2.shp, down in the table of contents so that it is just below the Rivers theme.

 e Click the Theme Properties button.

f Change the theme name to **Yearly Rain(mm)** and click OK.

g Turn off the Cities and City Names themes and turn on Yearly Rain (mm).

   (1) Which regions within South Asia get the least rainfall?

   (2) Which regions within South Asia get the most rainfall?

   (3) In step 5c you were comparing Calcutta, Herat, New Delhi, and Dhaka. Does the map of yearly rainfall that is on your screen now reflect the observation you made at that time? Explain.

h Turn off Yearly Rain (mm) and turn on Physical Features.

   What relationships do you see between South Asia's patterns of yearly rainfall and its physical features?

Step 7 Explore the monsoon's impact on agriculture and population density

a Turn on the Country Borders and the Country Names themes.

The rain patterns and physical features of an area have a significant impact on the way of life of the people who live there. Now you will look at those themes and determine the kind of impact they have on individual countries.

b Turn the Physical Features and Yearly Rain themes on and off to make your observations and to answer the questions below.

   (1) Which regions or countries of South Asia are suitable for agriculture and which are not? Explain.

   (2) In which regions of South Asia do you expect to see the lowest population density? Explain.

   (3) In which regions of South Asia do you expect to see the highest population density? Explain.

c Turn off Physical Features and Yearly Rain (mm) themes.

Now you will add agricultural data for the region and will see if your predictions are correct.

d   Click the Add Theme button.

e   Navigate to the data directory (**C:\esri\mapworld\mod3\data**).

f   Select **agri2.shp**. Click OK.

g   Click the Theme Properties button.

h   Change the theme name to **Agriculture** and click OK.

i   Make the table of contents window wider so that you can read the entire legend label. Do this by slowly moving your cursor to the line that separates the table of contents from the View window. When the cursor turns to a double arrow symbol, hold down the mouse button and drag the line to the right.

j   Drag Country Borders and Rivers to the top of the table of contents. Turn on the Agriculture theme.

  (1)   *Does the Agriculture theme reflect the predictions you made in step 7b? Explain.*

  (2)   *Why are grazing, herding, and oasis agriculture the major activities in Afghanistan?*

  (3)   *What do you know about rice cultivation that would help explain its distribution on the agriculture map?*

  (4)   *Is there any aspect of the agriculture map that surprised you? Explain.*

k   Turn off the Agriculture theme.

You will now examine population density in relation to precipitation and land use.

l   Click the Add Theme button.

m   Navigate to the data directory (**C:\esri\mapworld\mod3\data**).

n   Select **dens2.shp**. Click OK.

o   Change the theme name for Dens2.shp to **People/sq. km**.

p   Drag Country Borders and Rivers to the top of the table of contents. Turn on People/sq.km.

  (1)   *Does the People/sq km theme reflect the population predictions you made in step 7b? Explain.*

  (2)   *Why is Afghanistan's population density so low?*

  (3)   *Since most of Pakistan gets little to no rainfall, how do you explain the areas of high population density in that country?*

  (4)   *What is the relationship between population density and patterns of precipitation in South Asia?*

  (5)   *What is the relationship between population density and physical features in South Asia?*

Step 8   Exit ArcView

In this GIS Investigation, you explored the patterns of monsoon rainfall in South Asia. You used ArcView to compare monthly and annual patterns of precipitation in cities throughout the region and explore the relationship between those patterns and the region's physical features. After analyzing this data, you added themes reflecting patterns of agriculture and population density and analyzed the relationship between those human characteristics and the region's climate and landforms.

a   Ask your teacher for instructions on where to save this ArcView project and on how to rename the project.

b   If you are not going to save the project, exit ArcView by choosing Exit from the File menu. When asked if you want to save changes to region3.apr, click No.

NAME _____   DATE _____

Student answer sheet

Module 3
Physical Geography II:
Ecosystems, Climate, and Vegetation

Regional case study: Seasonal Differences

Step 3   Observe patterns of rainfall

a-1   Which month gets the most rainfall in Bombay? _____

a-2   Which months appear to get little or no rainfall in Bombay? _____

a-3   Approximately how much rainfall does Bombay get each year (in inches)? _____

a-4   Write a sentence summarizing the overall pattern of rainfall in Bombay in an average year.

_____

_____

_____

b-1   How did this change the map?

_____

_____

b-2   How did this change the charts?

_____

_____

c   Analyze the charts and fill in the Mangalore section of the table below.

CITY	MONTHS WITH RAINFALL	HIGHEST MONTHLY RAINFALL (INCHES)	TOTAL ANNUAL RAINFALL (INCHES)
Mangalore			
Bombay			
Ahmadabad			

d-1   Complete the rest of the table in step c above.

d-2   As you move northward along the subcontinent's west coast, how does the pattern of rainfall change?

_____

_____

d-3   Although the monthly rainfall amounts differ, what similarities do you see among the overall rainfall patterns of these three cities?

_____

_____

Step 4   Compare coastal and inland cities

a   Complete the table below.

CITY	MONTHS WITH RAINFALL	HIGHEST MONTHLY RAINFALL (INCHES)	TOTAL ANNUAL RAINFALL (INCHES)
Bangalore			

b   How does the rainfall pattern of Bangalore compare with that of Mangalore?
     Similarities: _____
     Differences: _____

d   What is the distance between the two cities? _____

e   How can this data help you explain the differences between patterns of rainfall in inland Bangalore and
     coastal Mangalore?

     _____

     _____

     _____

     _____

     _____

     _____

     _____

     _____

     _____

     _____

Step 5   Compare eastern and western Asian cities

a-1   Analyze the charts and complete the table below.

CITY	MONTHS WITH RAINFALL	HIGHEST MONTHLY RAINFALL (INCHES)	TOTAL ANNUAL RAINFALL (INCHES)
Kabul			
Herat			

a-2   Describe the pattern of rainfall in these two cities.

       _____

       _____

a-3   How do you think Afghanistan's rainfall pattern will affect the way of life in that country?

       _____

       _____

b-1   Analyze the charts and complete the table below.

CITY	MONTHS WITH RAINFALL	HIGHEST MONTHLY RAINFALL (INCHES)	TOTAL ANNUAL RAINFALL (INCHES)
Calcutta			
Dhaka			

b-2 Describe the pattern of rainfall in these two cities.

_____

_____

c   What is happening to the patterns of rainfall as you move from west to east across South Asia?

_____

_____

Step 6   Observe yearly precipitation

g-1 Which regions within South Asia get the least rainfall?

_____

_____

g-2 Which regions within South Asia get the most rainfall?

_____

_____

g-3 In step 5c you were comparing Calcutta, Herat, New Delhi, and Dhaka. Does the map of yearly rainfall that is on your screen now reflect the observation you made at that time? Explain.

_____

_____

_____

_____

h   What relationships do you see between South Asia's patterns of yearly rainfall and its physical features?

_____

_____

_____

_____

Step 7   Explore the monsoon's impact on agriculture and population density

b-1 Which regions or countries of South Asia are suitable for agriculture and which are not? Explain.

_____

_____

_____

_____

b-2 In which regions of South Asia do you expect to see the lowest population density? Explain.

_____

_____

_____

_____

b-3 In which regions of South Asia do you expect to see the highest population density? Explain.

_____

_____

_____

_____

j-1   Does the Agriculture theme reflect the predictions you made in step 7b? Explain.

j-2   Why are grazing, herding, and oasis agriculture the major activities in Afghanistan?

j-3   What do you know about rice cultivation that would help explain its distribution on the agriculture map?

j-4   Is there any aspect of the agriculture map that surprised you? Explain.

p-1   Does the People/sq km theme reflect the population predictions you made in step 7b? Explain.

p-2   Why is Afghanistan's population density so low?

p-3   Since most of Pakistan gets little to no rainfall, how do you explain the areas of high population density in that country?

p-4   What is the relationship between population density and patterns of precipitation in South Asia?

_____

_____

_____

_____

p-5   What is the relationship between population density and physical features in South Asia?

_____

_____

_____

_____

NAME _____   DATE _____

Seasonal Differences
Middle school assessment

For this activity, you are to assume the role of an American student who is spending a year living in South Asia as an exchange student. Your task is to write four letters to friends or family back home about your experiences and observations during your year in South Asia. Your four letters should be dated January 1, April 1, July 1, and October 1. Using the ArcView project as a guide, describe seasonal changes in your city and ways that your daily life and the lives of people around you reflect those changes. You may choose to spend your year in or near any one of the following locations: Bombay, Calcutta, or Dhaka. You may use additional sources such as your geography book, encyclopedias, and the Internet to help you develop your letters.

Use the space below to brainstorm for your essay.

January 1

April 1

July 1

October 1

Seasonal Differences

Assessment rubric

Middle school

STANDARD	EXEMPLARY	MASTERY	INTRODUCTORY	DOES NOT MEET REQUIREMENTS
The student understands how to use maps, graphs, and databases to analyze spatial distributions and patterns.	Uses GIS as a tool to analyze the patterns of monsoon rains in South Asia through mapping and creating original charts based on the data provided. Accurately depicts the seasonal weather conditions in a given city.	Uses GIS as a tool to analyze the patterns of monsoon rains in South Asia through mapping and viewing charts. Accurately depicts most of the seasonal weather conditions in a given city.	Identifies patterns of precipitation on maps and charts using GIS. Accurately depicts some of the seasonal weather conditions for a given city.	Has difficulty identifying patterns of precipitation using maps and charts in GIS. Has difficulty depicting any seasonal weather conditions for the given city.
The student understands how physical processes shape places.	Shows understanding of the effects of seasonal rains and other climate factors on the characteristics of many South Asian cities throughout an entire year. Creatively incorporates this into each letter.	Shows understanding of the effects of seasonal rains and other climate factors on the characteristics of one particular South Asian city throughout an entire year. Incorporates this into each letter.	Shows limited understanding of the effects of seasonal rains on the characteristics of one particular South Asian city throughout an entire year. Incorporates some of this information into each letter.	Attempts to describe seasonal climate changes in the region of South Asia. Has difficulty incorporating this information into each letter.
The student understands how variations within the physical environment produce spatial patterns that affect human adaptation.	Uses great detail and specific examples to illustrate the impact of seasonal climate changes on the daily life of people in a variety of South Asian cities throughout an entire year. Creatively incorporates this into each letter.	Clearly illustrates the impact of seasonal climate changes on the daily life of people in a particular South Asian city throughout an entire year. Incorporates this into each letter.	Shows limited understanding of the impact of seasonal climate changes on the daily life of people in a particular South Asian city throughout an entire year. Incorporates some of this information into each letter.	Attempts to describe the impact of seasonal climate changes on the daily life of people in South Asia. Has difficulty incorporating this information into each letter.

This is a four-point rubric based on the National Standards for Geographic Education. The "Mastery" level meets the target objective for grades 5–8.

NAME _____   DATE _____

Seasonal Differences
High school assessment

For this activity you are to assume the role of an American student who is spending a year traveling in South Asia. Your task is to write four letters to friends or family back home about your experiences and observations during your year abroad. Each letter should be written from a different South Asian city. Your four letters should be dated January 1, April 1, July 1, and October 1. Using the ArcView project as a guide, describe seasonal characteristics of each city on the date you are writing and ways that your daily life and the lives of people around you reflect those characteristics. You may use additional sources such as your geography book, encyclopedias, and the Internet to help you develop your letters.

Use the space below to brainstorm for your essay.

January 1

April 1

July 1

October 1

Seasonal Differences

Assessment rubric

High school

STANDARD	EXEMPLARY	MASTERY	INTRODUCTORY	DOES NOT MEET REQUIREMENTS
The student understands how to use technologies to represent and interpret the earth's physical and human systems.	Uses GIS as a tool to analyze the patterns of monsoon rains in South Asia through mapping and creating original charts based on the data provided. Accurately depicts the seasonal weather conditions in the four cities.	Uses GIS as a tool to analyze the patterns of monsoon rains in South Asia through mapping and viewing charts. Accurately depicts most of the seasonal weather conditions in the four cities.	Identifies patterns of precipitation on maps and charts using GIS. Accurately depicts some of the seasonal weather conditions for some of the cities.	Has difficulty identifying patterns of precipitation using maps and charts in GIS. Has difficulty depicting the seasonal weather conditions for the cities.
The student understands the changing human and physical characteristics of places.	Shows understanding of the effects of seasonal rains and other climate factors on the characteristics of four South Asian cities. Describes the seasonal changes for an entire year for each city. Creatively incorporates this into each letter.	Shows understanding of the effects of seasonal rains and other climate factors on the characteristics of four different South Asian cities throughout an entire year (one city for each season). Incorporates this into each letter.	Shows limited understanding of the effects of seasonal rains on the characteristics of one or two South Asian cities throughout an entire year. Incorporates some of this information into each letter.	Attempts to describe seasonal climate changes in the region of South Asia. Has difficulty incorporating this information into each letter.
The student understands how the characteristics of different physical environments provide opportunities for or place constraints on human activities.	Uses great detail and specific examples to illustrate the impact of seasonal climate changes on the daily life of people in four South Asian cities. Describes the seasonal changes for an entire year for each city. Creatively incorporates this into each letter.	Clearly illustrates the impact of seasonal climate changes on the daily life of people in four different South Asian cities throughout an entire year (one city for each season). Incorporates this into each letter.	Shows limited understanding of the impact of seasonal climate changes on the daily life of people in one or two South Asian cities throughout an entire year. Incorporates some of this information into each letter.	Attempts to describe the impact of seasonal climate changes on the daily life of people in South Asia. Has difficulty incorporating this information into each letter.

This is a four-point rubric based on the National Standards for Geographic Education. The "Mastery" level meets the target objective for grades 9–12.

Sibling Rivalry
An advanced investigation

Lesson overview

Students will study climatic phenomena El Niño and La Niña by downloading map images from the GLOBE (Global Learning and Observations to Benefit the Environment) Web site. They will incorporate these images into an ArcView project and identify patterns and characteristics of these phenomena, then assess the impact these anomalies have on the global and local environment.

Estimated time One to two 45-minute class periods

Materials
✔ Internet access
✔ Access to a graphics software program (e.g., Adobe Illustrator®, Jasc® Paint Shop Pro®, CorelDRAW®)
✔ Student handouts from this lesson to be copied:
- GIS Investigation sheets (pages 169 to 174)
- Student answer sheets (pages 175 to 178)

Standards and objectives

National geography standards

GEOGRAPHY STANDARD	MIDDLE SCHOOL	HIGH SCHOOL
7 The physical processes that shape the patterns of Earth's surface	The student understands how to predict the consequences of physical processes on Earth's surface.	The student understands the dynamics of the four basic components of Earth's physical systems: the atmosphere, biosphere, lithosphere, and hydrosphere.
15 How physical systems affect human systems	The student understands human responses to variations in physical systems.	The student understands strategies to respond to constraints placed on human systems by the physical environment.
18 How to apply geography to interpret the present and plan for the future	The student understands how to apply the geographic point of view to solve social and environmental problems by making geographically informed decisions.	The student understands how to use geographic knowledge, skills, and perspectives to analyze problems and make decisions.

Objectives

The student is able to:
- Create online maps using GLOBE data sets and download map images into ArcView for analysis.
- Compare and contrast various controls of climate to determine characteristics of El Niño and La Niña weather patterns.
- Describe the potential impact of El Niño at a global, regional, and local level.

GIS skills and tools
- Create maps using the GLOBE Web site and import them into ArcView
- Georeference a GLOBE map image by creating a world file
- Change the map projection
- Compare image and feature data sets

For more on geographic inquiry and these steps, see Geographic Inquiry: Thinking Geographically (pages xxi to xxiii).

Teacher notes

Lesson introduction

Begin the lesson with a brainstorming session on what El Niño is. Record your students' responses on the board. Discuss their ideas: identify correct impressions and important points. Be sure to record this information in a way and place that will allow your students to compare investigative findings with preconceptions. After the initial discussion, provide a brief overview of what El Niño and La Niña are. Use the Teacher Resource CD to locate additional information if you need it. Explain to the students that in the following GIS Investigation, they will be making observations and comparing El Niño to its countervariation, La Niña.

Student activity

 Before completing this lesson with students, we recommend that you complete it yourself. Doing so will allow you to modify the activity to suit the specific needs of your students. The lesson is designed for students working individually at the computer, but it can be modified to accommodate a variety of instructional settings.

Distribute the GIS Investigation sheets to the class. Explain that this investigation will have them create maps on the Internet using data from the GLOBE Web site. The GLOBE program is a joint effort between several U.S. agencies: NOAA (National Oceanic and Atmospheric Administration), NASA (National Aeronautics and Space Administration), NSF (National Science Foundation), and EPA (Environmental Protection Agency), along with more than 95 other countries, to bring real-world environmental science into the classroom. The data displayed at the GLOBE site is a combination of the work of students from around the world and scientists using techniques ranging from visual observation with the naked eye to analysis of data from satellites. GLOBE has a special series of maps dedicated to observation of El Niño and La Niña global events. Your students will create maps online, then download the images into ArcView with the GIS Investigation as their guide. They will then be free to download other maps and data that might help them to predict the overall impact of El Niño.

The GIS Investigation has the students saving images from the Internet into a folder named Image in the module 3 data directory. If you would like students to save to another location, be sure to provide them instruction on where and how to save.

Steps 7–9 can be done as part of the assessment. They include instructions on how to look at a sample metadata document and how to write a simple one.

Things to look for while the students are working on this activity:

- Are students selecting appropriate data sets to map to gain understanding of El Niño and La Niña?
- Are they saving their images into an appropriate directory?

Conclusion　　After students have identified a variety of characteristics of El Niño and La Niña, have them share their observations with the class. Compare those observations with notes from the original discussion.

Assessment　　*Middle school: Highlights skills appropriate to grades 5 through 8*

Students will write an essay that describes the effects of El Niño and La Niña years on their local area or an area you designate for research. Their observations should be focused on weather patterns (temperature and precipitation). They should provide data and printed maps to support their findings.

High school: Highlights skills appropriate to grades 9 through 12

Students will write an essay that describes the effects of El Niño and La Niña years on their local area or an area you designate for research. Their observations should include weather patterns (temperature and precipitation) and other factors that affect the local economy. For example, if the research area is highly agricultural, then greater than normal rainfall could cause damage to crops. They should provide data and printed maps to support their findings.

Extensions
- Download local precipitation and temperature data for the past 10 years. Identify patterns and attempt to predict future weather events based on historic patterns.
- Have students collect and enter their local climate data into ArcView for analysis.
- Select a region as unlike your own as possible, and have students study the consequences of El Niño in an unfamiliar setting.
- Check out the Resources by Module section of the Teacher Resource CD for print and media resources on the topics of El Niño, La Niña, and Southern Oscillation or visit *www.esri.com/mappingourworld* for Internet links.

NAME _____   DATE _____

Sibling Rivalry
An advanced investigation

Note: Due to the dynamic nature of the Internet, the URLs listed in this lesson may have changed, and the graphics shown below may be out of date. If the URLs do not work, refer to this book's Web site for an updated link: www.esri.com/mappingourworld.

ACQUIRE

ASK

EXPLORE

ACT

ANALYZE

Answer all questions on the student answer sheet handout

Every two to seven years, the climatic phenomenon El Niño occurs over the tropics just south of the equator and off the western coast of Central and South America. In this GIS Investigation, you will record your observations of this event and identify characteristics of both El Niño and its counterpart, La Niña.

Step 1   Visualize and acquire GLOBE data

The GLOBE program (Global Learning and Observations to Benefit the Environment) is a joint effort between several U.S. agencies: NOAA (National Oceanic and Atmospheric Administration), NASA (National Aeronautics and Space Administration), NSF (National Science Foundation), and EPA (Environmental Protection Agency), along with 95 other countries, to bring real-world environmental science into the classroom. The data that is displayed at the GLOBE site is a combination of the work of students and scientists from around the world using everything from the naked human eye to satellites to make their observations. GLOBE has a special series of maps dedicated to El Niño and La Niña. You will create your maps online and then download the images into ArcView for further analysis.

a   Go to the GLOBE home page on the Internet at www.globe.gov.

b   Under **GLOBE DATA**, click **Visualizations**.

c   Click the **GLOBE Maps** link.

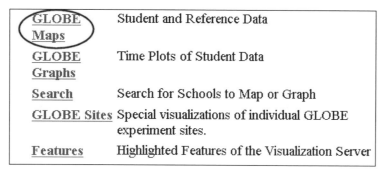

The GLOBE Maps page opens and you see options to click and create your own map.

d   Change the following parameters to create your map:

- Change the Map size to large.
- Change the Data Source to La Niña/El Niño/SO.

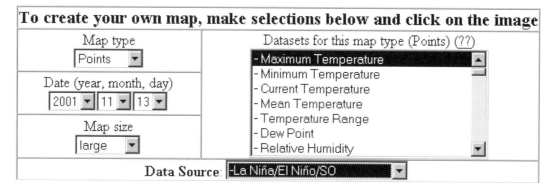

e   Click the Redraw Map button.   Redraw map

Now you see the default map for this particular data source. It is the El Niño Predicted Temperature Anomaly for the month ending March 31, 1999. This is the default map because the winter of 1998–1999 was determined to be an El Niño year. You will change the map parameters so you get a map for sea surface temperatures.

f   Scroll down and change the map parameters to match the settings in the graphic below.

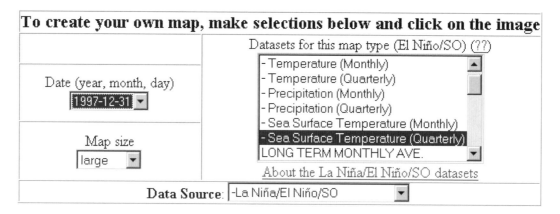

g   Click the Redraw map button. | Redraw map |

An anomaly means any significant variation from the norm. In this case, the map shows any deviation from the normal average quarterly sea surface temperatures. Any temperature that is within the normal range is identified by the color gray on the map. Temperatures that are above normal are identified in the color range of yellow–orange–red. Temperatures that are below normal are in the color range of aqua–blue–purple. A color legend is located at the top left-hand corner of the map image.

In order to use this map image in ArcView, you need to save it.

h   For Windows users, right-click on the map image. For Macintosh users, hold your cursor over the map image.

A menu or option box appears.

i   Click the option for saving the image (e.g., Save Picture As) and navigate to the module 3 data directory (**C:\esri\mapworld\mod3\data\image**) or a location your teacher designates.

j   Change the name of the image to **abc_sstemp1297.gif**, where abc are your initials.

k   Repeat steps 1f–1j to draw and save a map with the following parameters:
   • Sea Surface Temperature (Quarterly)
   • Date: 1999-03-31
   • Save image as: **abc_sstemp0399.gif** (abc are your initials)

l   Minimize your Web browser.

Step 2   Convert images to .tif

ArcView is unable to import .gif images. Therefore, you must use a graphics software program to convert the two .gif images you downloaded from the GLOBE Web site into a form that ArcView can import (.tif).

a   Open a graphics software program of your choice.

b   Navigate to the location of your saved GLOBE images (**C:\esri\mapworld\mod3\data\image**). Open **abc_sstemp1297.gif**.

c   From the File menu, click Save As. Keep the same file name, but change the Type of file to **Tagged Image File Format (.tif, .tiff)**. Be sure to save the map image in an *uncompressed form*.

d   Repeat this procedure for **abc_sstemp0399.gif**.

Now your images are saved in a file format that ArcView can read.

e   Close the graphics software program.

Step 3   Georeference the images

In order for ArcView to know where a map image is in relation to latitude and longitude, it needs a special file known as a world file. This file georeferences the image: it tells ArcView where the image needs to sit in relationship to latitude and longitude coordinates. Now you will georeference the images.

a   Use "My Computer" or "Windows Explorer" (on PCs) or the "Finder" (on Macintosh) to navigate to the Image folder (**C:\esri\gised\module3\data\image**).

b   Copy the file **master.flw** and paste it twice into the same Image folder. Now you have three copies of master.flw in the Image folder.

> *Note: If you did not save your downloaded images to this image folder, you need to paste the two copies of master.flw into the folder where the two images are saved.*

Master.flw is a master georeferencing file (also known as the world file) to use with all the large-format images you retrieve from the GLOBE home page. It will "tell" ArcView where to place the image.

c   Rename one of the world files to match the file name of the first image you downloaded. You also need to change the file extension from .flw to .tfw. The "t" tells ArcView to georeference the .tif file with the same name. For the image named abc_elnino1297.tif, you need to rename the world file **abc_sstemp1297.tfw**.

d   Change the file name of the other pasted world file to match the other downloaded image. It now reads **abc_sstemp0399.tfw**.

You still have a copy of the original master.flw file in the Images folder. When you need to download additional images later, you can copy this file again.

Step 4   Start ArcView and add the georeferenced images

a   Double-click the ArcView icon on your computer's desktop.

b   Navigate to the module 3 directory (**C:\esri\mapworld\mod3**) and open **adv3.apr**.

When the project opens, the View window appears with the following themes displayed: Latitude & Longitude, Climate, Countries, and Oceans.

c   Maximize the ArcView window and stretch the view so the map is bigger.

d   Click the Add Theme button and navigate to the folder where the georeferenced images are saved (**C:\esri\mapworld\mod3\data\images**).

e   Select Image Data Source and add both images (**abc_sstemp1297.tif** and **abc_sstemp0399.tif**).

> *Note: If you received an error message, check to make sure your .tif image files are uncompressed. You will need to go back to your graphics program to do this. If the files are compressed, delete them and go back to step 2 to repeat steps 2 and 3.*

f   Drag the Latitude & Longitude theme so that it is at the top of the table of contents. Turn on one of the new themes.

The two themes do not match up. This is because the images are in Mollweide projection and the view is set to the default decimal degree (unprojected) view. You will change it to Mollweide.

   g   From the View menu, select Change Map Projection. Click Mollweide and click OK. All the themes line up with the images.

      Note: If you cannot see the image properly in the view, it is because there was an error in georeferencing. Go back to step 3 and check that you changed the world file to match the file name of the image. Also, check that the extension is ".tfw"—if the file name or .tfw are incorrect by even one character, you will not see the image properly. After you correct the file names, replace the image themes in ArcView.

   h   Save your project according to your teacher's instructions.

   ?   *Record the new name of the project and its location on your answer sheet.*

Step 5   Analyze characteristics of El Niño and La Niña

El Niño and La Niña are climate anomalies that affect sea surface temperature and precipitation. The scientific term for El Niño/La Niña is the "Southern Oscillation." The image maps you are exploring are Predicted Anomaly Maps. They show and describe scientists' predictions for anomalies in sea surface temperature.

   a   Zoom in on the area of the Pacific Ocean just west of the coast of South America.

   b   Record your temperature observations in the table on your answer sheet, taking note of the legend in the upper left of the map image itself.

   c   Go back to the GLOBE Web site and download precipitation maps for the same time period. Follow the procedure in steps 1–4 to create and import the images into ArcView.

   ?   *Record your observations of precipitation characteristics for the same time period on your answer sheet.*

   d   Save your project.

The impact of El Niño and La Niña are felt at a global scale; it is not limited to the small region of the Pacific where it is born.

   e   If time permits, download other GLOBE map images that you think will help you create a detailed definition for the global climate events of El Niño and La Niña. Compare these patterns with the Climate theme already in the table of contents.

   ?   *f*   Synthesize the information you've recorded and develop a definition of El Niño and La Niña. Write these definitions on your answer sheet.

   g   Save your project.

Step 6   **Are El Niño and La Niña equal and opposite?**

According to Sir Isaac Newton's third law of motion, for every action there is an equal and opposite reaction. The same holds true for the weather—which is, after all, influenced by physical elements and properties. Is La Niña the "equal and opposite" reaction to El Niño? Is one event better than the other? Your answer depends on where you live.

a   Look at the maps you've downloaded and focus on different regions of the world. Complete the table on your answer sheet by filling in your observations for El Niño and La Niña's effect on the different regions of the world. Consult your teacher for what kinds of weather phenomena you should investigate under "Other."

b   Based on the data you recorded in the previous question, do you think La Niña is equal and opposite to El Niño? Explain your answer.

c   Analyze the climate in your region and local area during the last El Niño (1997) and La Niña (1999). Consult basic climate data (temperature, precipitation, and storms). Refer to the following questions to guide you, and record your answers on the answer sheet:

     (1)   Was one year better than the other for you and your community?

     (2)   If in one year you received greater than normal rainfall, did your town have problems with flooding?

     (3)   If weather was unseasonably warm and mild, did outdoor activities such as amusement parks have greater turnout?

     (4)   How did these things affect the local economy?

d   As an assessment for this GIS Investigation, write an essay that answers the questions above and create ArcView maps to support your conclusions. Check with your teacher for additional guidelines for this assignment.

Step 7   **Exit ArcView**

When you are finished analyzing the El Niño and La Niña data, you may exit ArcView.

a   Save your project.

b   Exit ArcView by choosing Exit from the File menu. When asked if you want to save changes to your project, click No.

NAME _____ DATE _____

Student answer sheet

Module 3
Physical Geography II:
Ecosystems, Climate, and Vegetation

Advanced investigation: Sibling Rivalry

Step 4   Start ArcView and add the georeferenced images

h   Write the new name you gave the project and where you saved it.

_____   _____
(Name of project.   **(Navigation path to where project is saved.**
For example: kzeregion5.apr)   **For example: C:\student\kze\mapworld)**

Step 5   Analyze characteristics of El Niño and La Niña

c   Record your observations of precipitation characteristics for the same time period.

YEAR / SO	TEMPERATURE CHARACTERISTICS	PRECIPITATION CHARACTERISTICS
1997 / El Niño		
1999 / La Niña		

f   Synthesize the information you've recorded and develop a definition of El Niño and La Niña. Write these definitions below.

El Niño:

_____

_____

_____

_____

La Niña:

_____

_____

_____

_____

Step 6   Are El Niño and La Niña equal and opposite?

a   Complete the table below by filling in your observations for El Niño and La Niña's effect on the different regions of the world.

WORLD REGION	TEMPERATURE		PRECIPITATION		OTHER		OTHER	
	DEC. 1997 EL NIÑO	MARCH 1999 LA NIÑA	DEC. 1997 EL NIÑO	MARCH 1999 LA NIÑA	DEC. 1997 EL NIÑO	MARCH 1999 LA NIÑA	DEC. 1997 EL NIÑO	MARCH 1999 LA NIÑA
North America								
South America								
Europe								
Africa								

Step 6   Are El Niño and La Niña equal and opposite? (continued)

a   (continued)

WORLD REGION	TEMPERATURE		PRECIPITATION		OTHER		OTHER	
	DEC. 1997 EL NIÑO	MARCH 1999 LA NIÑA	DEC. 1997 EL NIÑO	MARCH 1999 LA NIÑA	DEC. 1997 EL NIÑO	MARCH 1999 LA NIÑA	DEC. 1997 EL NIÑO	MARCH 1999 LA NIÑA
Asia								
Oceania								
Pacific Ocean								

b   Based on the data you recorded in the previous question, is La Niña equal and opposite to El Niño? Explain your answer.

_____

_____

_____

_____

c-1   Was one year better than the other for you and your community?

_____

_____

_____

_____

c-2   If in one year you received greater than normal rainfall, did your town have problems with flooding?

_____

_____

_____

_____

c-3   If weather was unseasonably warm and mild, did outdoor activities such as amusement parks have greater turnout?

_____

_____

_____

_____

c-4   How did these things affect the local economy?

_____

_____

_____

_____

_____

_____

_____

module 4

Human Geography I
Population Patterns and Processes

The distribution and size of the earth's population results from the interplay between the complex forces of growth and migration.

The March of Time: A global perspective
Students will observe and analyze the location and population of the world's largest cities from the year 100 C.E. through 2000 C.E. In this investigation, students will describe spatial patterns of growth and change among the world's largest urban centers during the past two thousand years and speculate on reasons for the patterns they observe.

Growing Pains: A regional case study of Europe and Africa
In this lesson, students will compare the processes and implications of population growth in the world's fastest and slowest growing regions: Sub-Saharan Africa and Europe. Through the analysis of standard of living indicators in these two regions, students will explore some of the social and economic implications of rapid population growth.

Generation Gaps: An advanced investigation
Students will use global data to investigate variations in population age structure and the relationship of age structure to a country's rate of natural increase. After exploring global patterns, they will download and map U.S. Census data to examine the patterns of age structure in their own community.

The March of Time
A global perspective

Lesson overview

Students will observe and analyze the location and population of the world's largest cities from the year 100 C.E. through 2000 C.E. In this investigation, students will describe spatial patterns of growth and change among the world's largest urban centers during the past two thousand years and speculate on reasons for the patterns they observe.

Estimated time　　One to two 45-minute class periods

Materials　　✔ Student handouts from this lesson to be photocopied:
- GIS Investigation sheets (pages 187 to 196)
- Student answer sheets (pages 197 to 200)
- Assessment(s) (pages 202 to 207)
- Photocopy of the transparency (page 201)

Standards and objectives

National geography standards

	GEOGRAPHY STANDARD	MIDDLE SCHOOL	HIGH SCHOOL
1	How to use maps and other geographic representations, tools, and technologies to acquire, process, and report information from a spatial perspective	The student knows and understands how to use maps to analyze spatial distributions and patterns.	The student knows and understands how to use geographic representations and tools to analyze and explain geographic problems.
12	The processes, patterns, and functions of human settlement	The student knows and understands the spatial patterns of settlement in different regions of the world.	The student knows and understands the functions, sizes, and spatial arrangements of urban areas.
17	How to apply geography to interpret the past	The student knows and understands how the spatial organization of a society changes over time.	The student knows and understands how processes of spatial change affect events and conditions.

Objectives
The student is able to:
- Describe the location and size of the world's largest cities over time.
- Identify historical events and periods that influenced the location of cities throughout history.
- Explain the exponential pattern of growth among the world's urban populations in the past thousand years.

GIS skills and tools

 Use the Identify tool to learn more about a selected record

 Open the attribute table for a theme

 Sort data in descending order

 Add themes to the view

 Use the Find tool to locate a specific feature

 Label selected features in the active theme

 Clear all selections made

 Use the Pointer tool to select a label or graphic in the view

    Zoom in and out of the view

- Turn themes on and off
- Add and view image data
- Change the order of the table of contents to change the map display
- Rename a theme

For more on geographic inquiry and these steps, see Geographic Inquiry: Thinking Geographically (pages xxi to xxiii).

Teacher notes

Lesson introduction

Begin the lesson by dividing the students into pairs. Challenge each group to name the ten most populated cities in the world today. After five minutes, each group should share their list with the rest of the class. Use the blackboard or an overhead projector to tally the cities mentioned as each group reports. Based on the tally, circle the cities that were listed most often. Tell the class that they are going to do a GIS Investigation that will use real data to identify the 10 most populated cities in the world from the last two thousand years.

Before beginning the GIS Investigation, engage students in a discussion about the cities that are circled on their list.

- What do they know about these cities?
- In what countries are these cities located?
- How many people live in these urban centers?
- Has anyone ever visited one of these cities?
- Can they think of any reasons why some cities grow to be so large?

Student activity

 Before completing this lesson with students, we recommend that you complete it as well. Doing so will allow you to modify the activity to accommodate the specific needs of your students.

After the initial discussion, have the students work on the computer component of the lesson. Ideally, each student should be at an individual computer, but the lesson can be modified to accommodate a variety of instructional settings.

Distribute the March of Time GIS Investigation to the students. Explain that in this activity, they will use GIS to observe and analyze the location and size of the world's 10 largest cities in seven different time periods from 100 C.E. to the year 2000. The activity sheets will provide them with detailed instructions for their investigations. As they investigate, they will identify changes in both location and size of the world's largest cities, and speculate on possible reasons for the patterns they describe.

In addition to instructions, the activity sheets include questions to help students focus on key concepts. Some questions will have specific answers while others will require creative thought.

Things to look for while the students are working on this activity:

- Are the students using a variety of GIS tools?
- Are the students answering the questions as they work through the procedure?
- Are students asking thoughtful questions throughout the investigation?

Conclusion

Use the March of Time transparency to compare student ideas and observations from the GIS Investigation. Summarize student observations on the transparency as they share and discuss their observations with the class. Use this discussion as a forum to elaborate on relevant themes in world history (such as the decline of the Roman Empire or the Industrial Revolution) and the value of using geography's spatial perspective to interpret the past. Ideally, this discussion should take place in the classroom with a projection device that displays the ArcView project as students discuss it. If this is not possible, teachers may choose to conduct the discussion while students are still working on the computer individually or in small groups.

Assessment

Middle school: Highlights skills appropriate to grades 5 through 8

The middle school March of Time assessment asks the students to create a line graph of the most populous cities for each time period studied. Students will use the graph as a reference when writing an essay that compares two major time periods they have mapped. The essay will illustrate their knowledge and understanding of the changes in spatial patterns of major population centers.

High school: Highlights skills appropriate to grades 9 through 12

The high school March of Time assessment asks the students to create a line graph of the most populous cities for each time period studied. Students will use the graph as a reference when writing an essay that compares and contrasts three or more time periods they have mapped. The essay will illustrate their knowledge and understanding of the changes in spatial patterns of major population centers. In addition, they will take the historical information from the map and their own research to make predictions about future locations of major population centers.

Extensions

- Explore the relationship between physical characteristics of the landscape and the location of the world's most populated cities by adding themes reflecting world climate and vegetation data.

- Assign students to conduct research on the historic cities and time periods reflected in the project.

- Ask students what questions this activity has raised in their minds. Questions generated could serve as a springboard for further research and spatial analysis.

- Create map layouts in ArcView of the historic time periods. Have students print these layouts and use them in reports about the ancient European cities.

- Check out the Resources by Module section of the Teacher Resource CD for print and media resources on the topics of population, historical time periods, and ancient cities or visit *www.esri.com/mappingourworld* for Internet links.

NAME _____   DATE _____

The March of Time
A GIS investigation

ACQUIRE

ASK

EXPLORE

ACT

ANALYZE

Answer all questions on the student answer sheet handout

Step 1   **Start ArcView**

   a   Double-click the ArcView icon on your computer's desktop.

   b   If the Welcome to ArcView dialog appears (pictured below), click **Open an Existing Project** and click OK. If it doesn't appear, proceed to step 2.

Step 2   Open the global4.apr file

a   In this exercise, a project file has been created for you. To open the file, go to the **File** menu and choose **Open Project**.

b   Navigate to the exercise data directory (**C:\esri\mapworld\mod4**) and choose **global4.apr** from the list.

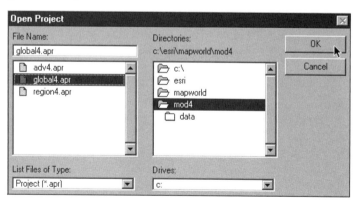

c   Click OK.

When the project opens, you see a world map in the View window with the following themes displayed: Geogrid, Major Rivers, Continents, and Ocean. The check mark next to the theme name tells you the theme is turned on and visible in the view.

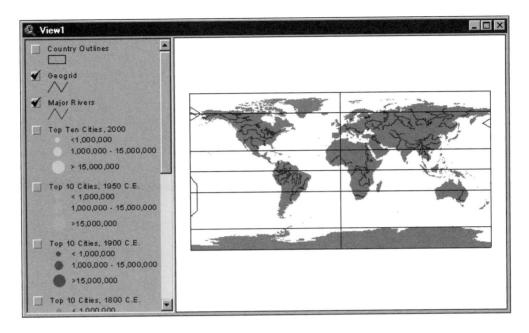

Step 3   Look at cities in 100 C.E.

a   The table of contents of the map is located on the left side of the ArcView window. Scroll down the table of contents until you see the theme called **Top 10 Cities, 100 C.E.** Click the box to the left of the name to turn on the theme.

This places a check mark in the box and adds a layer of points showing the locations of cities on the map.

b   On the answer sheet, write three observations about the location of the ten largest cities in the world in 100 C.E.

   (1)   Where are they located on the earth's surface?

   (2)   Where are they located in relation to each other?

   (3)   Where are they located in relation to physical features?

c   What are possible explanations for the patterns you see on this map?

Step 4   **Find historic cities and identify modern cities and countries**

   a   Make Top 10 Cities, 100 C.E. active by clicking its name in the table of contents. (Do not click the box with the check mark because the theme will no longer be displayed in the view.)

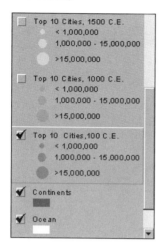

You know the theme is active because the box around it appears slightly raised above the other themes in the table of contents.

    b   You will use the Find tool to locate the historic cities on the map. Click the Find tool.

   c   Type **Carthage** in the Find Text in Attributes dialog box.

   d   Click OK.

The dot that represents Carthage turns yellow in the view (the map). In order to complete the table on the answer sheet, you need to identify the modern city located in the same location.

    e   Click the Identify tool and move your cursor over the view area. Notice how the cursor has a small "i" next to it.

   f   Click the yellow dot that represents Carthage.

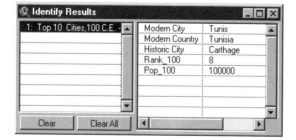

An Identify Results dialog appears with information in it that's needed to complete the table on the answer sheet.

?   *g*   Write the modern city name and modern country name for Carthage in the table on the answer sheet.

     h   When you're finished reading the information in the window, move it off the map by clicking its title bar and dragging it out of the way.

?   *i*   Use the Find tool and Identify tool to complete the information in the table on the answer sheet.

     j   Close the Identify Results window.

     k   Click the Clear All Selections button.

Step 5   Find the largest city of 100 C.E. and label it

?   *a*   What's your estimate of how many people lived in the world's largest city in 100 C.E.?

     b   Make sure Top 10 Cities, 100 C.E. is active.

     c   Click the Zoom to Active Themes button. This allows you to quickly zoom to the region of the world where the top 10 cities in 100 C.E. are located.

     d   Click the Open Theme Table button. The table shows you all the attribute data associated with the red dots on the map.

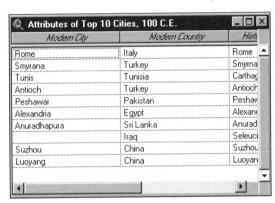

Attributes of Top 10 Cities, 100 C.E.

Modern City	Modern Country	Hist
Rome	Italy	Rome
Smyrana	Turkey	Smyrna
Tunis	Tunisia	Carthac
Antioch	Turkey	Antioch
Peshawar	Pakistan	Peshaw
Alexandria	Egypt	Alexand
Anuradhapura	Sri Lanka	Anurad
	Iraq	Seleuci
Suzhou	China	Suzhou
Luoyang	China	Luoyan

> *Note: Do not maximize this window. Doing so makes it difficult to get back to the view.*

e   Look at the table. Scroll right to see more entries. Remember: each entry in this table represents one of the points on the map.

f   Scroll right and locate the field name Pop_100. Click the name. It turns darker than the other field names when you click it.

Attributes of Top 10 Cities, 100 C.E.

Modern Country	Historic City	Rank_100	Pop_100
Italy	Rome	1	450000
Turkey	Smyrna	10	90000
Tunisia	Carthage	8	100000
Turkey	Antioch	5	150000
Pakistan	Peshawar	7	120000
Egypt	Alexandria	4	250000
Sri Lanka	Anuradhapura	6	130000
Iraq	Seleucia	3	250000
China	Suzhou	9	-99
China	Luoyang	2	420000

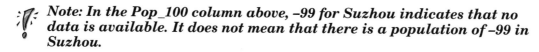

 Note: In the Pop_100 column above, –99 for Suzhou indicates that no data is available. It does not mean that there is a population of –99 in Suzhou.

 g Click the Sort Descending button to list the cities from largest to smallest in terms of population. Evaluate the table and answer the following questions:

? (1) *What was the largest city in 100 C.E.?*

? (2) *What was the population of the world's largest city in 100 C.E.?*

h Click the name of the largest city in the attribute table.

The selected record turns yellow in the attribute table and its corresponding dot on the map is yellow.

i Close the Attributes of Top 10 Cities, 100 C.E. table by clicking the × in the upper right of the window.

 j Click the Label tool. Click the selected city on the map to label it.

 k If you want to reposition the label, use the Pointer tool to click the text and drag it to where you want it to go.

 Hint: If you want to change the size of the text, choose Show Symbol Window from the Window menu, and change the size of the font by clicking the Font button. Change the font size. Close the Symbol window. The font size is changed.

 l Use the Clear Selection button to make all the cities in 100 C.E. red again. Click anywhere on the view to get rid of the black squares that surround the label.

 m Click the Zoom to Full Extent button to see the entire world in the view.

Step 6 Look at cities in 1000 C.E. and label the most populous city

a Turn on the Top Ten Cities, 1000 C.E. theme. Make the theme active (click the theme name).

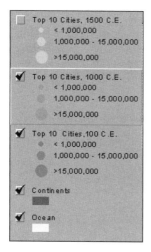

 b Click the Zoom to Active Theme button to zoom to the region of the world where the top 10 cities in 1000 C.E. are located.

c Look for similarities and differences between these points and the cities of 100 C.E. in the location and distribution of the world's largest cities.

(1)   What notable changes can you see from 100 C.E. to 1000 C.E.?

(2)   What similarities can you see between 100 C.E. and 1000 C.E.?

Now you will find, identify, and label the most populous city of 1000 C.E.

d First, make sure Top 10 Cities, 1000 C.E. is active.

e Click the Open Theme Table button. This shows you all the attribute data associated with the blue dots on the map.

> *Note: Do not maximize this window. Doing so makes it difficult to get back to the view.*

f Look at the table. Scroll right to see more entries. Remember: Each entry in this table represents one of the points on the map.

g Scroll right and locate the field name Pop_1000. Click the name. It turns darker than the other field names.

h Click the Sort Descending button to list the cities from largest to smallest in terms of population. Evaluate the table and answer the following two questions:

(1)   What was the largest city in 1000 C.E.?

(2)   What was the population of the world's largest city in 1000 C.E.?

i Click the name of the largest city in the Attributes Table. The selected record turns yellow in the attribute table and its corresponding dot on the map is yellow.

j Close the Attributes of Top 10 Cities, 1000 C.E. table.

k Click the Label tool. Click the selected city on the map to label it.

l If you want to reposition the label, use the Pointer tool to click the text and drag it to where you want it to go.

> *Hint: If you want to change the size of the text, choose Show Symbol Window from the Window menu, and change the size of the font by clicking on the Font button. Change the font size. Close the Symbol window. The font size is changed.*

m Use the Clear Selection button to make all the cities in 1000 C.E. blue again.

n Click the view to get rid of the black squares surrounding the label.

Step 7   Observe and compare other historic periods

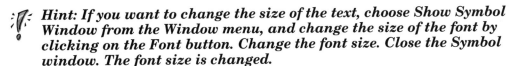

a Make the Continents theme active. Click Zoom to Active Theme. You see the entire world map.

b One at a time, turn on each of the remaining five themes representing the years 1500, 1800, 1900, 1950, and 2000.

As you do so, explore the map to complete the table on the answer sheet. Refer back to steps 5 and 6 to get information for 100 C.E. and 1000 C.E.

Hint: Turn themes on and off or change the order of the themes by moving them up or down in the table of contents to see each of the largest cities from all the time periods. Each tool works only on the active theme.

Step 8   State a hypothesis

 a   Make the Continents theme active. Click Zoom to Active Theme. You see the entire world map.

 b   Look back at the maps and at the chart you completed in step 7. In the table on the answer sheet, complete the left column by stating which periods in history are associated with the greatest changes. In the right column, state possible explanations for the changes that you see.

Step 9   Investigate cities in the present time

a   As a class, before you began the GIS Investigation, you made a guess about the top 10 cities in the world today. Turn on the Top Ten Cities, 2000 theme to see if any of your predictions are correct.

 (1)   How many of your original guesses are among the cities in Top Ten Cities, 2000?

(2)   Which cities did you successfully guess?

Most likely, there are cities on your list that are not in the Top Ten Cities, 2000 theme. In order to look at population data on these cities, you will add population data from 1998.

 b   Click the Add Theme button.

c   Navigate to the data directory (**C:\esri\mapworld\mod4\data**). Select **cities.shp** and click OK.

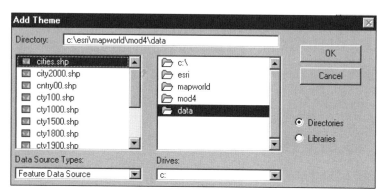

d   The major cities theme is added to your table of contents. Click the box next to cities.shp to turn on the theme.

e   Make Cities.shp the active theme by clicking on the theme name.

The box surrounding the name appears raised.

f From the Theme menu, select Properties. Change the theme name to **Major Cities, 1998**.

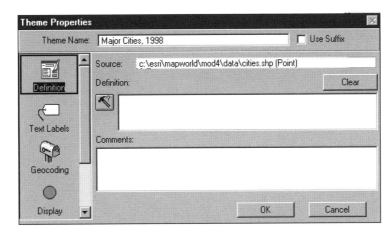

g Click OK.

The name changes in the table of contents.

h Use the Find tool to locate the other cities that you guessed in the beginning of this activity.

See example below:

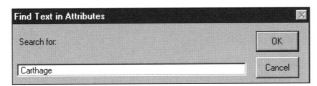

The city you selected is highlighted in yellow in the view. Because there are so many cities in this data set, you may not be able to see the yellow dot. If this is the case, click the Zoom to Selected button to focus on your selected city.

i Click the Identify tool and move your cursor over the view area.

j Click on the selected city in the view (the one yellow dot).

The Identify Results dialog pops up with information like that in the graphic below.

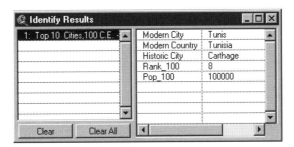

Note: This data is from 1998. It will give you a good but inexact comparison to the year 2000 data.

k Continue to find population data for other major cities that interest you. Use the Find tool and the Identify tool. When you're finished, close the Identify Results window.

? *(1) What are the rankings of the cities you chose that are not in the top 10?*

? *(2) How far are they from the top 10?*

Step 10 **Exit ArcView**

In this exercise, you explored population data from 100 C.E. through the year 2000. You used ArcView to find and identify and label the world's most populous cities throughout different time periods. After analyzing this data, you added data from 1998 and explored populations of your cities of interest.

a Ask your teacher for instructions on where to save this ArcView project and on how to rename the project.

b If you are not going to save the project, exit ArcView by choosing Exit from the File menu. When it asks if you want to save changes to global4.apr, click No.

NAME _____   DATE _____

Student answer sheet
Module 4
Human Geography I: Population Patterns and Processes

Global perspective: The March of Time

Step 3   Look at cities in 100 C.E.

b-1   Where are they located on the earth's surface?

_____

_____

_____

_____

b-2   Where are they located in relation to each other?

_____

_____

_____

_____

b-3   Where are they located in relation to physical features?

_____

_____

_____

_____

c   What are possible explanations for the patterns you see on this map?

_____

_____

_____

_____

_____

_____

_____

Step 4   Find historic cities and identify modern cities and countries

g, i   Use the Find tool and Identify tool to complete the information in the table below.

HISTORIC CITY NAME	MODERN CITY NAME	MODERN COUNTRY
Carthage		
Antioch		
Peshawar		

Step 5   Find the largest city of 100 C.E. and label it

a   What's your estimate of how many people lived in the world's largest city in 100 C.E.?

_____

g-1   What was the largest city in 100 C.E.?   _____

g-2   What was the population of the world's largest city in 100 C.E.?   _____

Step 6   Look at cities in 1000 C.E. and label the most populous city

c-1   What notable changes can you see from 100 C.E. to 1000 C.E.?

_____

_____

_____

_____

_____

c-2   What similarities can you see between 100 C.E. and 1000 C.E.?

_____

_____

_____

_____

_____

h-1   What was the largest city in 1000 C.E.?   _____

h-2   What was the population of the world's largest city in 1000 C.E.?   _____

Step 7 Observe and compare other historic periods

b   Explore the map to complete the table below. Refer back to steps 5 and 6 to get information for 100 C.E. and 1000 C.E.

YEAR C.E.	LARGEST CITY	POPULATION OF LARGEST CITY	MAJOR DIFFERENCES FROM PREVIOUS TIME PERIOD
100			
1000			
1500			
1800			
1900			
1950			
2000			

Step 8   State a hypothesis

b   In the table below, state in the left column which periods in history are associated with the greatest changes. In the right column, state possible explanations for the changes you see.

TIME PERIODS OF SIGNIFICANT CHANGE	EXPLANATION FOR CHANGE

Step 9   Investigate cities in the present time

a-1   How many of your original guesses are among the cities in Top Ten Cities, 2000?

_____

a-2   Which cities did you successfully guess?

_____

_____

_____

_____

_____

k-1   What are the rankings of the cities you chose that are not in the top 10?

_____

k-2   How far are they from the top 10? _____

Transparency master

YEAR C.E.	LARGEST CITY	POPULATION OF LARGEST CITY	MAJOR DIFFERENCES FROM PREVIOUS TIME PERIOD
100			
1000			
1500			
1800			
1900			
1950			
2000			

NAME _____   DATE _____

The March of Time
Middle school assessment

Part 1

Use the information you collected in steps 5, 6, and 7 of the GIS Investigation to complete the attached graph. Plot a line graph that shows the population of the largest city in each time period. Be sure to label the points and title the graph appropriately.

Part 2

Refer to the ArcView map you have created and write an essay that compares two of the time periods that you studied. You may use additional sources such as your history and geography books, encyclopedias, and the Internet to help you with your comparisons. On a separate piece of paper, address the following questions in your essay:

1   What was the primary means of transportation in each time period?
2   How did trade between various cities influence the location of the places?
3   How did advancements in technology affect the location of places?
4   How did distances between major cities change throughout time?
5   What physical features played important roles in the locations of cities (location relative to fresh water and waterways, elevation, etc.)?

Use the remainder of this page as a place to brainstorm for your essay.

The March of Time

Assessment rubric

Middle school

STANDARD	EXEMPLARY	MASTERY	INTRODUCTORY	DOES NOT MEET REQUIREMENTS
The student knows and understands how to use maps to analyze spatial distributions and patterns.	Creates an accurate and well-labeled graph of the most populous cities over time. Uses GIS to analyze population patterns and how they change throughout time, compares this with his/her original predictions, and makes predictions on future trends.	Creates an accurate graph of the most populous cities over time. Uses GIS to analyze population patterns and how they change throughout time, and compares this with his/her original predictions.	Creates a graph of the most populous cities over time that is inaccurate or mislabeled in some places. Uses GIS to view patterns of the location of major population centers.	Does not complete a graph of the most populous cities over time. Has difficulty identifying location patterns of major population centers.
The student knows and understands the spatial patterns of settlement in different regions of the world.	Describes in detail how major cities have changed and are influenced by their location relative to physical features, and elaborates on what these features are, and if they were a positive or negative force on the cities.	Identifies how major cities have changed and are influenced by their location, relative to physical features such as water and other major cities.	Identifies how the major cities (population centers) have spread throughout the various regions of the world.	Identifies the location of major cities, but does not identify a pattern of settlement.
The student knows and understands how the spatial organization of a society changes over time.	Compares the major cities of at least two time periods and makes predictions as to why changes occurred over time. Provides detailed evidence of how things such as technology and transportation influenced the societies that lived in these cities.	Compares the major cities of two time periods, and makes predictions as to why changes in these population centers occurred over time.	Identifies characteristics of the major population centers of two different time periods, but does not draw comparisons between the two eras.	Identifies a few characteristics of major population centers of two different time periods, or only describes characteristics of one time period's settlements.

This is a four-point rubric based on the National Standards for Geographic Education. The "Mastery" level meets the target objective for grades 5–8.

NAME _____   DATE _____

The March of Time
High school assessment

Part 1

Use the information you collected in steps 5, 6, and 7 of the GIS Investigation to complete the attached graph. Plot a line graph that shows the population of the largest city in each time period. Be sure to label the points and title the graph appropriately.

Part 2

Refer to the ArcView map you have created and write an essay that compares two of the time periods that you studied. Use at least three additional sources such as your history and geography books, encyclopedias, and the Internet to help you with your comparisons. On a separate piece of paper, address the following questions in your essay:

1   What was the primary means of transportation in each time period?

2   How did trade between various cities influence the location of the places?

3   What physical features played important roles in the locations of cities (location relative to fresh water and waterways, elevation, etc.)?

4   Are there any unusual shifts in population centers that you see from one period to the next? What do you think caused these changes?

5   How did advancements in technology affect the location of places?

6   Predict how future advancements in technology may affect the location of population centers in the next hundred years.

Use the remainder of the page as a place to brainstorm for your essay.

The March of Time

Assessment rubric

High school

STANDARD	EXEMPLARY	MASTERY	INTRODUCTORY	DOES NOT MEET REQUIREMENTS
The student knows and understands how to use geographic representations and tools to analyze and explain geographic problems.	Creates an accurate and well-labeled graph of the most populous cities over time. Uses GIS to analyze how population patterns change throughout time, compares this with his/her original thoughts, and makes predictions on future trends. Uses GIS to create an original map to illustrate these points.	Creates an accurate graph of the most populous cities over time. Uses GIS to analyze how population patterns change throughout time, compares this with his/her original thoughts, and makes predictions on future trends.	Creates a graph of the most populous cities over time that is inaccurate or mislabeled in some places. Uses GIS to analyze population patterns and how they change throughout time.	Does not complete a graph of the most populous cities over time. Uses GIS to analyze population patterns, but has difficulty identifying how they change throughout time.
The student knows and understands the functions, sizes, and spatial arrangements of urban areas.	Elaborates, by using visual and written products, on characteristics of major cities throughout time, and how factors such as trade routes and technology influenced the growth of these places.	Identifies characteristics of major cities throughout time and how factors such as trade routes and technology influenced the growth of these places.	Identifies characteristics of major cities throughout time.	Identifies major cities and a few characteristics of these places.
The student knows and understands how processes of spatial change affect events and conditions.	Cites specific examples of how changes in politics, transportation, and so on caused major population centers to shift from one location to another throughout time.	Identifies how changes in politics, transportation, and so on influenced the location of major population centers throughout time.	Lists one or two major cities that changed from one time period to the next and provides explanation for the change.	Lists one or two major cities that changed from one time period to the next.

This is a four-point rubric based on the National Standards for Geographic Education. The "Mastery" level meets the target objective for grades 9–12.

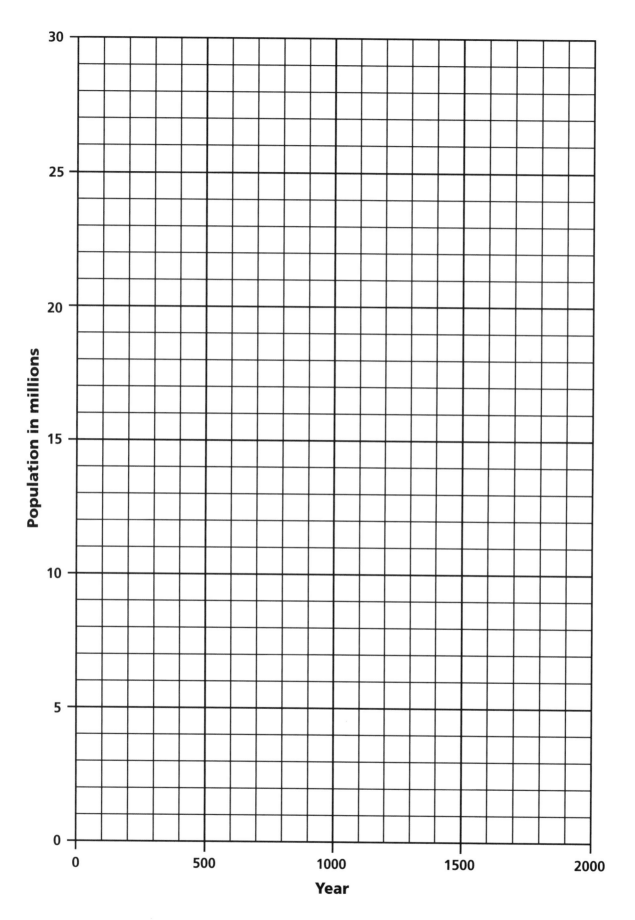

MODULE 4 • HUMAN GEOGRAPHY I: POPULATION PATTERNS AND PROCESSES

Answer key to graph

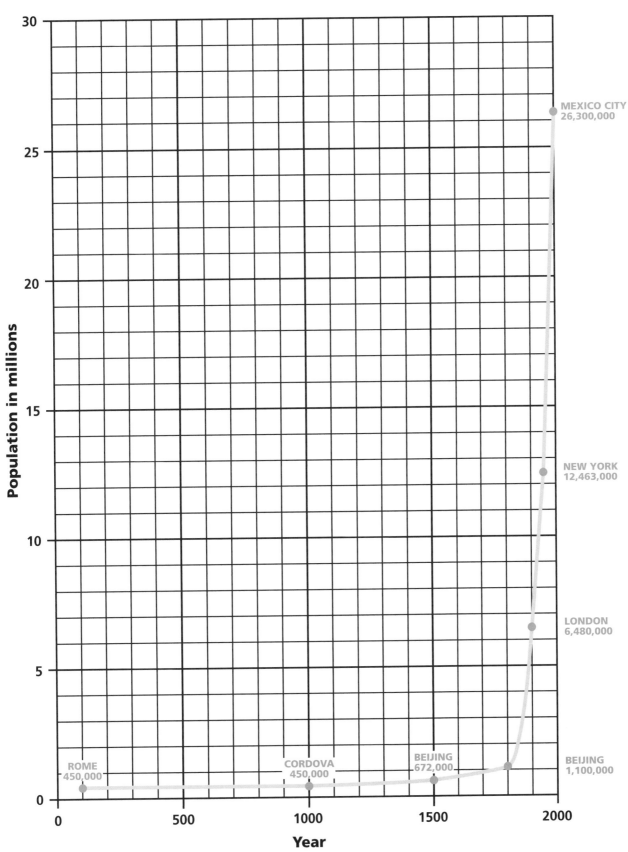

Population in millions (y-axis)

Year (x-axis)

MEXICO CITY
26,300,000

NEW YORK
12,463,000

LONDON
6,480,000

BEIJING
1,100,000

ROME
450,000

CORDOVA
450,000

BEIJING
672,000

Growing Pains
A regional case study of Europe and Africa

Lesson overview

In this lesson, students will compare the processes and implications of population growth in the world's fastest and slowest growing regions: sub-Saharan Africa and Europe. Through the analysis of standard of living indicators in these two regions, students will explore some of the social and economic implications of rapid population growth.

Estimated time   Two to three 45-minute class periods

Materials   ✔ Large sheets of construction paper or drawing paper and markers—one sheet of paper and one marker per student

         ✔ Student handouts from this lesson to be copied:
- GIS Investigation sheets (pages 213 to 224)
- Student answer sheets (pages 225 to 229)
- Assessment(s) (pages 230 to 233)

Standards and objectives

National geography standards

	GEOGRAPHY STANDARD	MIDDLE SCHOOL	HIGH SCHOOL
1	How to use maps and other geographic representations, tools, and technologies to acquire, process, and report information from a spatial perspective	The student knows and understands how to use maps to analyze spatial distributions and patterns.	The student knows and understands how to use geographic representations and tools to analyze and explain geographic problems.
9	The characteristics, distribution, and migration of human populations on Earth's surface	The student knows and understands the demographic structure of a population.	The student knows and understands trends in world population numbers and patterns.
18	How to apply geography to interpret the present and plan for the future	The student knows and understands how to apply the geographic point of view to social problems by making geographically informed decisions.	The student knows and understands how to use geographic knowledge, skills, and perspectives to analyze problems and make decisions.

Standards and objectives (continued)

Objectives

The student is able to:

- Describe the fundamentals of population growth by explaining the relationship between birth rate, death rate, and natural increase.
- Identify the fastest and slowest growing regions in the world today.
- Explain the socioeconomic implications of rapid population growth.
- Explain and understand the slow population growth in Europe and how standard of living indicators are affected.
- Explain and understand the rapid population growth in Africa and how standard of living indicators are affected.

GIS skills and tools

 Add themes to the view

 Change the properties of a theme

       Zoom in and out of the view

 Pan the view to see different areas of the map

 Use the Identify tool to learn more about a selected record

 Use the Text tool to add text to a layout of a map

 Resize or move text and graphics in a layout with the Pointer tool

 Change the theme name by using the Theme Properties button

- Turn themes on and off
- Change the order of the table of contents to change the map display
- Change theme names
- Create a layout and print a map

For more on geographic inquiry and these steps, see Geographic Inquiry: Thinking Geographically (pages xxi to xxiii).

Teacher notes

Lesson introduction Begin the lesson with a discussion of world population growth. Remind the students that the world's population reached six billion in 1999 and continues to grow.

Consider the following questions:

- Why is the earth's population growing?
- Are all regions growing at the same rate of speed?
- What does "overpopulation" mean and at what point would you characterize the world as overpopulated?

If your community or state is growing, that would be a good starting point for this discussion. How has population growth affected your community or state? Do students see this as a good thing or a bad thing?

Student activity   **Before completing this lesson with students, we recommend that you work through it yourself. Doing so will allow you to modify the activity to accommodate the specific needs of your students.**

After the initial discussion, have the students work on the computer component of the lesson. Ideally, each student should be at an individual computer, but the lesson can be modified to accommodate a variety of instructional settings.

Distribute the GIS Investigation handout to the students. Explain that in this activity, they will use GIS to identify the regions of the world where population is growing fastest and where it is growing slowest. In addition, they will use GIS to compare characteristics of regions with slow and rapid population growth. As they compare, they will gather information and form a hypothesis about the relationship between a region's rate of population growth and its standard of living.

The GIS activity will provide students with detailed instructions for their investigations. In addition to the instructions, the handout includes questions to help students focus on key concepts. Some questions will have specific answers while others call for speculation and require creative thought.

Things to look for while the students are working on this activity:

- Are the students using a variety of tools?
- Are the students answering the questions as they work through the procedure?
- Do students need help with the lesson's vocabulary?

Conclusion When the class has finished the GIS Investigation, give each student a piece of drawing paper on which to print the hypothesis they wrote in step 7 of the GIS Investigation. Students should tape their hypothesis to the wall or blackboard where they will provide the focus for a concluding discussion. Look for similarities and differences among the student hypotheses. Allow students to question each other to clarify confusing or contradictory statements. If possible, try to reach consensus about the relationship between a country's rate of natural increase and its standard of living based on evidence from the GIS Investigation.

Assessment

Middle school: Highlights skills appropriate to grades 5 through 8

The middle school assessment has the students taking the role of a special liaison to the United Nations, in charge of establishing partnerships between nations of slow and fast growth. The students will select two countries—one with fast growth and one with slow growth. They will use their work in the GIS Investigation to identify issues critical to each country and devise a way the countries can form a partnership to improve the standard of living in each place.

High school: Highlights skills appropriate to grades 9 through 12

The high school assessment has the students taking the role of a special liaison to the United Nations, in charge of establishing partnerships between nations of slow and fast growth. The students will select two groups of countries—one with fast growth and one with slow growth. They will use their work in the GIS Investigation to identify issues critical to each group and devise a way the countries can form a partnership to improve the standard of living for all countries involved.

Extensions

- Use the ArcView Chart document to create charts illustrating the relationship between population growth and indicators of standard of living.
- Explore and map additional data from the module 4 data directory. Look for additional social and economic implications of rapid population growth.
- Students can test their hypothesis about the relationship between standard of living, net migration, and the rate of natural increase, by investigating countries in other parts of the world.
- Assign an African country to each student and ask them to explore the population, birth rate, death rate, and standard of living indicators of the country. Create map layouts of the individual countries and use them in a report about the African country.
- Use the ArcView query function to identify countries that do not match your hypothesis. Try to figure out an explanation for these anomalies.
- Explore gender differences in standard of living in Europe and/or sub-Saharan Africa. Using the module data, map and analyze male and female life expectancy, male and female literacy rates, and male and female infant and child mortality.
- Conduct research on the impact of HIV/AIDS on death rates and population growth in sub-Saharan Africa.
- Check out the Resources by Module section of the Teacher Resource CD for print and media resources on the topics of Africa, Europe, demographics, and standard of living indicators or visit *www.esri.com / mappingourworld* for Internet links.

NAME _____   DATE _____

Growing Pains
A GIS investigation

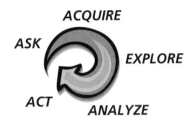

ACQUIRE

ASK

EXPLORE

ACT

ANALYZE

Answer all questions on the student answer sheet handout

In this GIS Investigation, you will observe and analyze population growth by looking at the natural increase of different countries. You will focus on Africa, the fastest growing region in the world, and Europe, the slowest growing region in the world. You will analyze the standard of living indicators for each region and form a hypothesis about the relationship between population growth and standard of living indicators.

Step 1   Start ArcView

a   Double-click the ArcView icon on your computer's desktop.

b   If the Welcome to ArcView dialog appears (pictured below), click **Open an Existing Project** and click OK. If it doesn't appear, proceed to step 2.

Step 2 Open the region4.apr file

a In this exercise, a project file has been created for you. To open the file, go to the **File** menu and choose **Open Project**.

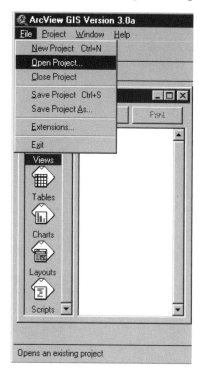

b Navigate to the exercise data directory (**C:\esri\mapworld\mod4**) and choose **region4.apr** from the list.

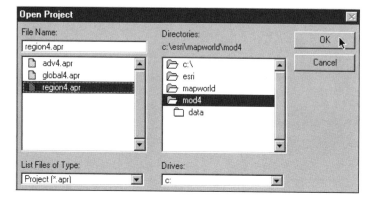

c Click OK.

d   When the project opens, click **Open** to open the Population Growth view.

You see a map in the window with two themes turned on (Countries and Ocean). The check mark next to the theme name tells you the theme is turned on and visible in the view. Two themes, Births/1000 and Deaths/1000, are listed in the table of contents, but are not turned on.

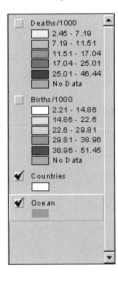

Step 3   Compare birth rate and death rate data

The world's population is growing because there are more births than deaths each year. This fact can be expressed as a simple formula:

Birth rate	= BR	BR − DR = N
Death rate	= − DR	
Natural increase	= NI	

In the first part of this activity, you will compare birth rates and death rates around the world to see if you can identify the regions that are growing fastest and slowest.

World Vital Events

World Vital Events Per Time Unit: 2001

(Figures may not add to totals due to rounding)

Time unit	Births	Deaths	Natural increase
Year	131,571,719	55,001,289	76,570,430
Month	10,964,310	4,583,441	6,380,869
Day	360,470	150,688	209,782
Hour	15,020	6,279	8,741
Minute	250	105	146
Second	4.2	1.7	2.4

Source: U.S. Bureau of the Census, International Data Base

This table from the U.S. Census Bureau shows the number of births, deaths, and rate of natural increase for the world population for every year, month, day, hour, and second in 2001.

a Turn on the Births/1000 theme by clicking the box to the left of the name in the table of contents. This places a check mark in the box and adds a layer displaying the number of births per thousand people.

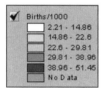

✔ Births/1000
- 2.21 - 14.86
- 14.86 - 22.6
- 22.6 - 29.81
- 29.81 - 38.96
- 38.96 - 51.45
- No Data

(1) Which world region or regions have the highest birth rates?

(2) Which world region or regions have the lowest birth rates?

b Turn on the Deaths/1000 theme by clicking the box to the left of its name.

(1) Which world region or regions have the highest death rates?

(2) Which world region or regions have the lowest death rates?

c Turn these themes on and off to compare them easily.

(1) If the overall rate of growth is based on the formula BR – DR = NI, which world regions do you think are growing the fastest?

(2) Which world regions do you think are growing the slowest?

d In order to learn more about the birth and death rates of specific countries, you can use the Identify tool. Click the Identify tool and move your cursor over the view area. Notice how the cursor has a small "i" next to it.

e Turn off the Deaths/1000 theme. Click the Births/1000 theme name to make it active. The box surrounding the theme name appears slightly raised.

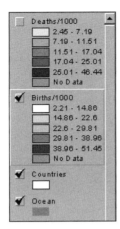

f Move the cursor over an African country where the birth rate is very high. Click on a country. You see an Identify Results window similar to the one below:

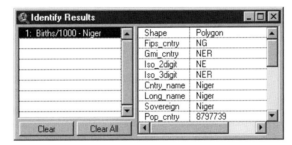

This database contains a lot of information about each country. As you scroll down the list, you see abbreviations in one column and numbers in another. The birth rate data is abbreviated as "Brthrate" and the death rate data is abbreviated as "Dthrate." In the example above, Niger has a birthrate of 51.45 births/1,000 living people, and a deathrate of 23.17 deaths/1,000 living people.

g Click the × in the upper right of the Identify Results window to close it.

h Choose two European countries and two African countries and record their birth and death rates in the table on the answer sheet. Use the Identify tool, as described above, to find information on your chosen countries.

i List three questions that these two maps raise in your mind.

Step 4 Add a natural increase theme

You can test your predictions of the fastest and slowest growing regions (step 3c) by adding the natural increase theme to your map. This theme shows the yearly increase in population that results from the difference between births and deaths in each country.

 a Click the Add Theme button.

b Navigate to the exercise data directory (**C:\esri\mapworld\mod4\data**). Scroll down and select **natr_inc.shp**. Click OK.

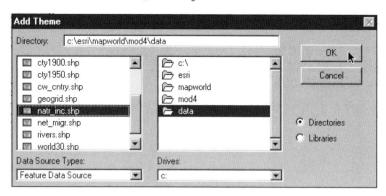

c Turn on the Natr_inc.shp theme. Turn off Births/1000 and Deaths/1000.

Like the birth and death rates, the numbers for Natural Increase are expressed as a rate per 1,000 population. This means that in all the countries colored dark blue on the map, there are between 20 and 39 people being added to the population each year for every 1,000 people already there.

> *Note: The actual growth rate of an individual country is based on its natural increase plus the net migration of people into or out of that country each year.*

(1) What is happening to the population in the countries that are red?

(2) Which world region is growing the fastest?

(3) Which world region is growing the slowest?

(4) Think about what it would mean for a country to have a population that is growing rapidly or one that is growing slowly. Which of these two possibilities (fast growth or slow growth) do you think would cause more problems within the country? On the answer sheet, briefly list some of the problems you would expect to see.

d Now you will change the name of the theme from Natr_inc.shp to Natural Increase. Click on the Natr_inc.shp theme to make it active.

e Click the Theme Properties button.

f Type **Natural Increase** in the Theme Name text box.

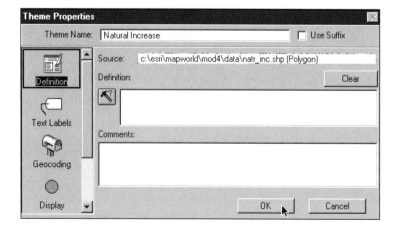

g   Click OK.

The theme name is changed in the table of contents. This descriptive name will be more useful than the abbreviated name.

Step 5   Look at standard of living indicators for Europe and Africa

Geographers look at certain key statistics when they want to compare the standard of living in different countries. They refer to these statistics as "indicators" because they typically reveal or provide some information about the quality of life in that country. The indicators that you will look at in this activity are the following:

- Population 60 years old and over: The percent of the total population that is in this age group.
- GDP per capita: The GDP (gross domestic product) divided by the total population for that year, where the GDP is expressed as a per-person figure.
- Infant mortality rate: The number of deaths of infants under one year old in a given year per 1,000 live births in the same year.
- Life expectancy: The number of years a newborn infant would live if prevailing conditions of mortality at the time of birth continue.
- Literacy rate: The percent of the adult population that can read and write.
- Services: The percent of the workforce that is employed in the service sector of the economy.

a   Click on the title bar of the Project window.

Notice that the Population Growth view goes "behind" the project window.

b   Click on Standard of Living Indicators to highlight it and click Open.

The Standard of Living Indicators view is focused on Europe and Africa. Europe is the slowest growing region and sub-Saharan Africa is the fastest growing region in the world.

c   Scroll down the table of contents to see each of the eight standard of living indicators theme names. The first indicator, Population >60, shows the percentage of each country's population over 60 years old.

In order to complete the chart on the answer sheet, you need to explore each of the eight standard of living indicators. Remember:

- A theme will cover the one beneath it when it is turned on. You will need to turn themes on and off to see all the indicators.
- You can also reposition the themes by dragging them to a new position in the table of contents.

? *d* Now, complete the table on the answer sheet.

Step 6   Add the net migration theme

The net rate of migration is a statistic that indicates the number of people per 1,000 gained or lost each year as a result of migration. A negative number indicates that more people are leaving the country than coming in. A positive number means more people are coming to the country than leaving it.

? *a* In step 5 you compared standard of living indicators in Europe and sub-Saharan Africa. Based on your observations of those indicators, which region would you expect to have a negative net migration? A positive net migration? Why?

b Click the Add Theme button.

c Navigate to the data directory (**C:\esri\mapworld\mod4\data**). Scroll down and select **net_migr.shp**.

d In the table of contents, click the box next to net_migr.shp to turn on the theme. Click the theme name to make it active.

You will now give the theme a more descriptive name: Net Migration.

 e Click the Theme Properties button.

f Type **Net Migration** in the Theme Name text box.

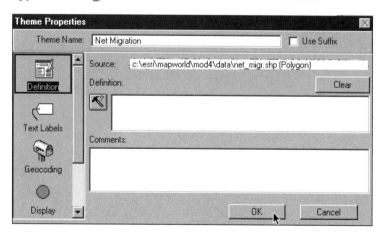

g Click OK.

? *h* Summarize the overall patterns of net migration in Europe and sub-Saharan Africa in the table on the answer sheet.

? *i* What are possible political or social conditions or events that could explain any of the migration patterns you see on the map?

Step 7   Draw conclusions

a   Click the Population Growth view title bar.

> *:⚡:* *Note: If you cannot see the title bar, click the Standard of Living title bar and move that view out of the way.*

b   Click the Zoom In tool and draw a box around Africa and Europe.

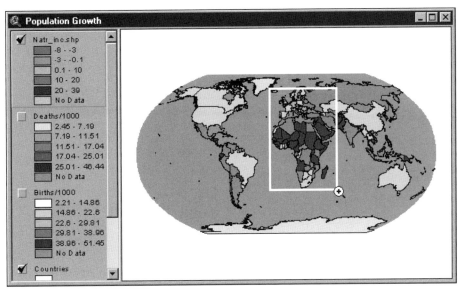

c   Make sure the Natr_inc.shp theme is turned on and active.

d   Toggle back and forth between the two views so you can see the Rate of Natural Increase for each country compared to its Net Migration. You can do this by clicking the title bar for each view.

Think about the following question: What correlation, if any, exists between countries with high rates of natural increase and high rates of net migration?

e   Based on your map investigations, write a hypothesis about how a country's rate of natural increase affects its standard of living and its net rate of migration.

f   In the table on the answer sheet, illustrate your hypothesis with data from one European country and one sub-Saharan African country. Use the Identify tool to see the data for an individual country. (Remember: You must make a theme active and click the Identify Tool before clicking on a country in the view.)

Step 8   Create a layout

a   Make sure the Population Growth view is active by clicking its title bar. Turn off all themes except Natural Increase and Ocean.

b   From the View menu, select Layout.

c   Select Landscape from the Template Manager and click OK.

d   You see a layout of the view, its map legend, a north arrow, and scale.

Quite often, it's easier to work with the map layout if it's larger.

e   Click the bottom right margin of the Layout window and drag the corner outward to make the window larger (see below for an example).

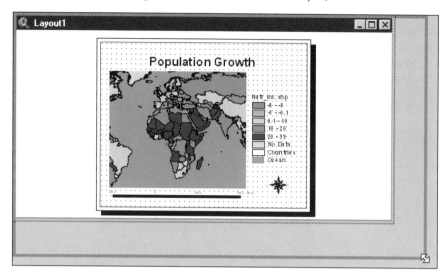

The layout redraws so it's larger.

f   Double-click the north arrow that's in the bottom right of the layout.

g   Choose a north arrow that you like from the North Arrow Template Manager by double-clicking it.

The layout automatically updates.

h   Resize the north arrow. Click on one of the black squares at the corners of the graphic and drag the square to make the graphic smaller or larger.

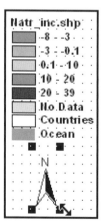

i   Click any blank area on the layout to complete the resize.

The black squares disappear from around the north arrow.

j   Double-click the scale bar to see its properties.

What are the units of measurement?

By default, ArcView assigns the correct units of measurement and intervals based on the view.

k   Because the displayed values are correct, click Cancel to return to the layout.

Step 9   Label your map and print it

When making maps, it's important to include the cartographer's name and date that the map was created. In order to make room for that in the bottom right corner of the map, you may need to move the map legend and north arrow up slightly.

a   Click the map legend once so you can see the black squares at the corners. Click the legend again and hold the cursor down. Drag the map legend to a suitable location closer to the top of the map.

b   Do the same for the north arrow. If necessary, move the title "Population Growth" over to the left.

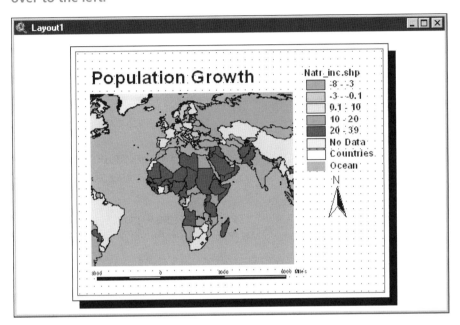

c   To add the name of the cartographer (that's you!) and date, click the Text tool.

d   Click the layout where the cartographer's name will go.

A Text Properties dialog appears.

e   In the dialog, type your name and the date of the project on the line below it. Click OK.

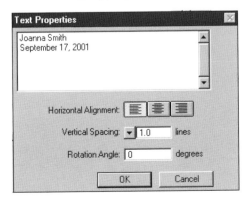

f   Click the Pointer tool. Click the text box with your name and date to resize it or move it.

Your layout of Population Growth is now ready to print. If you need to make any final adjustments to your map, do them now. Otherwise, proceed to the next step.

g   From the File menu, click Print. Click OK if the printer you want to use is selected. Otherwise, consult your teacher for instructions on how to print to the correct computer.

Step 10   Make a standard of living indicators map and print it

a   Close the Layout1 window.

b   Click the title bar of the Standard of Living Indicators view so you can see it.

c   Turn off all themes except Net Migration and Oceans. Move Net Migration to the top of the table of contents.

d   From the View menu, select Layout.

e   Select Landscape from the Template Manager and click OK.

f   Click OK to select a New Layout. By default, ArcView names this Layout2.

g   Follow the procedures outlined in steps 8 and 9 to design your presentation map and print it.

Step 11   Save your project and exit ArcView

In this exercise, you explored world population growth and analyzed standard of living indicators in the fastest and slowest growing regions of the world (Africa and Europe). You added themes and worked in two views. You created a layout for each view and printed your maps.

a   Ask your teacher for instructions on where to save this ArcView project and on how to rename the project. Save it according to your teacher's instructions.

b   If you are not going to save the project, exit ArcView by choosing Exit from the File menu. When asked if you want to save changes to region4.apr, click No.

NAME _____   DATE _____

Student answer sheet
Module 4
Human Geography I: Population Patterns and Processes

Regional case study: Growing Pains

Step 3   Compare birth rate and death rate data

a-1   Which world region or regions have the highest birth rates?

_____

a-2   Which world region or regions have the lowest birth rates?

_____

b-1   Which world region or regions have the highest death rates?

_____

b-2   Which world region or regions have the lowest death rates?

_____

c-1   If the overall rate of growth is based on the formula BR – DR = NI, which world regions do you think are
      growing the fastest?

_____

c-2   Which world regions do you think are growing the slowest?

_____

h   Choose two European and two African countries and record their birth and death rates below.

COUNTRY AND CONTINENT NAME	BIRTH RATE/1,000	DEATH RATE/1,000
Niger (Africa)	51.45	23.17

i   List three questions that these two maps raise in your mind.

_____

_____

_____

_____

_____

Step 4   Add a natural increase theme

c-1   What is happening to the population in the countries that are red?

_____

c-2   Which world region is growing the fastest?

_____

c-3   Which world region is growing the slowest?

_____

c-4   Think about what it would mean for a country to have a population that is growing rapidly or one that is growing slowly. Which of these two possibilities (fast growth or slow growth) do you think would cause more problems within the country? Briefly list some of the problems you would expect to see.

Most problems would occur in countries with how much growth?

_____

Problems there would include:

_____

_____

_____

_____

_____

Step 5   Look at standard of living indicators for Europe and Africa

d   Complete the following table:

INDICATOR	COMPARE SUB-SAHARAN AFRICA AND EUROPE	WHAT DOES THIS "INDICATE" ABOUT THE STANDARD OF LIVING IN THESE REGIONS?
Population > 60 years		
GDP per capita		
Infant mortality rate		
Life expectancy		
Literacy rate		
Percent of workforce in service sector		

Step 6   Add the net migration theme

a   In step 5 you compared standard of living indicators in Europe and sub-Saharan Africa. Based on your observations of those indicators, which region would you expect to have a negative net migration?

_____

A positive net migration?

_____

Explain your answer.

_____

_____

_____

_____

_____

h   Summarize the overall patterns of net migration in Europe and sub-Saharan Africa.

NET MIGRATION IN SUB-SAHARAN AFRICA	NET MIGRATION IN EUROPE

i   What are possible political or social conditions or events that could explain any of the migration patterns you see on the map?

_____

_____

_____

_____

_____

Step 7 Draw conclusions

e   Based on your map investigations, write a hypothesis below about how a country's rate of natural increase affects its standard of living and its net rate of migration.

_____

_____

_____

_____

_____

f   In the table below, illustrate your hypothesis with data from one European country and one sub-Saharan African country.

EUROPE	DATA	AFRICA
	Country name	
	Natural increase	
	Net migration	

Step 8 Create a layout

j   What are the units of measurement?

_____

NAME _____   DATE _____

Growing Pains
Middle school assessment

You have been selected by the United Nations to establish a model partnership between nations. Your job is to select two countries—one country with a slow growth rate from Europe and one with a fast growth rate from Africa. Refer to your initial GIS Investigation if you need help in identifying growth rate in a country. Select the countries from the list below:

Belgium	Italy	Norway
Germany	Nigeria	Poland
Mozambique	Madagascar	Tanzania
Somalia	Zambia	
United Kingdom	Denmark	

Once you have selected your two countries, you need to write a report that addresses the following points:

1 Identify issues critical to each country in regard to growth and standard of living.

2 How can these countries work together to help improve the standard of living in their countries and address their respective problems?

3 Do current relationships (such as trade agreements) exist between these countries and are these positive or negative in nature? How can these current partnerships be improved upon?

Support this report with maps and data from the GIS Investigation. You may also want to refer to additional sources such as textbooks, encyclopedias, and the Internet to find out additional details about your countries.

Growing Pains

Assessment rubric

Middle school

STANDARD	EXEMPLARY	MASTERY	INTRODUCTORY	DOES NOT MEET REQUIREMENTS
The student knows and understands how to use maps to analyze spatial distributions and patterns.	Uses GIS to compare and analyze growth and demographic trends in countries throughout the world. Makes predictions from the data on future population trends. Creates original printed/electronic maps to support findings.	Uses GIS to compare and analyze growth and demographic trends in countries throughout the world. Creates views that isolate regions of slow and fast growth. Creates printed/electronic maps to support findings.	Uses GIS to identify regions and countries with slow and fast growth rates. Printed maps do not support findings.	Has difficulty identifying patterns in population and demographics. No printed maps are included.
The student knows and understands the demographic structure of a population.	Analyzes demographic data of countries with fast and slow rates of natural increase and draws conclusions on how the structure of these populations came to exist.	Identifies and analyzes data that illustrates the demographic makeup of countries with slow and fast rates of natural increase.	Identifies demographic characteristics of countries with slow and fast rates of natural increase.	Has difficulty distinguishing differences between countries with slow and fast rates of natural increase.
The student knows and understands how to apply the geographic point of view to social problems by making geographically informed decisions.	Identifies critical growth issues for a country of fast growth and a country of slow growth. Creates a detailed program by which the countries could establish a partnership of nations to solve growth-related issues. Elaborates on how this program could be replicated by other countries.	Identifies critical growth issues for a country of fast growth and a country of slow growth. Determines several ways that the countries could form a partnership for the mutual benefit of all in regard to growth issues.	Identifies critical growth issues for a country of fast growth and another of slow growth. Determines one or two ways the countries could form a partnership.	Identifies critical growth issues for two countries, but does not address the issue of establishing a partnership between nations.

This is a four-point rubric based on the National Standards for Geographic Education. The "Mastery" level meets the target objective for grades 5–8.

NAME _____   DATE _____

Growing Pains
High school assessment

You have been selected by the United Nations to establish a model partnership among nations. Your job is to select two groups of countries—one group with a slow growth rate and one group with a fast growth rate. Each group of countries should consist of four to five nations from Europe or sub-Saharan Africa. The countries within each group need to share similar growth characteristics. Refer to your initial GIS investigation if you need help identifying growth rate in a particular country.

Once you have selected your two groups, you will write a report that addresses the following points:

1 Identify issues related to growth and standard of living that are critical to each group's welfare.

2 How can these countries work together to help improve the standard of living in their countries and address their respective problems?

3 Do current relationships (such as trade agreements) exist between these countries and are these positive or negative in nature? How can these current partnerships be improved upon?

Support this report with maps and data from the GIS Investigation. You may also want to refer to additional sources such as textbooks, encyclopedias, and the Internet to find out additional details about your countries.

Growing Pains

Assessment rubric

High school

STANDARD	EXEMPLARY	MASTERY	INTRODUCTORY	DOES NOT MEET REQUIREMENTS
The student knows and understands how to use geographic representations and tools to analyze and explain geographic problems.	Uses GIS to compare and analyze growth and demographic trends in countries throughout the world. Makes predictions from the data provided and additional sources on future population trends. Creates original printed/electronic maps to support findings.	Uses GIS to compare and analyze growth and demographic trends in countries throughout the world. Makes predictions from the data on future population trends. Creates printed/electronic maps to support findings.	Uses GIS to compare and analyze growth and demographic trends in countries throughout the world. Makes predictions from the data on future population trends. Printed maps do not support findings.	Attempts to make comparisons between countries using demographic data in a GIS. No printed maps are included.
The student knows and understands trends in world population numbers and patterns.	Creates two groups of countries, each with similar demographic trends, including standard of living indicators. One group represents fast-growth countries and one represents slow-growth. Makes predictions on how these trends will change through time.	Creates two groups of countries, each with similar demographic trends, including standard of living indicators. One group represents fast-growth countries and one represents slow-growth.	Creates two groups of countries with similar growth characteristics. One group represents fast-growth countries and one represents slow-growth.	Identifies one or two countries that represent either slow or fast growth.
The student knows and understands how to use geographic knowledge, skills, and perspectives to analyze problems and make decisions.	Creates a report that establishes a coalition of slow- and fast-growth countries that work together for the mutual benefit of all involved. The report takes into account issues critical to the countries that participate and elaborates on possible solutions.	Creates a report that establishes a coalition of slow- and fast-growth countries that work together for the mutual benefit of all involved. The report takes into account issues critical to the countries that participate.	Lists several issues that are critical to fast- and slow-growth countries. Attempts to find ways in which each group of countries can partner.	Identifies one or two issues critical to countries with slow growth and those with fast growth.

This is a four-point rubric based on the National Standards for Geographic Education. The "Mastery" level meets the target objective for grades 9–12.

Generation Gaps
An advanced investigation

Lesson overview

Students will use global data to investigate variations in population age structure and the relationship of age structure to a country's rate of natural increase. After exploring global patterns, they will download and map U.S. census data to examine the patterns of age structure in their own community.

Estimated time   Two to three 45-minute class periods

Materials   ✔ Student handouts from this lesson to be copied:
- GIS Investigation sheets (pages 239 to 244)
- Student answer sheets (pages 245 to 246)

Standards and objectives

National geography standards

GEOGRAPHY STANDARD	MIDDLE SCHOOL	HIGH SCHOOL
1 How to use maps and other geographic representations, tools, and technologies to acquire, process, and report information from a spatial perspective	The student knows and understands how to use maps to analyze spatial distributions and patterns.	The student knows and understands how to use geographic representations and tools to analyze and explain geographic problems.
9 The characteristics, distribution, and migration of human populations on the earth's surface	The student knows and understands the demographic structure of a population.	The student knows and understands trends in world population numbers and patterns.
18 How to apply geography to interpret the present and plan for the future	The student knows and understands how to apply the geographic point of view to social problems by making geographically informed decisions.	The student knows and understands how to use geographic knowledge, skills, and perspectives to analyze problems and make decisions.

Objectives

The student is able to:
- Explain how the age structure of a population is affected by its rate of natural increase.
- Describe age structure trends in different world regions.
- Use the Internet to locate U.S. Census Bureau data for their own community.
- Use GIS to map acquired data from the Internet.
- Describe patterns of age structure in their local area.

GIS skills and tools

- Obtain census data from the Internet and prepare database tables for GIS implementation
- Add tables to a project
- Copy themes
- Join tables in a project
- Use the Legend Editor to map thematic data
- Create a new view

For more on geographic inquiry and these steps, see Geographic Inquiry: Thinking Geographically (pages xxi to xxiii).

Teacher notes

Lesson introduction

Introduce the concept of population age structure by displaying and discussing several examples of population pyramids. You can obtain population pyramids for any world country at the U.S. Census Bureau's International Database (IDB) (www.census.gov/ipc/www/idbpyr.html). Compare a population pyramid for a country that is growing rapidly with one for a country that is growing very slowly. Initially, just display the pyramids without telling the students what countries they represent—refer to them as Country A and Country B. Ask students to consider the following questions:

- How is the makeup of population different in these two countries?
- One of these countries is growing rapidly and one is growing slowly—can you determine which is which? How do you know?
- What specific issues does each country face because of its age structure?

Explain to students that in this lesson they will investigate age structure trends around the world and then download U.S. Census data to examine patterns of age structure in their own community.

Student activity Prior to beginning this activity with students, you should complete the Geography Network Registration Form (step 4 in the GIS Investigation). Designate one e-mail address for your students to use when they complete the GIS Investigation.

Distribute the Generation Gaps GIS Investigation. Students should follow the GIS Investigation guidelines to analyze global and local patterns of age structure within specific populations. After looking at global patterns, they will download census data for their own county and examine local age distribution patterns.

 Note: When downloading the census data from the Internet, each student needs to have a folder or a computer directory that they can save the data to. Students are instructed to consult their teacher on how to name the data that's downloaded and where to save it. Without this information, students will have difficulty accessing the data at a later time.

 Teacher Tip: In step 5b, students are directed to unzip the census data files. If this is a task that your students are not familiar with, you will need to explain the process to them.

Conclusion To conclude this lesson, engage the class in a discussion of the implications of population age structure. Be sure to bring out the concept that the age structure of a population in a country or a community affects its key socioeconomic issues. Countries and communities with young populations (high percentage under age 15) need to invest more in schools while countries and communities with older populations (high percentage ages 60 and over) might need to invest more in the health sector. The age structure can also be used to help predict potential political issues. For example, the rapid growth of a young adult population unable to find employment can lead to unrest. Large numbers of the elderly can lead to political concerns about health benefits and pension plans.

Challenge the class to identify local social and political issues that are related to the age structure of the community. Bring in newspaper clippings about stories related to these issues for further discussion and display in the classroom.

Assessment Students will create a layout that illustrates the distribution of one particular age group in their county. They will submit a written report summarizing their observations and explaining their data analysis. The report should address all or some of these points:

- Summarize the spatial distribution of the selected age group in your county.
- Analyze the patterns that you observe by relating those patterns to other features of the county such as:
  - Location of age-based institutions (colleges, retirement communities, etc.)
  - Proximity to major roads and transportation systems
  - Economic activities
  - Housing characteristics
  - Income and educational attainment of the population

 Note: Due to the independent nature of this lesson, there is no supplied assessment rubric. You are free to design assessment rubrics that meet the needs of your specific adaptation.

Extensions
- Do an Internet search on "population pyramid" to find directions for using a spreadsheet program to create population pyramids. Make a pyramid of your own community and incorporate it into an ArcView project, displaying demographic data about your community.
- Download TIGER data from the U.S. Census Bureau for a college town and a town with a large retirement population. Map and compare the age distributions in these communities with your own community.
- Analyze your county data and make a prediction on population growth for the future. Create map layouts that support your prediction. Download Census 2000 data for your county and enhance your claim by using that data to support your prediction.
- Check out additional resources available online at *www.esri.com/ mappingourworld.*

NAME _____   DATE _____

Generation Gaps
An advanced investigation

 Note: Due to the dynamic nature of the Internet, the URLs listed in this lesson may have changed, and the graphics shown below may be out of date. If the URLs do not work, refer to this book's Web site for an updated link: www.esri.com/mappingourworld.

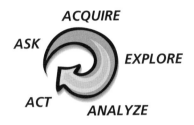

Answer all questions on the student answer sheet handout

Step 1   Start ArcView and open the adv4.apr file

a   Start ArcView.

b   Open the **adv4.apr** project from your exercise directory (**C:\esri\mapworld\mod4 \adv4.apr**).

When the project opens, you see the map in the view with two themes turned on (Ocean and Natural Increase).

Step 2   Create new themes

The world's population is growing because there are more births than deaths each year. This fact can be expressed as a simple formula, BR (birth rate) – DR (death rate) = NI (natural increase). The spatial patterns displayed on the Natural Increase map in View 1 clearly indicate where population is growing rapidly and slowly in the world today. In this part of the activity, you will see if there are comparable spatial patterns in the world's age structure.

World Vital Events

World Vital Events Per Time Unit: 2001

(Figures may not add to totals due to rounding)

Time unit	Births	Deaths	Natural increase
Year	131,571,719	55,001,289	76,570,430
Month	10,964,310	4,583,441	6,380,869
Day	360,470	150,688	209,782
Hour	15,020	6,279	8,741
Minute	250	105	146
Second	4.2	1.7	2.4

Source: U.S. Bureau of the Census, International Data Base

?   *(1)   What world regions have the highest rates of natural increase today?*

?   *(2)   What world regions have the lowest rates of natural increase today?*

a   From the Edit menu, select Copy Themes.

b   Paste two copies of the theme back into the view by selecting Paste from the Edit menu two times.

c   Now you will rename each of the themes. Make one of the pasted themes active.

   d   Click the Theme Properties button.

e   Name it **% Population <15 Yrs** and click OK.

f   Follow steps 2d through 2e above. Name the second theme **% Population 60+ Yrs**.

Step 3   Thematically map world age structure

a   Double-click the % Population <15 Yrs theme to open its Legend Editor.

b   For Classification Field, select **Pop0_14**.

   c   Click the Null Value button. The Null Values dialog appears.

d   Make sure the Field is Pop0_14. Set the Null Value to –99.

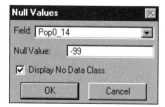

e   Click OK. You see that the –99 Null Value has been incorporated into the No Data class.

f   From the Color Ramps drop-down list, choose **Green monochromatic**. Click Apply. Close the Legend Editor.

g   Repeat these steps with the theme you named **% Population 60+ Yrs**. For Classification Field, select Pop_60_9.

h   From the Color Ramps drop-down list, choose **Magenta monochromatic**. Click Apply. Close the Legend Editor.

i   Turn off the Natural Increase theme. Toggle back and forth between the % Population 60+ Yrs theme and the % Population <15 Yrs theme to see them both.

     (1)   *What regions of the world have populations with a high percent below age 15?*

     (2)   *What regions of the world have populations with a relatively low percent below age 15?*

     (3)   *Why do you think there is a much higher percent of children in some populations than in others?*

     (4)   *Why are there far more people 60 years of age and older in some populations than in others?*

     (5)   *Based on the maps in Generation Gaps View 1, how do you think natural increase is related to age structure?*

j   Minimize the ArcView window.

Step 4   **Download census data for your county**

a   Before beginning this part of the exercise, create a folder to store the census data you will download later. Ask your teacher where to put the folder and how to name it. One suggestion is to put the folder in your computer's C:Temp\ drive and name it yournamecensusdata. For example:

joesmithcensusdata

b   Use your Web browser to go to the Geography Network site at www.geographynetwork.com/data/tiger2000. (At this ESRI site, you will be able to download 2000 U.S. Census data for your county.)

c   Type in the e-mail address your teacher gives you and click Enter. A new page opens.

d   Select your state from the drop-down list or click on your state on the map.

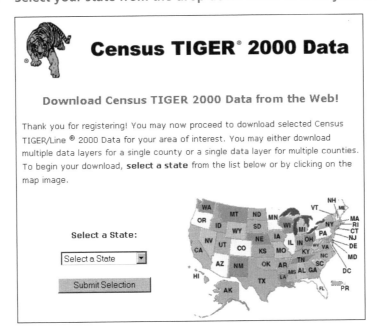

e Click Submit Selection.

f Select your county from the Select by County drop-down list. Click Submit Selection.

A list of available data for your county appears.

g Check off all of the following data layers:

- Block Groups 2000
- Line Features - Roads
- Water Polygons
- Census Block Group Demographics (SF1)

h Scroll to the bottom of the page and click Proceed to Download.

"Your Data File is Ready" appears.

i Click the Download File Now button at the bottom of the page.

j Select Save this File to Disk and navigate to the folder that you created in step 5a. Change the file name to **your state abbreviation countyname.zip**. (Eliminate all spaces from the file name.)

For example:

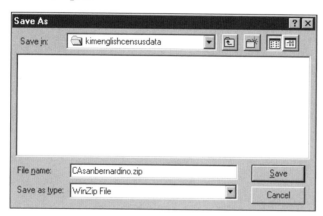

k Click Save. When the download is complete, select Close and quit your Web browser.

Step 5   Unzip your data and add it to the project

a Locate your census data in the designated directory.

The downloaded file (stateabbreviationcountyname.zip) contains the data layers you selected. Each of those layers is, in turn, a zipped file.

b Unzip the individual layers and save the unzipped data in a location that your teacher specifies. Open and read the readme.html file that's included with the data.

> *Note: It is very important that your data is saved to a directory that you can access later. If you intend to save this project and open it later, you must have access to the data.*

c Maximize the ArcView project.

d From the Window menu, select adv4.apr to return to the Project window.

e   Click on the Views icon and click New.

A blank new view opens.

f   From the View menu, select Properties and rename the view **Generation Gaps View 2**.

g   Click the Add Theme button to add the downloaded themes from the Census Bureau. Navigate to the location of the census data you downloaded and add all of the themes. These themes end with the following suffixes: grp00.shp, lka.shp, and wat.shp.

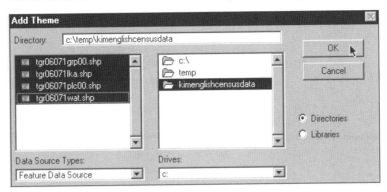

 Note: The number included in the name of the file is the census number for your county.

h   Arrange the themes so that the streets theme (lka.shp) is on top, followed by water (wat.shp), and Census Block Groups (grp00.shp) is on the bottom.

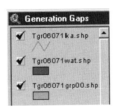

Step 6   **Join census tables to the map**

a   Make the Project window active, click on the Tables icon, and click Add.

b   Navigate to where you saved your census data and select the file ending with **sf1grp.dbf**.

c   Click on the field name, Stfid.

d   Go back to View 2 and make the grp00.shp theme active.

e   Click the Open Theme Table button to open the Attributes of grp00.shp table.

f   Click on the field name, Stfid.

g   Click the Join button to join the two tables.

The census data now appears in the Attributes of grp00.shp table.

h   Close the attribute table. View 2 becomes active.

Step 7   **Analyze census data for your county**

a   Double-click grp00.shp to open the Legend Editor.

b   Select Graduated Color for the Legend Type. Select Age_under5 for the Classification Field. Click Apply and close the Legend Editor.

The map displays the distribution of children between birth and five years of age in your county.

c   Change the name of the theme to **Age under 5 yrs**.

d   If it's hard to see color patterns with the roads theme (lka.shp) on, turn it off.

? *Are there identifiable concentrations of children between birth and five years old in your county? How do you explain these concentrations?*

e   Copy and paste the Age under 5 yrs. theme. Use the Legend Editor to map the classification field, Age_65_up. Choose a different color ramp.

The map displays the distribution of adults over the age of 65 in your county.

? *Where is the greatest number of this age cohort found?*

f   Change the name of this theme to Age over 65 yrs. Repeat the process of copying a theme and mapping a new age group to explore other age group populations.

? *(1)   Do you have any colleges in your county? Any retirement communities? Are these institutions reflected in the census data?*

? *(2)   On the answer sheet, identify patterns of age distribution in your county and suggest explanations for those patterns.*

Step 8   **Save your project and exit ArcView**

You can further explore population in your county by investigating the TIGER data you downloaded. For information on how to read all the abbreviations and interpret the assigned codes, visit the Census Bureau's Web site on TIGER data at www.census.gov/geo/www/tiger or read the Readme file that was downloaded from the Geography Network.

In this exercise, you analyzed the natural increase and population by age of different countries. You searched for your county data on the Internet, downloaded it, and incorporated it into an ArcView project. Now the data is available for analysis and presentation.

a   When you're finished working with this project, save it according to your teacher's instructions.

b   Exit ArcView.

NAME _____   DATE _____

Student answer sheet
Module 4
Human Geography I: Population Patterns and Processes

Advanced investigation: Generation Gaps

Step 2   Create new themes

2-1   What world regions have the highest rates of natural increase today?

_____

_____

_____

2-2   What world regions have the lowest rates of natural increase today?

_____

_____

_____

Step 3   Thematically map world age structure

i-1   What regions of the world have populations with a high percent below age 15?

_____

_____

_____

i-2   What regions of the world have populations with a relatively low percent below age 15?

_____

_____

_____

i-3   Why do you think there is a much higher percent of children in some populations than in others?

_____

_____

_____

_____

i-4   Why are there far more people 60 years of age and older in some populations than in others?

_____

_____

_____

_____

i-5　Based on the maps in Generation Gaps View 1, how do you think natural increase is related to age structure?

_____

_____

_____

_____

_____

Step 7　Analyze census data for your county

d　Are there identifiable concentrations of children between birth and five years old in your county? How do you explain these concentrations?

_____

_____

_____

_____

_____

e　Where is the greatest number of this age cohort found? _____

f-1　Do you have any colleges in your county? Any retirement communities? Are these institutions reflected in the census data?

_____

_____

_____

_____

_____

f-2　In the space below, identify patterns of age distribution in your county and suggest explanations for those patterns.

_____

_____

_____

_____

_____

_____

_____

_____

_____

_____

Human Geography II
Political Geography

Invisible boundary lines on the earth's surface divide our world into discrete political entities and have significant influence on the stability and cohesiveness of the countries they define.

Crossing the Line: A global perspective
Students will explore the nature and significance of international political boundaries. Through an investigation of contemporary political boundaries, they will identify boundary forms, compare patterns of size and shape, and explore the influence of boundaries on national cohesiveness and economic potential. By comparing world political boundaries in 1992 and 2000, students will observe the evolution of boundary characteristics over time.

A Line in the Sand: A regional case study of Saudi Arabia and Yemen
Students will study the creation of a new border between Yemen and Saudi Arabia on the Arabian Peninsula. Using data included in the June 2000 Treaty of Jeddah, they will draw the new boundary described in the treaty and analyze the underlying physiographic and cultural forces that influenced the location of that boundary. In the process they will come to understand how any map of the world must be considered a tentative one, as nations struggle and cooperate with each other.

Starting from Scratch: An advanced investigation
Students will use physiographic (physical features) and anthropographic (cultural features) data to redraw some of the world's international boundaries, thereby creating states characterized by internal cohesiveness and economic parity. By comparing their maps to ones reflecting contemporary political boundaries, students will identify world regions where political boundaries are in conflict with physical and cultural imperatives.

Crossing the Line
A global perspective

Lesson overview

Students will explore the nature and significance of international political boundaries. Through an investigation of contemporary political boundaries, they will identify boundary forms, compare patterns of size and shape, and explore the influence of boundaries on national cohesiveness and economic potential. By comparing world political boundaries in 1992 and 2000, students will observe the evolution of boundaries over time.

Estimated time   Two to three 45-minute class periods

Materials   ✔ Student handouts from this lesson to be copied:
- GIS Investigation sheets (pages 255 to 264)
- Student answer sheets (pages 265 to 270)
- Assessment(s) (pages 272 to 275)

Standards and objectives   *National geography standards*

GEOGRAPHY STANDARD	MIDDLE SCHOOL	HIGH SCHOOL
3 How to analyze the spatial organization of people, places, and environments on Earth's surface	The student knows and understands how to use the elements of space to describe spatial patterns.	The student knows and understands how to apply concepts and models of spatial organization to make decisions.
13 How the forces of cooperation and conflict among people influence the division and control of Earth's surface	The student knows and understands the multiple territorial divisions of the student's own world.	The student knows and understands why and how cooperation and conflict are involved in shaping the distribution of social, political, and economic spaces on Earth at different scales.
18 How to apply geography to interpret the present and plan for the future	The student knows and understands how varying points of view on geographic context influence plans for change.	The student knows and understands contemporary issues in the context of spatial and environmental perspectives.

Objectives
The student is able to:
- Define and give examples of physiographic, geometric, and anthropographic boundaries.
- Describe the political and economic implications of a country's size and shape.
- Explain the relationship between boundary characteristics and national cohesiveness.
- Give examples and explain the nature of international boundary changes in the late twentieth century.

GIS skills and tools

 Zoom to a specific area in the view

 Draw a line in the view

 Identify a feature in the view

 Zoom to the full extent

 Add a theme to the view

 Pan to a different section of the map

 Find a feature in a theme

 Zoom to the selected feature

 Change the theme name

ACQUIRE geographic resources

ASK geographic questions

EXPLORE geographic data

ACT on geographic knowledge

ANALYZE geographic information

For more on geographic inquiry and these steps, see Geographic Inquiry: Thinking Geographically (pages xxi to xxiii).

Teacher notes

Lesson introduction

Write the following quotation on the board or on a transparency:

"When you go around the Earth in an hour and a half . . . you look down there and you can't imagine how many borders and boundaries you cross, again and again and again, and you don't even see them . . . from where you see it, the thing is a whole, and it's so beautiful."
Russell L. Schweickart
Apollo 9, March 3–13, 1969
(As published in *The Overview Effect*, 1998)

Use this quotation as a springboard to discuss the following questions:

- What are boundaries?
- Who draws the boundary lines?
- What purpose do boundaries serve?
- If boundaries are invisible lines, how do you know when you've crossed one?
- Once you've crossed one of these invisible lines, what has changed?
- Can you think of any problems that boundaries may cause?

Throughout the discussion, emphasize that although most boundaries are unmarked and invisible, they determine our perception of spaces and places on earth. Boundaries between countries help maintain order in the world because they define internationally recognized and sovereign political entities. Conflict can result when boundary lines are disputed. Boundaries are also potential sources of conflict because they are the point of contact between neighboring people.

Challenge students to identify places in the world where international boundaries have changed or are in conflict. What do students know about the reasons for those boundary changes and conflicts? Tell the class that they are going to do a GIS Investigation that will explore the characteristics of modern international boundaries and investigate recent boundary changes.

Student activity

 Before completing this lesson with students, we recommend that you complete it as well. Doing so will allow you to modify the activity to accommodate the specific needs of your students.

After the initial discussion, have the students work on the computer component of the lesson. Ideally, each student should be at an individual computer, but the lesson can be modified to accommodate a variety of instructional settings.

Distribute the GIS Investigation sheets to the students. Explain that in this activity, they will use GIS to investigate different types of international boundaries, explore the implications of various boundary configurations, and observe boundary changes in recent years. The activity sheets will provide them with detailed instructions for their investigations.

In addition to instructions, the handout includes questions to help students focus on key concepts. Some questions have specific answers while others call for speculation and have a range of possible responses. In addition, answers to many questions will vary with student knowledge of current events and contemporary world political issues.

Things to look for while the students are working on this activity:
- Are the students using a variety of GIS tools?
- Are the students answering the questions as they work through the procedure?
- Are students asking thoughtful questions throughout the investigation?

 Teacher Tip: This GIS Investigation contains instructions for students to periodically stop and save their work. These are good spots to stop the class for the day and to pick up the investigation the next day. Be sure to inform your students as to how they should rename their project and where to save it.

Conclusion

Use a projection device to display the global5.apr in the classroom. As a group, compare student observations and conclusions from the lesson. Students can take turns being the "driver" on the computer to highlight boundaries and observations that are identified by members of the class. Focus on the following aspects of the GIS Investigation in your discussion:
- Where did students observe the coincidence of political and physiographic boundaries?
- What examples of territorial morphology (countries representing categories of different shapes and sizes) did the students find? Ask students to speculate on ways that a country's size and shape could influence its sense of unity or cohesiveness.
- How can a country's boundaries influence its economic strength and advantage?

- What kinds of problems are likely to arise when political and anthropographic (cultural) boundaries do not coincide?
- What is the nature of the boundary changes that occurred between 1992 and 2000? Based on student response in step 8, which of the new countries do students believe are in the strongest position today in terms of cohesiveness?

Assessment

Middle school: Highlights skills appropriate to grades 5 through 8

In the middle school assessment, students are asked to identify an international boundary in 2000 that they predict could change in the next 25 years. They are asked to prepare a map of the projected boundary change and describe its impact in terms of the lesson's concepts.

- What types of boundaries are involved in the projected change?
- How will the territorial morphology of the countries involved be affected by the projected change?
- What will the economic impact of the projected change be?
- How will the projected change affect the internal cohesiveness of all countries involved?

High school: Highlights skills appropriate to grades 9 through 12

In the high school assessment, students are asked to identify two international boundaries in 2000 that they predict could change in the next 25 years. One of their predictions should involve splitting a current country into two or more smaller countries, and one of their predictions should involve merging two or more countries into one larger one. They are asked to prepare a map of the projected boundary changes and compare them in terms of the lesson's concepts.

- What types of boundaries are involved in the changes?
- How will the territorial morphology of the countries involved be affected by the projected changes?
- What will the economic impact of the projected changes be?
- How will the projected changes affect the internal cohesiveness of all countries involved?

Extensions

- Assign students to conduct research on world boundary changes during the twentieth century. Use ArcView to prepare a sequence of layouts reflecting those changes.
- Explore the nature of political boundaries in your own community and state. What kind of boundaries are they, how does the shape of your town or state affect its cohesiveness, what are the economic advantages and disadvantages of your town or state's boundary configuration, how have your town or state's boundaries changed over time?
- Search newspapers and magazines (both on- and offline) for coverage of border conflicts and related issues around the world. Use ArcView to illustrate the nature of these conflicts.
- Check out the Resources by Module section of the Teacher Resource CD for print and media resources on the topic of political boundaries or visit *www.esri.com/mappingourworld* for Internet links.

NAME _____   DATE _____

Crossing the Line
A GIS investigation

ACQUIRE

ASK

EXPLORE

ACT

ANALYZE

Answer all questions on the student answer sheet handout

Boundaries are invisible lines on the earth's surface. They divide the surface area into distinct separate political entities. In this activity, you will use GIS to investigate different types of international boundaries, explore the implications of various boundary configurations, and observe boundary changes in recent years. When you have completed the activity, you will use your knowledge of boundary dynamics to speculate on world boundaries that are likely to change in the future.

Step 1   Start ArcView

   a   Double-click the ArcView icon on your computer's desktop.

   b   If the Welcome to ArcView dialog appears (pictured below), click **Open an Existing Project** and click OK. If it doesn't appear, proceed to step 2.

Step 2   Open the global5.apr file

a   In this exercise, a project file has been created for you. To open the file, go to the **File** menu and choose **Open Project**.

b   Navigate to the module 5 directory (**C:\esri\mapworld\mod5**) and choose **global5.apr** from the list.

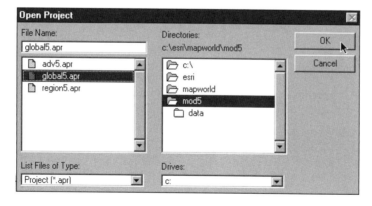

c   Click OK.

The project opens and you see a composite satellite image of the world. The check mark next to the theme name tells you the theme is turned on and visible in the view.

Step 3   **Explore mountain ranges as physiographic boundaries**

As astronaut Russell L. Schweickart said, if you could view the world from space, you would see no boundary lines. Boundaries are human-made lines that define the world's political entities.

There are several types of boundaries between countries. One type of boundary is called a *physiographic* boundary. They are based upon natural features on the landscape such as mountain ranges or rivers.

a   Click the box next to the Boundaries 2000 theme to turn it on. A check mark appears and the red lines display the international boundaries for the year 2000.

 b   Click the Zoom In tool. Click on the map and drag a box around Europe. The view is now centered on Europe.

c   Drag another box around Europe to zoom in more so you can see the physical features in greater detail.

d   Turn off Boundaries 2000 by clicking the check mark next to the theme name.

Locate Europe's mountain ranges in the satellite image. Notice that mountains such as the Pyrenees Mountains in northeastern Spain form a natural boundary. You will use the Draw Line tool to draw lines where you see a mountain range forming a natural boundary between different parts of the continent.

First, you will select a symbol type and color for drawing.

e   From the Window menu, select Show Symbol Window.

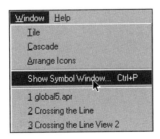

f   Click the Pen button to display the Pen Palette window.

g Select the solid line if it isn't already selected. Change the size to 2.

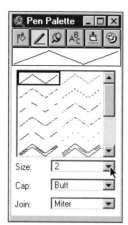

 h Click the Paintbrush button to see the Color Palette. Click yellow. Close the Color Palette by clicking the "X" in the upper right corner of the window.

Now you are ready to draw a physiographic boundary in Europe.

 *i* Select the Draw Line tool from the Draw tools menu.

j Click the westernmost edge of the Pyrenees Mountains to start your line. Continue clicking along the path of the mountain range until you reach its easternmost edge. Double-click to end the line. The yellow line is displayed in the view.

> *Note: Sometimes, ArcView leaves colored boxes around the line. If this happens, click the Pointer tool and then click once on the map. The colored boxes disappear.*

k Turn on Boundaries 2000 and make it active (click its name). You see that your line corresponds to a border between two countries.

 l Click the Identify tool. Click the country that borders the Pyrenees Mountains and that is to the north.

An Identify Results window appears.

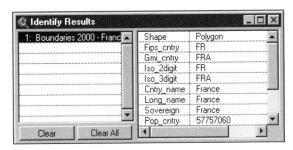

The left side of the window shows the theme name and the beginning of the country name. You can see the complete country name by looking at the right side of the window. The field "Cntry_name" reveals the full name.

m   Click the country on the other side of the border.

?   *The Pyrenees Mountains are the border between which two countries?*

n   Use the Identify tool to find other Western European countries where physiographic boundaries created by mountains correspond to actual political boundaries. Click a country to see its information in the Identify Results window.

?   *Complete the table on the answer sheet.*

o   Close the Identify Results window by clicking the "✕" in the upper right corner of the window.

Step 4   **Explore bodies of water as physiographic boundaries**

a   Turn on the Rivers, Lakes, and Countries 2000 themes.

Wherever countries have physiographic boundaries based on rivers, the red boundary line disappears beneath the blue river in the view. Look closely at Europe to see if you can find any boundaries that are rivers. The different colored countries will help you find these places.

b   Make sure Boundaries 2000 is the active theme.

c   Click the Identify tool. Click a country that has a river as all or part of a boundary.

?   *In the table on the answer sheet, record the names of three sets of countries that share a boundary that's a river.*

In order to identify the names of the rivers, you must make Rivers the active theme and then use the Identify tool.

Coastlines are also physiographic boundaries. Countries that do not have a coastline are said to be landlocked.

?   *d*   Look at the view and name three landlocked countries in Western Europe. Use the Identify tool if you don't know the name of a specific country.

Step 5 Explore geometric boundaries

Another type of boundary is a *geometric* boundary. Geometric boundaries consist of straight or curved lines that do not correspond to physical features on the earth's surface.

 a Click the Zoom to Full Extent button to see all the continents. Turn off Satellite Image.

 b Click the Zoom In tool. Use it to zoom in on Africa.

You see many rivers that overlap boundaries throughout the African continent.

c Make sure Boundaries 2000 is the active theme.

d Look at the view and locate countries that have geometric boundaries (boundaries that are not physiographic).

 e Click the Identify tool. Use your Identify tool to identify African countries that are separated by geometric boundaries.

? *Record three sets of countries in the table on the answer sheet.*

f Close the Identify Results window.

 g Click Zoom to Full Extent to see the entire world in the view.

h Ask your teacher if you should stop here and save this ArcView project. Follow your teacher's instructions on how to rename the project and where to save it. If you do not need to save the project, proceed to step 6.

? *Write the new name you gave the project and where you saved it.*

Step 6 Explore anthropographic boundaries based on language and religion

A third type of boundary is an *anthropographic* boundary. This boundary marks the transition between cultural characteristics on the landscape. Anthropographic boundaries are based on characteristics such as language, religion, or ethnicity.

a Turn off Boundaries 2000, Countries 2000, and Rivers.

b Click the Add Theme button.

c Navigate to the data directory (**C:\esri\mapworld\mod5\data**).

d Select **language.shp**. Hold down the Shift key and scroll down to select **religion.shp**. Click OK.

e   Drag Boundaries 2000 to the top of the table of contents.

f   Turn on Language.shp. It displays the distribution of major language groups on the earth. Scroll down the table of contents so you can read the Language.shp legend. Drag the right edge of the table of contents to widen it.

g   Observe the pattern of anthropographic boundaries based on language in the world.

h   Use the Identify tool to determine the principal language groups in South America and Western Europe. (Don't forget to make Language.shp the active theme.) Record your answers on the answer sheet.

i   Turn on Boundaries 2000.

j   Use the Identify and Zoom tools to locate three examples in the world where political boundaries coincide with anthropographic boundaries based on language. Record them on the answer sheet.

k   Click Zoom to Full Extent. Turn off Boundaries 2000 and Language.shp.

l   Turn on Religion.shp. The distribution of major religions is displayed.

m   Observe the pattern of anthropographic boundaries based on religion throughout the world.

n   Use the Identify and Zoom tools to determine the principal religions in North America and Africa. Record them on the answer sheet.

o   Turn on Boundaries 2000.

p   Use the Identify and Zoom tools to locate three examples in the world where political boundaries coincide with anthropographic boundaries based on religion.

Step 7   Review physiographic, geometric, and anthropographic boundaries

Use your Zoom and Identify tools to find additional examples of physiographic, geometric, and anthropographic boundaries between countries and record your findings in the table on the answer sheet.

Step 8   **Explore the impact of boundary shape, cultural diversity, and access to natural resources**

Boundaries determine the size and shape, or territorial morphology, of countries. Size and shape can exert a powerful influence on the cohesiveness of a country. Small compact nations or circular/hexagonal ones, for example, are more easily united than ones that are elongated or fragmented.

   a   Click the Zoom to Full Extent button to see the whole world again. Turn off all themes except for Countries 2000 and Ocean. Make Countries 2000 the active theme.

   b   Click the Find tool.

     c   When the Find window appears, type **Chile** and click OK. Chile is highlighted yellow in the view, but may be small and difficult to see.

   d   Click Zoom to Selected Feature to get a closer look at Chile's shape.

The table on the answer sheet illustrates six types of countries based on shape and gives an example of each. Chile is listed as an example of an elongated country.

   e   Use the Find, Zoom, and Pan tools to locate another example of each type of country. Record them in the table on the answer sheet in the Example 2 column. Remember, you can use the Identify tool to find the names of countries that you do not know.

Another factor that influences cohesiveness is the extent of cultural diversity.

   f   Click the Zoom to Full Extent button. Turn on Language.shp and Boundaries 2000. Turn off Countries 2000.

     (1)   *By using language groups as an indicator of cultural uniformity, identify three countries that reflect cultural uniformity.*

     (2)   *By using language groups as an indicator of cultural diversity, identify three countries that reflect cultural diversity.*

Boundaries also influence economic activities. Earlier in this GIS Investigation, you identified landlocked countries in Western Europe. Historically, these countries were limited in their ability to trade directly with other nations because imports and exports had to pass through other countries en route to their destination.

   g   Click the Zoom to Full Extent button to see the whole world again. Turn off Boundaries 2000 and Language.shp. Turn on Countries 2000.

     h   Use the ArcView tools and buttons you've learned in this investigation to find an example of a landlocked country on each continent listed in the table on the answer sheet. For a continent that does not have a landlocked country, write "none."

Boundaries also influence economic activities by establishing a country's access to natural resources.

     i   Click the Zoom to Full Extent button. Make sure Rivers and Lakes are turned off.

     j   Click the Add Theme button. Navigate to the data directory (**C:\esri\mapworld \mod5\data**) and double-click **oil_gas.shp**.

     k   Turn on Oil_gas.shp. The locations of oil and gas sources around the world are displayed.

     l   Use the Zoom In tool to focus on Southeast Asia.

m   Use the Pan, Zoom, and Identify tools to help you answer the following questions.

   (1)   *Name two Southeast Asian countries that do not have any oil and gas resources within their borders.*

   (2)   *Name two Southeast Asian countries that have oil and gas resources within their borders.*

n   Turn off Oil_gas.shp.

o   Click the Zoom to Full Extent button. Turn off Countries 2000 and Ocean, and turn on Satellite Image.

Step 9   Explore boundary changes in the 1990s

Political boundaries can change in numerous ways. Large countries may split into several smaller ones, small countries may combine to produce larger ones, territories that were once part of one country may be incorporated into another.

a   Click the Add Theme button. Navigate to the module 5 data directory (**C:\esri \mapworld\mod5\data**) and select **cntry92.shp**.

b   Make Cntry92.shp the active theme and turn it on. International boundaries from 1992 display as yellow lines.

c   Click the Theme Properties button. Type **Boundaries 1992** as the new theme name. Click OK.

d   Turn on Boundaries 2000.

> *Note: Because 1992 boundaries cover the 2000 boundaries in the view, the 2000 boundary lines are not visible if they are the same as they were in 1992. The only 2000 boundary lines that are visible are those that did not exist in 1992.*

e   Observe the map closely to see the differences between 1992 and 2000. What kind of changes do you see? (Use your Zoom and Pan tools to get a good look at these changes.)

   (1)   *Describe three political boundary changes you see between 1992 and 2000.*

   (2)   *Name two countries that existed in 1992 but do not exist in 2000.*

f   Ask your teacher if you need to save this ArcView project. If you do, follow your teacher's instructions on how to name the project and where to save it. If you do not need to save the project, continue to the next page.

Write the new name you gave the project and where you saved it.

Countries in groups A and B below are new countries that have emerged since 1992.

Group A	**Group B**
Czech Republic	Russia
Slovakia	Belarus
Slovenia	Ukraine
Croatia	Moldova
Bosnia and Herzegovina	Armenia
Serbia	Azerbaijan
Montenegro	Georgia
Macedonia	Kazakhstan
Eritrea	Uzbekistan
	Tajikistan
	Turkmenistan
	Kyrgyzstan

g Select three countries from group A and three from group B and complete the table on the answer sheet. Use the information and skills you learned in this GIS investigation to answer the questions.

Step 10   Exit ArcView

In this exercise, you used ArcView to explore the various types of political boundary and their impact on the countries they define. You added themes and used the Find, Identify, Zoom, and Pan tools to investigate the maps. You observed and analyzed boundary changes between 1992 and 2000.

a Ask your teacher for instructions on where to save this ArcView project and on how to rename the project.

b If you are not going to save the project, exit ArcView by choosing Exit from the File menu. When asked if you want to save changes to global5.apr, click No.

NAME _____   DATE _____

Student answer sheet
Module 5
Human Geography II: Political Geography

Global perspective: Crossing the Line

Step 3 Explore mountain ranges as physiographic boundaries

m The Pyrenees Mountains are the border between which two countries?

_____ **and** _____

n Complete the following table (consult an atlas to find the names of unknown mountain ranges):

COUNTRIES THAT HAVE MOUNTAIN RANGES AS POLITICAL BOUNDARIES	MOUNTAINS THAT FORM THE BOUNDARY
and	.
and	
and	

Step 4 Explore bodies of water as physiographic boundaries

c In the following table, record the names of three sets of countries that share a boundary that's a river:

COUNTRIES THAT HAVE RIVERS AS BOUNDARIES	RIVER THAT FORMS THE BOUNDARY
and	
and	
and	

d Look at the view and name three landlocked countries in Western Europe. Use the Identify tool if you don't know the name of a specific country.

_____

_____

_____

Step 5 Explore geometric boundaries

e   Record three sets of countries in the following table:

COUNTRIES THAT ARE SEPARATED BY GEOMETRIC BOUNDARIES
and
and
and

h   Write the new name you gave the project and where you saved it.

(Name of project. **For example: kzeregion5.apr)**	**(Navigation path to where project is saved.** **For example: C:\student\kze\mapworld)**

Step 6 Explore anthropographic boundaries based on language and religion

h   Determine the principal language groups in the following regions. (Don't forget to make Language.shp the active theme.)

South America: _____

Western Europe: _____

j   Locate three examples in the world where political boundaries coincide with anthropographic boundaries based on language.

ANTHROPOGRAPHIC BOUNDARIES BASED ON LANGUAGE COINCIDE WITH POLITICAL BOUNDARIES BETWEEN
and
and
and

n   Determine the principal religions in the following regions:

North America: _____

Africa: _____

p   Locate three examples in the world where political boundaries coincide with anthropographic boundaries based on religion.

ANTHROPOGRAPHIC BOUNDARIES BASED ON RELIGION COINCIDE WITH POLITICAL BOUNDARIES BETWEEN
and
and
and

Step 7   Review physiographic, geometric, and anthropographic boundaries

Record your findings in the following table:

CONTINENT	PHYSIOGRAPHIC BOUNDARIES SEPARATE THE FOLLOWING COUNTRIES	GEOMETRIC BOUNDARIES SEPARATE THE FOLLOWING COUNTRIES	ANTHROPOGRAPHIC BOUNDARIES SEPARATE THE FOLLOWING COUNTRIES
North and Central America	_____ and _____ Boundary formed by (circle one): Mountains Rivers Lakes	_____ and _____	_____ and _____
South America and the Caribbean	_____ and _____ Boundary formed by (circle one): Mountains Rivers Lakes	_____ and _____	_____ and _____
Europe	_____ and _____ Boundary formed by (circle one): Mountains Rivers Lakes	_____ and _____	_____ and _____
Africa	_____ and _____ Boundary formed by (circle one): Mountains Rivers Lakes	_____ and _____	_____ and _____
Asia	_____ and _____ Boundary formed by (circle one): Mountains Rivers Lakes	_____ and _____	_____ and _____

Step 8 Explore the impact of boundary shape, cultural diversity, and access to natural resources

e Locate another example of each type of country. Record them in the following table in the Example 2 column. Remember, you can use the Identify tool to find the names of countries that you do not know.

TYPE OF COUNTRY	EXAMPLE	EXAMPLE 2
Elongated	Chile	
Fragmented	Philippines	
Circular/Hexagonal	France	
Small/Compact	Bulgaria	
Perforated	South Africa	
Prorupted	Namibia	

f-1 By using language groups as an indicator of cultural uniformity, identify three countries that reflect cultural uniformity.

_____

_____

_____

f-2 By using language groups as an indicator of cultural diversity, identify three countries that reflect cultural diversity.

_____

_____

_____

h Use the ArcView tools and buttons you've learned in this investigation to find an example of a landlocked country on each of the following continents. For a continent that does not have a landlocked country, write "none."

CONTINENT	LANDLOCKED COUNTRY
North America (including Central America)	
South America	
Africa	
Asia	

m-1 Name two Southeast Asian countries that do not have any oil and gas resources within their borders.

_____

_____

m-2 Name two Southeast Asian countries that have oil and gas resources within their borders.

_____

_____

Step 9 Explore boundary changes in the 1990s

e-1   Describe three political boundary changes you see between 1992 and 2000.

_____

_____

_____

e-2   Name two countries that existed in 1992 but do not exist in 2000.

_____

_____

f   Write the new name you gave the project and where you saved it.

_____   _____
(Name of project.   **(Navigation path to where project is saved.**
For example: kzeregion5.apr)   **For example: C:\student\kze\mapworld)**

g   Turn over.

g Select three countries from group A and three from group B and complete the following table. Use the information and skills you learned in this GIS investigation to answer the questions.

GROUP	COUNTRY	WHAT TYPE OF BOUNDARIES DOES IT HAVE?	HOW WOULD YOU CHARACTERIZE ITS SHAPE?	WHAT ECONOMIC ADVANTAGES OR DISADVANTAGES DO YOU SEE?	DO THE CURRENT CHARACTERISTICS OF THE COUNTRY PROMOTE COHESIVENESS OR SPLITTING APART?
A					
B					

NAME _____   DATE _____

Crossing the Line
Middle school assessment

In this assessment activity, you must predict the future! Instead of using a crystal ball, you will use a GIS to see 25 years into the future of the world.

1 Use the information you learned in this GIS Investigation to identify a current international boundary that you think will change in the next 25 years.

2 Draw a map (in ArcView or with paper and pencil) to illustrate what this boundary will look like in 25 years. Be sure to include a legend, north arrow, scale, and date of creation on the map.

3 Write an essay that describes the consequences of the change you predict. Address the following questions in your essay:

   • What types of boundaries are involved in the projected change?
   • How will the territorial morphology of the countries involved be affected by the projected change?
   • What will the economic impact of the projected change be?
   • How will the projected change affect the internal cohesiveness of all countries involved?

Crossing the Line

Assessment rubric

Middle school

STANDARD	EXEMPLARY	MASTERY	INTRODUCTORY	DOES NOT MEET REQUIREMENTS
The student knows and understands how to use the elements of space to describe spatial patterns.	Creates a digital map using GIS to illustrate a predicted international boundary change. The student takes into account a variety of data when developing his/her new map.	Creates a map to illustrate a predicted international boundary change. Takes into account a variety of data when developing his/her new map.	Creates a map to illustrate a predicted international boundary change. Provides some data to support ideas.	Describes a change in a border, but does not provide a map and has little or no data to support ideas.
How cooperation and conflict among people contribute to economic and social divisions of Earth's surface.	Uses maps and written description to illustrate how the proposed boundary changes will affect the countries involved, including their role in the global economy. Provides sufficient data to support ideas.	Describes how the proposed boundary changes will affect the countries involved, including their role in the global economy. Provides sufficient data to support ideas.	Attempts to describe how the proposed boundary changes will affect the countries involved, but does not address effect on global economy. Provides some data to support ideas.	Does not address economic issues for the countries involved in the boundary changes.
The student knows and understands how varying points of view on geographic context influence plans for change.	Describes the perspectives of the countries involved in the proposed boundary change. Addresses a variety of issues including political, cultural, and so on. Also addresses how this will affect the global community.	Describes the perspectives of the countries involved in the proposed boundary change. Addresses a variety of issues including political, cultural, and so on.	Attempts to describe the perspectives of the countries involved in the proposed boundary change. Addresses only one issue in their description.	Describes only one country's perspective on the proposed boundary change.

This is a four-point rubric based on the National Standards for Geographic Education. The "Mastery" level meets the target objective for grades 5–8.

NAME _____   DATE _____

Crossing the Line
High school assessment

In this assessment activity, you must predict the future! Instead of using a crystal ball, you will use a GIS to see 25 years into the future of the world.

1 Use the information you learned in this GIS Investigation to identify two current international boundaries that you think will change in the next 25 years. One prediction must involve splitting a current country into two or more smaller countries, and the other must merge two or more countries into one larger one.

2 Draw a map (in ArcView or with paper and pencil) to illustrate what these boundaries will look like in 25 years. Be sure to include a legend, north arrow, scale, and date of creation on the map.

3 Write an essay that describes potential consequences of the change you predict. Address the following questions in your essay:

- What types of boundaries are involved in the changes?
- How will the territorial morphology of the countries involved be affected by the projected changes?
- What will the economic impact of the projected changes be?
- How will the projected changes affect the internal cohesiveness of all countries involved?

Crossing the Line

Assessment rubric

High school

STANDARD	EXEMPLARY	MASTERY	INTRODUCTORY	DOES NOT MEET REQUIREMENTS
The student knows and understands how to apply concepts and models of spatial organization to make decisions.	Creates a digital map using GIS to illustrate a predicted international boundary change. Takes into account a variety of data when developing his/her new map.	Creates a map to illustrate a predicted international boundary change. The map illustrates two types of changes: the merging of countries together, and the splitting of countries apart. Takes into account a variety of data when developing his/her new map.	Creates a map to illustrate a predicted international boundary change. The map shows only one type of change. Provides some data to support ideas.	Describes a change in a border, but does not provide a map and has little or no data to support ideas.
The student knows and understands why and how cooperation and conflict are involved in shaping the distribution of social, political, and economic spaces on Earth at different scales.	Uses maps and written description to illustrate how the proposed boundary changes will affect the countries involved, including their role in the global network. Provides sufficient data to support ideas.	Describes how the proposed boundary changes will affect the countries involved and their role in the global network. Provides sufficient data to support ideas.	Attempts to describe how the proposed boundary changes will affect the countries involved, but does not address effect on global network. Provides some data to support ideas.	Does not address any specific issues for the countries involved in the boundary changes.
The student knows and understands contemporary issues in the context of spatial and environmental perspectives.	Describes the perspectives of the countries involved in the proposed boundary change. Addresses a variety of issues including political, cultural, and so on. Also addresses how this will affect the perspective of the global community.	Describes the perspectives of the countries involved in the proposed boundary change. Addresses a variety of issues including political, cultural, and so on.	Attempts to describe the perspectives of the countries involved in the proposed boundary change. Addresses only one issue in their description.	Describes only one country's perspective on the proposed boundary change.

This is a four-point rubric based on the National Standards for Geographic Education. The "Mastery" level meets the target objective for grades 9–12.

A Line in the Sand
A regional case study of Saudi Arabia and Yemen

Lesson overview

Students will study the creation of a new border between Saudi Arabia and Yemen on the Arabian Peninsula. Using data included in the June 2000 Treaty of Jeddah, they will draw the new boundary described in the treaty and analyze the underlying physiographic and cultural forces that influenced the location of that boundary. In the process they will come to understand how any map of the world must be considered a tentative one, as nations struggle and cooperate with each other.

Estimated time   Three 45-minute class periods

Materials   ✔ Student handouts from this lesson to be copied:
- GIS Investigation sheets (pages 283 to 295)
- Student answer sheets (pages 297 to 301)
- Assessment(s) (pages 302 to 305)

Standards and objectives   *National geography standards*

GEOGRAPHY STANDARD	MIDDLE SCHOOL	HIGH SCHOOL
1 How to use maps and other geographic representations, tools, and technologies to acquire, process, and report information from a spatial perspective	The student understands how to use maps to analyze spatial distributions and patterns.	The student understands how to use geographic representations and tools to analyze and explain geographic problems.
4 The physical and human characteristics of places	The student understands how physical processes shape places and how different human groups change places.	The student understands the changing human and physical characteristics of places.
13 How the forces of cooperation and conflict among people influence the division and control of Earth's surface	The student understands the multiple territorial divisions of the student's own world.	The student understands why and how cooperation and conflict are involved in shaping the distribution of social, political, and economic spaces on Earth at different scales.
18 How to apply geography to interpret the present and plan for the future	The student understands how various points of view on geographic context influence plans for change.	The student understands contemporary issues in the context of spatial and environmental perspectives.

Standards and objectives (continued)

Objectives

The student is able to:

- Describe the physical and human characteristics of the Arabian Peninsula.

- Define and describe the Empty Quarter.

- Explain major elements of the Treaty of Jeddah boundary agreement between Saudi Arabia and Yemen.

- Identify physical and cultural characteristics of the Arabian Peninsula that are reflected in the new Saudi–Yemeni border agreement.

GIS skills and tools

 Zoom in on the map

 Identify attributes of a feature

 Zoom to the previous extent

 Zoom to the active theme

 Add themes to the view

 Add a point at a specific latitude/longitude coordinate intersection

 Draw a line

 Edit the vertices of a line or polygon

 Clear selected features

 Select a feature from the active theme

 Select a graphic

- Create a new theme
- Print a map
- Create a buffer

GIS skills and tools (continued)

For more on geographic inquiry and these steps, see Geographic Inquiry: Thinking Geographically (pages xxi to xxiii).

Teacher notes

Lesson introduction

Divide your class into small groups. Explain to your students that the lesson they will begin today is about drawing boundary lines. In order to identify some of the important considerations in the demarcation of boundaries, each group will take five minutes to consider the following hypothetical scenario:

Size limitations in the school building require that their classroom be divided to create two new, smaller classrooms. Other than the wall dividing the two classrooms, there will be no new construction. Each group is charged with two tasks:

- Identify the features of the present classroom which are valuable to the teachers and students who use that room (windows, for example).
- Suggest a possible boundary line to divide the present classroom and identify the features from the previous step that each of the new classrooms will have.

When five minutes have passed, make a list on the blackboard or an overhead projector of the valuable classroom features that students identified in the first step. Let each group report on the boundary they propose. Use this activity as a springboard for a discussion of the issues involved in the creation of national boundaries. Be sure to include the following points in the discussion:

- Certain features of the physical environment have greater value than others to the people who will occupy and use that space.
- The human uses of a place influence the perceived value of its physical features.
- When a boundary line is drawn, it may not be possible to divide the valuable features evenly between the parties involved.

Tell the class that this activity will explore a twenty-first-century case of the demarcation of a boundary between two countries. Although at a much different scale, this decision involved many of the same issues they faced in drawing a hypothetical boundary in their classroom.

Student activity

 Before completing this lesson with students, we recommend that you work through it yourself. Doing so will allow you to modify the activity to accommodate the specific needs of your students.

After the initial discussion, have the students work on the computer component of the lesson. Ideally each student should be at an individual computer, but the lesson can be modified to accommodate a variety of instructional settings.

Distribute the GIS Investigation sheets to your students. Explain that in this activity they will use GIS to explore a region of the world where a recent boundary dispute has been settled after 65 years of conflict: the Arabian Peninsula. They will explore alternatives for boundaries between the countries involved and analyze the underlying physiographic and cultural considerations that played a part in the resolution of that conflict.

The GIS activity will provide students with detailed instructions for their investigations. In addition to the instructions, the handout includes questions to help them focus on key concepts. Some questions will have specific answers while others require creative thought.

Things to look for while the students are working on this activity:

- Are the students using a variety of tools?
- Are they answering the questions as they work through the procedure?
- Are they experiencing any difficulty managing the display of information in their view as they turn themes on and off?

Conclusion

Refer your students to the activity that introduced this lesson: the creation of a hypothetical boundary line that divides their classroom into two new rooms. Review their conclusions and ask them to identify parallel issues in the settlement of the boundary dispute between Saudi Arabia and Yemen.

- Certain features of the physical environment have greater value than others to the people who will use that space. On the Arabian Peninsula, areas that get enough precipitation for agriculture, areas of grassland for grazing, sources of water, and areas with the strategic advantage of mountain peaks have greater value.

- The human uses of a place influence the perceived value of its physical features. On the Arabian Peninsula, livestock herding and farming are examples of traditional human uses.

- When a boundary line is drawn, it may not be possible to divide the valuable features evenly between the parties involved. On the Arabian Peninsula, most of the areas that get enough precipitation for agriculture, areas of grassland for grazing, and sources of water went to Yemen.

Ask students to identify issues that played a role in the Saudi–Yemeni border conflict that were not present in the classroom boundary scenario. For example:

- Historic boundaries and patterns of political control in the region played an important part in the Saudi–Yemeni border conflict. Discuss important historical events and periods such as the Ottoman Empire, the consequences of World War I in this region, international interest in the region during the twentieth century, and the British Protectorate of Aden.

- Nomadism and strong identity with regional tribal traditions are at odds with the delineation of a fixed boundary in this region. Discuss the various factors that influence a community's or a region's sense of itself.

Note: Students may wonder why the Saudis were willing to yield so much territory to Yemen. Ask them to speculate on possible reasons for this apparent generosity. The Treaty of Jeddah states that the two countries will negotiate if sources of "shared natural wealth" are discovered in the border region. This means that Saudi Arabia reserves the right to reopen negotiations in the event that something that they value very highly—oil or gas, for example—is discovered near the new boundary. Also, Saudi Arabia has long been interested in constructing a pipeline to the Arabian Sea across the southern part of the peninsula. The Saudis may have been hoping that a generous settlement with Yemen on the border issue could make the Yemenis more willing to agree to a Saudi pipeline across their territory.

Assessment

Middle school: Highlights skills appropriate to grades 5 through 8

In the middle school assessment, students will write a newspaper article reporting on the settlement of the Saudi–Yemeni border dispute by the Treaty of Jeddah, which was agreed to in June 2000. The article, written from either a Saudi or a Yemeni perspective, should describe the new boundary established by the treaty and analyze underlying physiographic and cultural considerations that influenced the location of that boundary. They will also prepare a map to go with the article.

High school: Highlights skills appropriate to grades 9 through 12

In the high school assessment, students will write a newspaper article reporting on the settlement of the Saudi–Yemeni border dispute by the Treaty of Jeddah, which was agreed to in June 2000. The article, written from either a Saudi or a Yemeni perspective, should describe the new boundary established by the treaty and analyze underlying physiographic and cultural considerations that influenced the location of that boundary. The article should also include information about historical factors that contributed to this 65-year-old boundary conflict. A map should accompany the article.

Extensions

- In the introductory activity, have student groups negotiate with each other to arrive at a mutually satisfactory boundary for the two new classrooms.
- Use the Internet to identify other areas of the world where international boundaries are in dispute.
- Research the events of World War I on the Arabian Peninsula. Create an ArcView project illustrating these events.
- Use ArcView to compare the countries of the Arabian Peninsula by mapping and analyzing relevant economic and demographic data.
- Check out the Resources by Module section of the Teacher Resource CD for print and media resources on the topics of Saudi Arabia, Yemen, and the Treaty of Jeddah or visit *www.esri.com / mappingourworld* for Internet links.

NAME _____   DATE _____

A Line in the Sand
A GIS investigation

ACQUIRE

ASK

EXPLORE

ACT

ANALYZE

Answer all questions on the student answer sheet handout

The ever-changing map of the world reflects the forces of conflict and cooperation among nations and peoples of the world. In this GIS Investigation, you will explore one of the first boundary changes of the twenty-first century—the creation of a new border between Yemen and Saudi Arabia on the Arabian Peninsula. After more than 60 years of conflict, the two nations signed the historic boundary agreement in June 2000. Using data provided in the Treaty of Jeddah, you will create a map reflecting the treaty's territory, and analyze underlying physiographic and cultural considerations that influenced the location of the boundary.

Step 1   Start ArcView

a   Double-click the ArcView icon on your computer's desktop.

b   If the Welcome to ArcView dialog appears (pictured below), click **Open an Existing Project** and click OK. If it doesn't appear, proceed to step 2.

Step 2   Open the region5.apr file

a   In this exercise, a project file has been created for you. To open the file, go to the **File** menu and choose **Open Project**.

b   Navigate to the exercise data directory (**C:\esri\mapworld\mod5**) and choose **region5.apr** from the list.

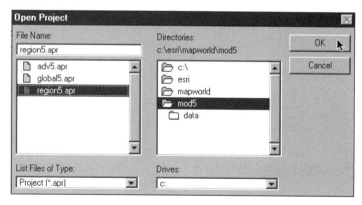

c   Click OK.

When the project opens, you will see a shaded relief map of the Middle East and North Africa. A red box locates the Arabian Peninsula.

d   Maximize your ArcView window. Stretch your View window so it's bigger.

Step 3   Identify countries that border the Arabian Peninsula

 a   Click the Zoom to Active Theme button. Now the red box that surrounds the Arabian Peninsula fills the view.

b   Scroll down the table of contents until you see a theme called Neighbors: Outline. Click the box to the left of the theme name to turn it on.

c   Make Neighbors: Outline the active theme. You can tell that it is the active theme because it looks like a raised button.

d   Click the Identify tool. Click the countries on the map that border the Arabian Peninsula to the north.

?   *Record their names on your answer sheet.*

Step 4   Investigate the physical characteristics of the Arabian Peninsula

a   The map on your screen is a shaded relief map. It depicts landforms and relief.

?   *(1)   Is any part of the Arabian Peninsula mountainous?*

?   *(2)   If so, where are the mountains located?*

b   Turn on the Bodies of Water and Streams themes.

Most of the streams you see on your map are intermittent, which means that they are dry during some parts of the year.

?   *(1)   Are there any parts of the Arabian Peninsula that do not have any water at all? If so, where are these regions?*

?   *(2)   Do you see any relationship between landforms and the availability of water?*

c   Turn off the Streams theme and observe the distribution of permanent bodies of water on the Arabian Peninsula.

d   Click the Zoom In tool. Click and drag a small box around an area of blue dots.

Now you can see the bodies of water more closely.

?   *Describe the bodies of water on your answer sheet.*

e   Click the Zoom to Previous Extent button to return to your view of the entire peninsula.

f   Turn off the following themes: Bodies of Water, Neighbors: Outline, and Shaded Relief.

Your view should now display a map that looks like this:

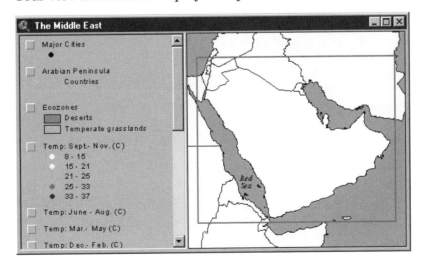

g   Turn on the Annual Precipitation theme.

Amounts of rainfall are given in millimeters. Here is a conversion table that compares millimeters to inches (25.4 mm. = 1 in.).

MM.	100	200	300	400	500	600	700
IN.	3.9	7.9	11.7	15.6	19.5	23.4	27.3

?   *(1)   A desert is defined as a place that gets less than 10 inches of rain per year. How many millimeters equal 10 inches?*

?   *(2)   Based on the amounts of rainfall displayed on the map, do you think there is much farming on the Arabian Peninsula? Explain.*

?   *(3)   Approximately what percent of the Arabian Peninsula is desert?*

h   Turn off Annual Precipitation and turn on Temp: Sept. - Nov. (C).

Use this conversion table to help you answer the next questions.

°C	0°	5°	10°	15°	20°	25°	30°	35°
°F	32°	41°	50°	59°	68°	77°	87°	95°

?   *What is the approximate range of temperatures across the Arabian Peninsula during this period?*

The three themes below Temp: Sept.- Nov. display temperature information for the periods December–February, June–August, and March–May.

i   Turn the temperature themes on and off one at a time to see the change of temperatures on the Arabian Peninsula through the four seasons.

?   *(1)   Which season is the hottest?*

?   *(2)   What is the approximate range of temperatures across the Arabian Peninsula during this period?*

j   Turn off any temperature themes that are still on. Turn on the Ecozones theme.

?   *(1)   What relationship do you see between the Arabian Peninsula's ecozones as displayed on this map and its patterns of landforms, precipitation, and temperature?*

?   *(2)   Use your answers from previous questions and turn different themes on and off to complete the Physical Characteristics of the Arabian Peninsula table on your answer sheet. List three observations for each physical characteristic.*

?   *(3)   In your opinion, which of the region's physical characteristics would be considered "valuable" in a boundary decision? Explain.*

Step 5   Investigate the human characteristics of the Arabian Peninsula

The population of the Arabian Peninsula is approximately 45 million. The majority of this population lives in Saudi Arabia (22 million) and Yemen (17.5 million). The remaining 5.5 million can be found in Oman, the United Arab Emirates, and Qatar.

a   Turn off Ecozones. Turn on the Arabian Peninsula Countries theme. This theme names and locates the countries, but does not show their borders. You will explore the borders of these countries in the second part of the lesson.

b   Turn on the Major Cities theme.

c   Click the Add Theme button. Navigate to the module 5 data directory (**C:\esri \mapworld\mod5**). Hold down the Shift key and click once on each of the following file names to select all of them at the same time: **agri.shp**, **pop_dens.shp**, **roads.shp**, and **springs.shp**. Click OK.

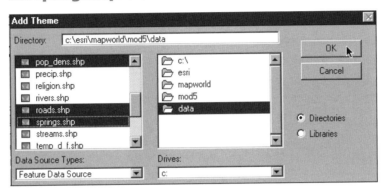

d   Turn on the Agri.shp theme. This theme shows the agricultural activity on the Arabian Peninsula.

(1)   What is the principal agricultural activity on the peninsula?

(2)   Based on what you now know about the physical characteristics of the region, why do you think the agricultural activity is so limited?

e   Turn on the Pop_dens theme. Drag the Major Cities theme above the Pop_dens theme. The human population around major cities and throughout the Arabian Peninsula is displayed as number of people per square kilometer.

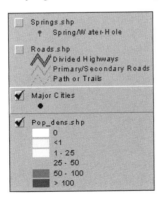

(1)   How does Yemen compare to the rest of the Arabian Peninsula in population density?

(2)   Describe the overall population density of the Arabian Peninsula.

f Turn off Pop_dens.shp. Make sure Agri.shp and Springs.shp are turned on.

Springs.shp shows the location of springs and water holes.

? *g* Look at the map. On your answer sheet, speculate about the ways water is most commonly and frequently used at these springs and water holes.

? *h* Use your answers from previous step 5 questions and analysis of the maps to complete the table on your answer sheet. List three observations for each human characteristic.

? *If an international boundary were to be drawn across some part of the Arabian Peninsula, how would these characteristics influence the perception of certain regions as being more "valuable" than others?*

Step 6 **Locate and describe the Empty Quarter**

a Turn off Springs.shp and turn on Roads.shp.

Take note of the large area with virtually no roads in the southern part of the peninsula. This region is called the Rub´ al-Khali and is also known as the Empty Quarter. The Empty Quarter is important to this lesson because most Saudi Arabian borders with its southern neighbors cross this region.

b Turn the following themes on and off so you can observe the characteristics of the Empty Quarter: Ecozones, Streams, Precipitation, Temp, Pop_dens, and Agri.shp.

? (1) *Complete the table on your answer sheet. List three observations in each column.*

? (2) *What difficulties would an area like this present if an international boundary must cross it?*

c Turn off all themes except Major Cities, Arabian Peninsula Countries, Arabian Peninsula Area, Neighbors, Arabian Peninsula, and Ocean.

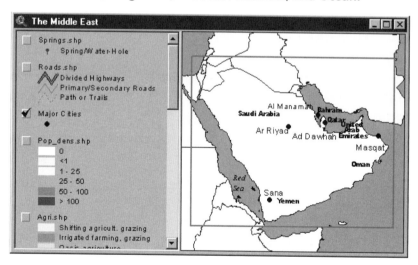

? *d* Ask your teacher for instructions on where to save this ArcView project and on how to rename the project. If you are not going to save the project, proceed to step 7b now. Otherwise, record the project's new name and where you saved it on your answer sheet.

e From the File menu, click Exit.

f Click No when asked if you want to save changes.

Step 7 Explore Saudi Arabia's southern boundaries

a Start ArcView. Navigate to the directory where you renamed and saved region5.apr. Open the project.

b Close the Middle East view.

c In the Project window, click the Saudi Arabia Southern Boundaries view and click Open.

This view looks very similar to the Middle East view. This theme reflects the boundary agreements Saudi Arabia made with most of its southern neighbors at the end of the twentieth century.

d Turn on 20th Century Boundary.

? *(1) Are the boundaries what you expected them to be?*

? *(2) Which boundary remained unsettled?*

As Richard Schofield, an international boundary expert put it, this boundary was "the last missing fence in the desert." The only part of the boundary that was mutually agreed upon was the western area adjacent to the Red Sea. Over the years, the boundary has shifted. Now you will add themes that reflect some of the major boundary changes.

e Click the Add Theme button. Navigate to the data directory (**C:\esri\mapworld\mod5 \data**). Scroll to the bottom of the list, hold down the Shift key, and click once on each of the following file names: **Yemen1.shp**, **Yemen2.shp**, and **Yemen3.shp**. Click OK.

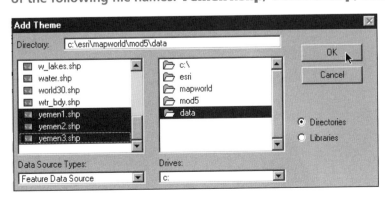

f   **Turn on Yemen1.shp.**

This red line represents the boundary between Yemen and Saudi Arabia established by the Treaty of Ta'if in 1934. It is the only part of the boundary that both countries recognized at the turn of the twenty-first century.

g   **Turn on Yemen2.shp.**

This green line represents the Saudi–Yemeni border recognized by Yemen at the end of the century. It is based on lines established when Yemen (then the Aden Protectorate) was under British control in the early twentieth century. Most maps used these lines to delineate the extent of Yemen prior to 2000. This boundary was not recognized by Saudi Arabia.

h   **Turn on Yemen3.shp.**

This purple line represents the Saudi–Yemeni border claimed by Saudi Arabia at the end of the twentieth century. It is based on lines established by the Saudis in the mid-1930s. This line was still being used on Saudi Arabian maps to represent the boundary in the 1990s.

i   **Click the Zoom In tool. On your map, click the label Yemen twice. Now you are zoomed into the country of Yemen.**

?   *What does the area between the green and purple lines represent?*

j   **Turn on Agri.shp.**

?   *What is the principal economic activity of the regions in dispute?*

k   **Turn off Agri.shp and turn on Pop_dens.shp.**

?   *Describe the population distribution in the disputed territory.*

If you were asked to settle the disputed boundary between Saudi Arabia and Yemen, where would you draw the line? In this next step, you will draw a proposed boundary line between Saudi Arabia and Yemen, using the Draw Line tool.

l   **From the Graphics Tools drop-down menu, select the Draw Line tool.**

m   **On your map, click the eastern end point of the red boundary line. (This is the boundary that both countries agree on.) Proceed eastward (to your right) and click a proposed boundary line. Double-click when you get to the end of your boundary. Now you have an additional black line that extends from the Red Sea to Oman.**

In the next step, you will view the new boundary actually agreed upon by Saudi Arabia and Yemen in 2000.

MODULE 5 • HUMAN GEOGRAPHY II: POLITICAL GEOGRAPHY

Step 8   **Draw the Saudi–Yemeni boundary**

In June of 2000, Saudi Arabia and Yemen signed the Treaty of Jeddah, which settled their 65-year-long boundary dispute. The boundary agreement had three parts. The first part of the treaty reaffirmed agreement on the 1934 Ta'if line (Yemen1.shp). The agreement did say, however, that the line would be amended in any place where it cuts through villages.

a   Click the Zoom In tool. Zoom to the area of the Ta'if line (red line) by dragging a box around it.

b   Click the Add Theme button. Navigate to the module 5 data directory (**C:\esri \mapworld\mod5\data**). Select **cit_town.shp**. Click OK.

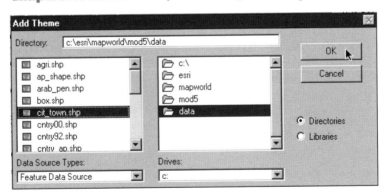

?   *(1)*   *Does the red line go through any cities or towns? (Hint: you may need to zoom in again to answer the question.) If yes, approximately how many does the boundary pass through?*

?   *(2)*   *How would you decide which side of the town to put the boundary on? Remember, this decision would determine whether the residents of that village would be citizens of Saudi Arabia or Yemen.*

The second part of the Treaty of Jeddah determined the new boundary from the end of the red line to the border with Oman, 500 miles to the east. The treaty did not actually draw the line, but gave its starting and ending points and points in between as latitude/longitude grid coordinates. You will now plot these points on your map.

c   Turn off Cit_town.shp and Pop_dens.shp. Make Arabian Peninsula Countries the active theme. Click the Zoom to Active Theme button.

d From the Window menu, select Show Symbol Window. Select the Marker Palette (it looks like a pushpin). Change the size to 8.

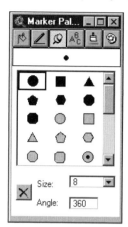

e Click the Color Palette button. Change the color to red.

f Click the Add Lat/Long point button. In the dialog, type **19.00** for latitude and **52.00** for longitude. Click OK.

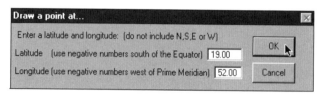

You see the red point appear at the border with Oman.

g To enter the remaining latitude and longitude points determined in 2000, consult the table below (points 2–17). For each point, you must first click the Add Lat/Long point button and then enter the coordinates in the table. Enter all the points now.

Point	Latitude	Longitude
1	19.00	52.00
2	18.78	50.78
3	18.61	49.12
4	18.17	48.18
5	17.45	47.60
6	17.12	47.47
7	16.95	47.18
8	16.95	47.00
9	17.28	46.75

Point	Latitude	Longitude
10	17.23	46.37
11	17.25	46.10
12	17.33	45.40
13	17.43	45.22
14	17.43	44.65
15	17.40	44.57
16	17.43	44.47
17	17.43	44.37

Does the new line seem to favor Yemen or Saudi Arabia? Explain.

The third and final part of the Treaty of Jeddah clarified the maritime boundary between Saudi Arabia and Yemen. A maritime boundary defines the offshore limits of a country. It too was defined by a series of latitude/longitude grid coordinates.

h Follow the same procedure as steps 8f and 8g to map the maritime boundary between Saudi Arabia and Yemen.

Point	Latitude	Longitude
1	16.40	42.77
2	16.40	42.15
3	16.29	41.78

? *Look at your map. What body of water does the maritime boundary traverse?*

Now you will use the points you have mapped to draw the new international boundary between Saudi Arabia and Yemen.

i From the View menu, select New theme. For Feature type, select line. Click OK.

j Ask your teacher where you should save your new theme. Navigate to the location your teacher directed you to and name the new theme **ABCYemen00**, where ABC are your initials. Click OK.

The new theme is now added to your table of contents. (You may need to scroll up to see it.) It has a dotted line around the theme's check box. This indicates that the theme is in "Edit" mode and can be created and modified. Because ArcView randomly selects a color for a new theme, you need to change it.

k Double-click the new Yemen00 theme to open its Legend Editor.

l Double-click the symbol box to bring up the Symbol window.

m Click the Color Palette button. Click a blue color. Click Apply and close the Legend Editor and the Color Palette window.

Now you will use the Draw Line tool and create the boundary line by "connecting the dots" of the latitude and longitude points you plotted earlier.

n Click the Draw Line tool. Click the red dot at the Oman boundary. Click each red dot to create the boundary line. When you get to the red line (Yemen 1.shp), click along its path to duplicate it. Double-click to end the line when you've completed the boundary into the Red Sea.

o Click the Zoom In tool. Zoom in on the boundary. Check to see if your line goes precisely through each dot and follows the Yemen1.shp theme exactly.

p If you find points on the line that need adjusting, click the Vertex Edit tool. Click the boundary line where it needs modification. Small boxes called vertices are visible.

q Click a vertex and drag it to its correct point on the map. Continue this process until the entire line is in the correct place.

r Once you are satisfied with your line, select Stop Editing from the Theme menu. When asked if you want to save your edits, click Yes.

s Click the Clear Selected Feature button. You can now see the blue line that represents the new territorial boundary between Saudi Arabia and Yemen.

t From the Theme menu, select Properties. Rename the theme **2000 Boundary**.

? *How does the actual boundary established by the Treaty of Jeddah compare with the boundary you drew earlier?*

u Turn on Agri.shp and Pop_dens.shp.

? *Write three observations about the boundary line created by the Treaty of Jeddah.*

v Save your project.

Step 9 Define the pastoral area

The Treaty of Jeddah included additional provisions about the new Saudi–Yemeni boundary. One of these was the creation of a "pastoral area" on either side of the boundary. Shepherds from either Yemen or Saudi Arabia are allowed to use the pastoral area and water sources on both sides of the border according to tribal traditions. The treaty declared that the pastoral area extend 20 kilometers on either side of the border. In this step, you will map the 20-kilometer pastoral area.

? *a* How many miles is 20 kilometers? (Hint: 1 kilometer = .6214 miles.)

b Make 2000 Boundary the active theme.

c Click the Select Feature tool. Click the 2000 Boundary line. It turns yellow in the view. From the Theme menu, select Buffer.

d Type **20** as the buffer distance. Click OK.

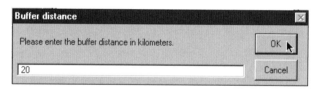

? *(1) In which part of the Saudi–Yemeni border will the pastoral area be most significant? Explain.*

? *(2) Why do you think the Treaty of Jeddah created a pastoral area?*

e Click Clear Selected Features. The 2000 Boundary line turns back to blue.

Step 10 Create a map of the Arabian Peninsula

Before you print a map of the Arabian Peninsula, you need to clean up the map.

a Click the Pointer tool. Click the boundary line you first drew (it's black and doesn't have the buffer around it). Press the Delete key. The line disappears from the view.

b Make the Arabian Peninsula active and click the Zoom to the Active Theme button. The entire Arabian Peninsula fills the view.

c   Make sure that the following themes are turned on:

- 2000 Boundary
- 20th Century Boundary
- Major Cities
- Arabian Peninsula Countries
- Neighbors
- Arabian Peninsula
- Ocean

    d   From the File menu, select Print to print a copy of your map.

Step 11   **Exit ArcView**

In this exercise, you explored physical and human characteristics of the Arabian Peninsula. After analyzing this data, you explored boundary issues in this region and plotted the new Yemeni–Saudi boundary established by the 2000 Treaty of Jeddah.

    a   Save your project.

    b   From the File menu, click Exit. When it asks if you want to save your changes, click No.

NAME _____   DATE _____

Student answer sheet

Module 5
Human Geography II: Political Geography

Regional case study: A Line in the Sand

Step 3   Identify countries that border the Arabian Peninsula

d   Record the names of the countries on the map that border the Arabian Peninsula to the north.

_____   _____   _____

Step 4   Investigate the physical characteristics of the Arabian Peninsula

a-1   Is any part of the Arabian Peninsula mountainous? _____

a-2   If so, where are the mountains located?

    _____

    _____

b-1   Are there any parts of the Arabian Peninsula that do not have any water at all? If so, where are these regions?

    _____

    _____

b-2   Do you see any relationship between landforms and the availability of water?

    _____

    _____

d   Describe the bodies of water:

    _____

    _____

g-1   How many millimeters equal 10 inches?

    _____

g-2   Based on the amounts of rainfall displayed on the map, do you think there is much farming on the Arabian Peninsula? Explain.

    _____

    _____

    _____

g-3   Approximately what percent of the Arabian Peninsula is desert? _____

h   What is the approximate range of temperatures across the Arabian Peninsula during this period?

    ° C: _____   ° F: _____

i-1   Which season is the hottest? _____

i-2   What is the approximate range of temperatures across the Arabian Peninsula during this period?
          _____ °C to _____ °C   _____ °F to _____ °F

j-1   What relationship do you see between the Arabian Peninsula's ecozones as displayed on this map and
       its patterns of landforms, precipitation, and temperature?

       _____

       _____

j-2   Complete the table. List three observations for each physical characteristic.

PHYSICAL CHARACTERISTICS OF THE ARABIAN PENINSULA	
Landforms and bodies of water	
Climate	
Ecozones	

j-3   In your opinion, which of the region's physical characteristics would be considered "valuable" in a
       boundary decision? Explain.

       _____

       _____

       _____

Step 5   Investigate the human characteristics of the Arabian Peninsula

d-1   What is the principal agricultural activity on the peninsula?

       _____

d-2   Based on what you now know about the physical characteristics of the region, why do you think the
       agricultural activity is so limited?

       _____

       _____

       _____

e-1   How does Yemen compare to the rest of the Arabian Peninsula in population density?

       _____

       _____

       _____

e-2   Describe the overall population density of the Arabian Peninsula.

_____

_____

_____

g   Speculate on the most frequent use of the water at these springs and water holes.

_____

_____

_____

h   Use your answers from previous step 5 questions and analysis of the maps to complete the table. List
      three observations for each human characteristic.

HUMAN CHARACTERISTICS OF THE ARABIAN PENINSULA	
Agricultural activities	
Population density and distribution	

If an international boundary were to be drawn across some part of the Arabian Peninsula, how would
these characteristics influence the perception of certain regions as being more "valuable" than others?

_____

_____

_____

Step 6   Locate and describe the Empty Quarter

b-1   Complete the table. List three observations in each column.

THE EMPTY QUARTER	
PHYSICAL CHARACTERISTICS	HUMAN CHARACTERISTICS

b-2   What difficulties would an area like this present if an international boundary must cross it?

_____

_____

_____

d   If you are going to save the project, record the project's new name and where you saved it.

_____
**(Name of project.
For example: Ksregion5.apr)**

_____
**(Navigation path to where project is saved.
For example: C:\student\kimsmith
\ksregion5.apr)**

Step 7   Explore Saudi Arabia's southern boundaries

d-1   Are the boundaries what you expected them to be?

_____

d-2   Which boundary remained unsettled?

_____

i   What does the area between the green and purple lines represent?

_____

_____

j   What is the principal economic activity of the regions in dispute?

_____

_____

k   Describe the population distribution in the disputed territory.

_____

_____

_____

Step 8   Draw the Saudi–Yemeni boundary

b-1   Does the red line go through any cities or towns? (Hint: you may need to zoom in again to answer the question.) If yes, approximately how many does the boundary pass through?

_____

_____

_____

b-2   How would you decide which side of the town to put the boundary on? Remember, this decision would determine whether the residents of that village would be citizens of Saudi Arabia or Yemen.

_____

_____

_____

g   Does the new line seem to favor Yemen or Saudi Arabia? Explain.

_____

_____

_____

h   What body of water does the maritime boundary traverse?

_____

t   How does the actual boundary established by the Treaty of Jeddah compare with the boundary you
     drew earlier?

     _____

     _____

     _____

     _____

u   Write three observations about the boundary line created by the Treaty of Jeddah.

     _____

     _____

     _____

     _____

     _____

     _____

Step 9   Define the pastoral area

a   How many miles is 20 kilometers? (Hint: 1 kilometer = .6214 miles.) _____

d-1 In which part of the Saudi–Yemeni border will the pastoral area be most significant? Explain.

     _____

     _____

     _____

     _____

d-2 Why do you think the Treaty of Jeddah created a pastoral area?

     _____

     _____

     _____

     _____

NAME _____   DATE _____

A Line in the Sand
Middle school assessment

You are a newspaper reporter assigned to cover the Treaty of Jeddah, signed on June 12, 2000, which settled the 65-year-old border dispute between Saudi Arabia and Yemen. You must choose to be a reporter for a newspaper in either Saudi Arabia or Yemen and write your article from that country's perspective. In preparing your article, you may use the Line in the Sand ArcView project as well as additional resources such as your history and geography books, encyclopedias, and the Internet. Your article should include the following:

- A map showing the new boundary line and a relevant physical or cultural characteristic discussed in your article
- A description of the physical and cultural characteristics of the region affected by the boundary change
- A description of the new boundary established by the treaty and its implications for people living in the affected areas

Use the remainder of the page as a place to brainstorm for your article.

MODULE 5 • HUMAN GEOGRAPHY II: POLITICAL GEOGRAPHY

A Line in the Sand

Assessment rubric

Middle school

STANDARD	EXEMPLARY	MASTERY	INTRODUCTORY	DOES NOT MEET REQUIREMENTS
The student knows and understands how to use maps to analyze spatial distributions and patterns.	Creates a detailed map using GIS that shows the new boundary lines and relevant physical or cultural characteristics.	Creates a map showing the new boundary line and most relevant physical or cultural characteristics.	Creates a map showing the new boundary line, but does not include any relevant physical or cultural characteristics.	Creates a map of the Middle East, but does not focus on the boundary issue between Saudi Arabia and Yemen.
The student knows and understands how physical processes shape places and how different human groups change places.	Writes a detailed description of the physical and cultural characteristics that will be affected by the boundary change and explains any possible ramifications.	Writes a description of the physical and cultural characteristics that will be affected by the boundary change.	Writes a description of physical or cultural characteristics that will be affected by the boundary change.	Describes physical or cultural characteristics of the region, but does not explain how these things are affected by the boundary change.
The student knows and understands the multiple territorial divisions of the student's own world.	Describes the implications of the new boundary for the people living in the affected areas. Includes quotes or stories from individuals in the area (these could be real or fictional and derived from research).	Describes the implications of the new boundary for the people living in the affected areas.	Describes characteristics of people living in the affected region, but does not relate it specifically to the boundary change.	Gives little description on the characteristics of the people living in the affected areas.
The student knows and understands how various points of view on geographic context influence plans for change.	Writes two clear and coherent news articles, one from the perspective of Saudi Arabia and one from the perspective of Yemen on the boundary issue.	Writes a clear and coherent news article from the perspective of either country involved in the boundary issue.	Writes a news article on the boundary issue, but does not offer the perspective of either country on the boundary issue.	Writes an essay on the boundary issue, but does not offer any geographic perspective, and it is not in the form of a news article.

This is a four-point rubric based on the National Standards for Geographic Education. The "Mastery" level meets the target objective for grades 5–8.

NAME _____   DATE _____

A Line in the Sand
High school assessment

You are a newspaper reporter assigned to cover the Treaty of Jeddah, signed on June 12, 2000, which settled the 65-year-old border dispute between Saudi Arabia and Yemen. You must choose to be a reporter for a newspaper in either Saudi Arabia or Yemen and write your article from that country's perspective. In preparing your article, you may use the Line in the Sand ArcView project as well as additional resources such as your history and geography books, encyclopedias, and the Internet. Your article should include the following:

- A map showing the new boundary line, the boundaries claimed by Yemen and Saudi Arabia prior to the settlement, and a relevant physical or cultural characteristic discussed in your article

- A description of the physical and cultural characteristics of the region affected by the boundary change

- A description of the historical factors that contributed to this long-standing conflict

- A description of the new boundary established by the treaty and its implications for people living in the affected areas

Use the remainder of the page as a place to brainstorm for your article.

A Line in the Sand

Assessment rubric

High school

STANDARD	EXEMPLARY	MASTERY	INTRODUCTORY	DOES NOT MEET REQUIREMENTS
The student knows and understands how to use geographic representations and tools to analyze and explain geographic problems.	Creates a detailed map using GIS showing the new boundary line and relevant physical and cultural characteristics.	Creates a map showing the new boundary line and most relevant physical or cultural characteristics.	Creates a map showing the new boundary line and relevant physical or cultural characteristics.	Creates a map showing the new boundary line, but does not include any relevant physical or cultural characteristics.
The student knows and understands the changing human and physical characteristics of places.	Writes a detailed description of the physical and cultural characteristics that will be affected by the boundary change and explains any possible ramifications.	Writes a description of the physical and cultural characteristics that will be affected by the boundary change.	Writes a description of physical or cultural characteristics that will be affected by the boundary change.	Describes physical or cultural characteristics of the region, but does not explain how these things are affected by the boundary change.
The student knows and understands why and how cooperation and conflict are involved in shaping the distribution of social, political, and economic spaces on Earth at different scales.	Describes the implications of the new boundary for the people living in the affected areas in relationship to social issues, politics, and the economy. Includes quotes or stories from individuals in the area (these could be real or fictional and derived from research).	Describes the implications of the new boundary for the people living in the affected areas in relationship to social issues, politics, and the economy.	Describes the implications of the new boundary for the people living in the affected areas.	Describes characteristics of people living in the affected region, but does not relate it specifically to the boundary change.
The student knows and understands contemporary issues in the context of spatial and environmental perspectives.	Writes two clear and coherent news articles, one from the perspective of Saudi Arabia and one from the perspective of Yemen, on the boundary issue. It includes historical factors that contributed to the conflict.	Writes a clear and coherent news article from the perspective of either country involved in the boundary issue. It includes historical factors that contributed to the boundary issue.	Writes a news article on the boundary issue, but does not offer the perspective of either country on the boundary issue. The article may include one or two historical factors.	Writes an essay on the boundary issue, but does not offer any geographic perspective, and it is not in the form of a news article.

This is a four-point rubric based on the National Standards for Geographic Education. The "Mastery" level meets the target objective for grades 9–12.

Starting from Scratch
An advanced investigation

Lesson overview

Students will use physiographic (physical features) and anthropographic (cultural features) data to redraw some of the world's international boundaries, thereby creating states characterized by internal cohesiveness and economic parity. By comparing their maps to ones reflecting contemporary political boundaries, students will identify world regions where political boundaries are in conflict with physical and cultural imperatives.

Estimated time   Three to four 45-minute class periods

Materials   ✔ Student handouts from this lesson to be copied:
  • GIS Investigation sheets (pages 311 to 316)
  • Student answer sheet (page 317)

Standards and objectives   *National geography standards*

	GEOGRAPHY STANDARD	MIDDLE SCHOOL	HIGH SCHOOL
1	How to use maps and other geographic representations, tools, and technologies to acquire, process, and report information from a spatial perspective	The student knows and understands how to use maps to analyze spatial distributions and patterns.	The student knows and understands how to use technologies to represent and interpret Earth's physical and human systems.
3	How to analyze the spatial organization of people, places, and environments on Earth's surface	The student knows and understands how to use the elements of space to describe spatial patterns.	The student knows and understands how to apply concepts and models of spatial organization to make decisions.
10	The characteristics, distribution, and complexity of Earth's cultural mosaics	The student knows and understands how to read elements of the landscape as a mirror of culture.	The student knows and understands how cultures shape the character of a region.
13	How the forces of cooperation and conflict among people influence the division and control of Earth's surface	The student knows and understands the multiple territorial divisions of the student's own world.	The student knows and understands the impact of multiple spatial divisions on people's daily lives.

Objectives

The student is able to:
• Describe the physical features that form natural boundaries between major regions on earth.
• Describe the distribution of major language groups and religions on earth.
• Explain how political and cultural boundaries influence a country's internal cohesiveness and opportunity for economic parity with other nations.

GIS skills and tools
- Change map symbolization with the Legend Editor
- Create new themes
- Digitize new polygons
- Populate the attribute table for a new theme
- Design a presentation layout and print it
- Analyze data and make decisions based on spatial patterns

For more on geographic inquiry and these steps, see Geographic Inquiry: Thinking Geographically (pages xxi to xxiii).

Teacher notes

Lesson introduction

Introduce the lesson by challenging students to identify regions in the world that are characterized by instability or long-term economic hardship. After generating a list, ask students if such problems in these regions are the result of boundary issues. Use this discussion to raise questions about boundary issues that influence stability, cohesiveness, and economic opportunity in a country.

- How can the size and shape of a country promote cohesiveness or instability?
- How do cultural (anthropographic) boundaries differ from political boundaries?
- How do cultural (anthropographic) boundaries contribute to cohesiveness or instability?
- How could the political boundaries of a country foster or hinder economic advantage?

Explain that in this lesson, students will have an opportunity to redraw the boundaries of the world. The purpose in this exercise will be to create a more stable and peaceful world by drawing boundaries that foster cohesiveness and economic parity among nations.

Student activity

 Before completing this lesson with students, we recommend that you complete it as well. Doing so will allow you to modify the activity to accommodate the specific needs of your students.

After the initial discussion, have the students work on the computer component of the lesson. This lesson is particularly well suited to students working in pairs, but the lesson can be modified to accommodate a variety of instructional settings.

Distribute and explain the Starting from Scratch GIS Investigation. It is important that students understand that the ultimate purpose of the activity is to create a more stable and peaceful world.

Explain that in this activity they will use GIS to observe and analyze cultural and physiographic data in order to determine the world's new boundaries. The GIS Investigation sheets will provide them with detailed instructions for the analysis of data and drawing of borders for their assigned continent.

Things to look for while students are working on this investigation:

- Are your students taking both physiographic and anthropographic boundaries into account as they draw their boundaries?
- Do they understand that there are many more factors that influence boundary decisions and that this lesson only uses simple data sets?
- Are they applying concepts of territorial morphology (size and shape) as they draw their boundaries?

 Teacher Tip: Students will need access to a computer and ArcView for the equivalent of two class periods to draw their new boundaries. Decide ahead of time how and where you want your students to save their projects. Encourage them to save their projects frequently.

Conclusion

Each student will present a New World map to the class, describing all boundary changes, and explaining how the decisions to make those changes were arrived at. Ideally, students will use a projection device to display their maps to the class; printed copies from an ArcView layout will also work. Be sure that presenters provide cultural or physiographic justifications for their boundary decisions. Use these presentations as a springboard to highlight present-day areas of instability, characterized by boundaries that conflict with physical and cultural imperatives in the area. Engage the students in a conversation that analyzes the data they used. Guide the students toward understanding that although language and religion are important in determining boundaries, there are many other factors that influence boundaries: access to natural resources, economics, and infrastructure are just a few. Using a GIS can help people deal with these complicated issues.

Assessments for middle- and high-school students

In the assessment, students will prepare a map and a report focusing on a country where political boundaries are causing cultural, political, or economic instability. The map will compare the country's present boundaries with new boundaries proposed by the student. The report will explain how boundary issues contribute to instability in the country today and why the new boundaries will foster cohesiveness and economic stability in the future.

In preparing their report, students should consider the influence that each of the following factors has on cohesiveness and economic strength:

- Size, shape, and relative location of countries
- Cultural characteristics (language and religion)
- Distribution of natural resources
- Physiographic connections and barriers between places

Extensions
- Assign students to use the Internet to collect and map additional data about the regions they have characterized as unstable.
- Assign students to use the Internet to locate historic maps of these regions and to prepare an ArcView layout reflecting boundary changes over time.
- Add climate, land-use, population density, and natural resource data to the project and revise boundaries based on this data.
- Have students create boundaries for all the continents and compare them to present-day boundaries.
- Ask students to use the Chart function to present religious and linguistic data that supports their boundary decisions.
- Check out the Resources by Module section of the Teacher Resource CD for print and media resources on the topic of international boundaries or visit *www.esri.com/mappingourworld* for Internet links.

NAME _____ DATE _____

Starting from Scratch
An advanced investigation

You have been selected to serve on a special commission charged with redrawing the boundaries of world nations. The goal of this commission is to defuse threats to world peace by creating political boundaries that foster stability, harmony, and economic parity among nations. You will create countries on one continent.

You may create anywhere from three to 15 countries on your continent. You will use this GIS Investigation to identify and analyze key variables that have a bearing on this important decision and to create a map reflecting your New World boundaries.

Step 1 Start ArcView

a Double-click the ArcView icon on your computer's desktop.

b If the Welcome to ArcView dialog appears (pictured below), click **Open an Existing Project** and click OK. If it doesn't appear, proceed to step 2.

Step 2 **Open the adv5.apr file**

a In this exercise, a project file has been created for you. To open the file, go to the **File** menu and choose **Open Project**.

b Navigate to the exercise data directory (**C:\esri\mapworld\mod5**) and choose **adv5.apr** from the list.

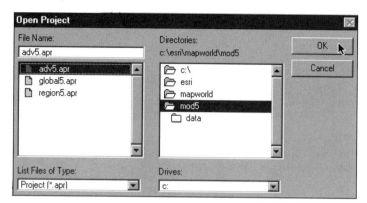

c Click OK. The project opens and you see a composite satellite image of the world in the view. Religion, language, rivers, lakes, continents, and ocean shapefiles are in the table of contents.

Step 3 **Explore project themes**

As you prepare to draw new country boundaries, you will use the Satellite Image and Rivers themes to evaluate physiographic boundaries, and you will use the Religion and Language themes to determine anthropographic boundaries.

> *Please note: Religion and language are not the only anthropographic factors that influence boundary decisions. For the purposes of this investigation, we decided to keep the data set small and simple.*

? *List other important factors that influence boundary decisions.*

a Zoom to South Asia. Turn on Religion.

? *b* Use the Religion legend to determine three principal religions of South Asia. Record them on the answer sheet.

c Turn off Religion and turn on Religion Transparent. Now you are able to see the religion boundaries and the physical features of the earth.

? *The boundary between which two religions corresponds to a physiographic boundary visible in the Satellite Image?*

? *d* Turn off Religion Transparent. Turn on Language. Identify the principal language groups in South Asia.

? *e* Turn off Language and turn on Language Transparent. The boundary between which two language groups corresponds to a physiographic boundary visible in the Satellite Image?

Step 4   **Create a new theme for new world boundaries**

    a   Zoom to Full Extent. From the View menu, select View Properties.

     b   Change the name of the view to **My New World**. Click OK.

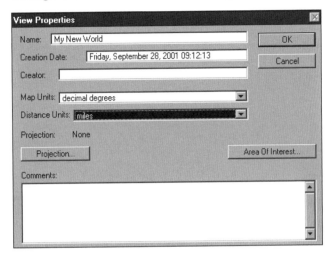

     c   Select New Theme from the View menu. Choose Polygon as the feature type.

     d   Ask your teacher how to name your new theme and where to save it. We suggest naming your theme **abnewtheme**, where ab are your initials.

     e   Open the Legend Editor for the new theme you created. Change the fill pattern from solid to transparent by clicking the first white box.

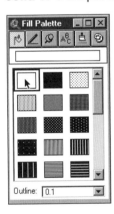

     f   Click the Paintbrush button. Select Outline from the drop-down menu. Click red. Apply your changes. Close the Color Palette and Legend Editor.

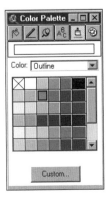

g Change the theme name to **New World**.

Step 5 Add language and religion fields to the New World table

 a Open the attribute table for New World.

b From the Edit menu, select Add Field.

c Complete the Field Definition dialog with the following parameters:

FIELD NAME	TYPE	WIDTH
Language	String	16
Religion	String	16

d From the Table menu, choose Stop Editing. Save your edits.

e Position the view and the attribute table so they are both visible. Make the view active by clicking its title bar.

f Save the project and rename it according to your teacher's instructions. Record the new name of the project and where you saved it.

Step 6 Draw the boundaries of new countries

It is now time to begin creating your new map of the world. You will select one continent and use the ArcView drawing tools to draw the boundaries (outlines) of the countries you create on that continent. It will be helpful to turn themes on and off so you can see the physiographic and anthropographic boundaries that exist. As you create new countries, identify the principal religion and language group of each in an attribute table.

Remember that your goal is to stabilize hot spots and foster world peace with your new boundary configuration. Consider the following points:
- Countries that are culturally uniform are generally more stable.
- Countries that have natural or physiographic boundaries as political boundaries tend to be more stable.
- Landlocked countries are at an economic disadvantage if they do not have some access to the sea.

a Choose the continent you will be working on and record it on the answer sheet.

b Zoom to your continent so you can see major landforms.

 c Select the Draw Polygon tool from the drawing tools drop-down menu.

d Make sure the New World theme is active. From the Theme menu, choose Start Editing.

e When you have chosen the first boundary to draw, click the cursor along the proposed boundary. You may need to click many times on a line that curves. Double-click to complete the polygon. If you have made a mistake, click the delete key and start again.

Now you need to add language and religion data to the new country.

f Make the New World Attribute Table active.

g   Click the Edit tool. In the ID field, change the number from zero to **1** and press Enter.

h   In the Language field, type the name of the major language group in the country you just created and press Enter. If there isn't one major language group, type **Mixed**.

i   In the Religion field, type the principal religion for the country you just created and press Enter. If there isn't one major religion, type **Mixed**.

j   From the Table menu, choose Save Edits. Make the view active.

Look at your boundary and determine whether there are any parts of it you need to edit. If you don't need to edit the boundary, proceed to step 6n. Otherwise, continue with the editing instructions below.

k   Zoom to a section of the boundary that you would like to edit.

l   Click the Vertex Edit tool. Click once on the new polygon and all the vertices display as small squares.

m   Click and drag these vertices to modify your boundary line. Add new vertexes to the line by clicking on it.

n   Repeat the process of making a new boundary line, adding its attribute data, and editing it to create countries in the remainder of your continent.

o   Pan your view over the continent you're working on and create country boundaries in the same manner. Save your edits and save the project periodically.

p   From the Theme menu, select Stop Editing and save your edits. Save your project.

Step 7   Label your map

a   Make sure New World is active and open its Theme Properties.

b   Click the Text Labels icon on the left side. From the Label Field drop-down list, choose ID. Click OK.

c   From the Window menu, select Show Symbol Window.

d   Click the Font Palette button. Make the text size **14** and the style **Bold**.

e   Click the Paintbrush button to display the Color Palette.

f   Choose Text from the Color drop-down menu. Click red. Close the Color Palette.

g   Click the Clear Selected Features button.

h   From the Theme menu, choose Auto-Label. Make sure the dialog has the following options selected:

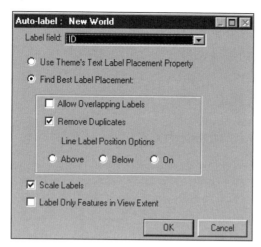

i   Click OK. The labels appear in the view.

Step 8   Create and print a layout of your New World boundaries

a   Turn off all themes except New World, Continents, and Ocean.

b   Zoom out so you can see your continent in its entirety.

c   Create a landscape layout from the View menu.

d   Use the Pointer tool to reposition any titles or graphics so there is room for your name and date in the layout.

e   Use the Text tool to add your name and the date to the map.

f   Reposition any graphics or text until you are satisfied with how the layout looks.

g   Print your map.

You should be prepared to show your New World map to your classmates and explain the cultural and physical features upon which you based your decisions.

Step 9   Exit ArcView

In this exercise, you used ArcView to explore patterns in the world's physiographic and anthropographic borders. Based on your observations, you drew new international boundaries to create countries that you feel would be more culturally and physiographically unified than those in the real world today.

a   Save your project.

b   From the File menu, click Exit.

NAME _____   DATE _____

Student answer sheet
Module 5
Human Geography II: Political Geography

Advanced investigation: Starting from Scratch

Step 3   Explore project themes
List other important factors that influence boundary decisions.

_____

_____

_____

_____

b   Use the Religion legend to determine three principal religions of South Asia. Record them here.

_____

_____

_____

c   The boundary between which two religions corresponds to a physiographic boundary visible in the Satellite Image?

_____

_____

d   Turn on Language. Identify the principal language groups in South Asia.

_____

_____

_____

e   The boundary between which two language groups corresponds to a physiographic boundary visible in the Satellite Image?

_____

_____

Step 5   Add language and religion fields to the New World table
f   Save the project and rename it according to your teacher's instructions. Record the new name of the project and where you saved it.

_____   _____
(Name of project.   **(Navigation path to where project is saved.**
For example: kzenewtheme.apr)   **For example: C:\student\kze\mapworld)**

Step 6   Draw the boundaries of new countries
a   Choose the continent you will be working on and record it here.

_____

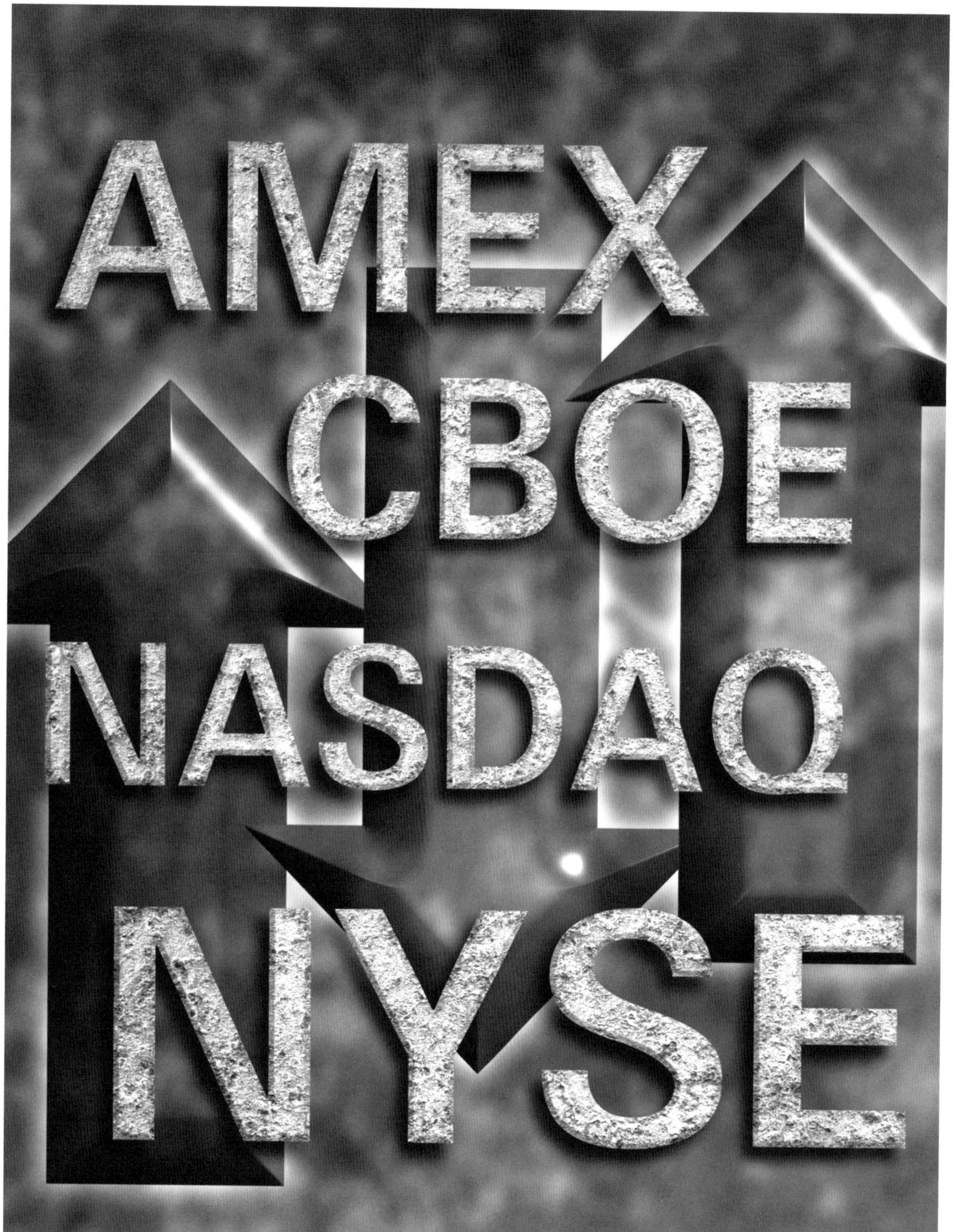
AMEX
CBOE
NASDAQ
NYSE

Human Geography III
Economic Geography

Economic development, modernization, and trade illustrate the interrelatedness of the global community.

The Wealth of Nations: A global perspective

In this lesson students will be presented with the three modes of economic production—agriculture, industry, and services—as the initial criteria for a country's developed or developing status. They will select layers of data from a group of economic indicators to determine patterns in developed and developing countries. They will be challenged to draw their own conclusions as to a country's economic development status and support their conclusions with data.

Share and Share Alike: A regional case study of North America and NAFTA

Students will explore trade in North America focusing on the three trading partners in the North American Free Trade Agreement (NAFTA)—Canada, Mexico, and the United States. They will study export data for the past 10 years from each of the NAFTA countries, then use this information to identify trading trends before and after NAFTA, and to assess its effectiveness. Finally, students will make an ArcView layout containing a map and charts that support their opinions.

Live, Work, and Play: An advanced investigation

Students will use the Internet to acquire the most recent data on job classifications in California. By exploring this data through a GIS, students will construct a thematic map for California counties that displays population, number of jobs in a given field, and average wage data. They will perform a complex query and create a map that illustrates their results.

The Wealth of Nations
A global perspective

Lesson overview

In this lesson students will be presented with the three modes of economic production—agriculture, industry, and services—as the initial criteria for a country's developed or developing status. They will select layers of data from a group of economic indicators to determine patterns in developed and developing countries. They will then be asked to draw their own conclusions as to a country's economic status, and to support those conclusions with data.

Estimated time   Two 45-minute class periods

Materials   ✔ Student handouts from this lesson to be copied:
- GIS Investigation sheets (pages 325 to 334)
- Student answer sheet (pages 335 to 337)
- Assessment(s) (pages 338 to 342)

Standards and objectives

National geography standards

GEOGRAPHY STANDARD	MIDDLE SCHOOL	HIGH SCHOOL
1 How to use maps and other geographic representations, tools, and technologies to acquire, process, and report information from a spatial perspective	The student understands how to make and use maps, globes, graphs, charts, models, and databases to analyze spatial distributions and patterns.	The student understands how to use geographic representations and tools to analyze, explain, and solve geographic problems.
11 The patterns and networks of economic interdependence on Earth's surface	The student understands ways to classify economic activity.	The student understands the classification, characteristics, and spatial distribution of economic systems.
18 How to apply geography to interpret the present and plan for the future	The student understands how varying points of view about geographic context influence plans for change.	The student understands how to use geographic knowledge, skills, and perspectives to analyze problems and make decisions.

Objectives

The student is able to:

- Define the three economic production criteria traditionally used to determine economic development status.
- Compare and contrast these criteria.
- Evaluate them as suitable measurements for developed or developing status.
- Understand additional economic indicators used to classify a country as developed or developing.
- Develop a definition of developed and developing.
- Predict a country's or region's economic status within the next 20 years.

GIS skills and tools

 Add themes to the view

 Find features

 Identify attributes of a feature

 Zoom to full extent of the view

 Zoom in to a specific area of the view

 Clear all selections made

 Specify a null data value to use in a theme legend

 Change a theme name

- Add a new view
- Modify a theme legend

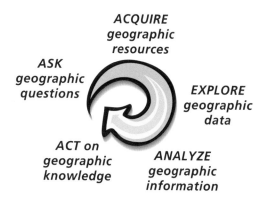

For more on geographic inquiry and these steps, see Geographic Inquiry: Thinking Geographically (pages xxi to xxiii).

Teacher notes

Lesson introduction

Introduce this lesson to your students with a discussion of the three economic production criteria: agriculture, industry, and services. Ask them what they know about these three methods of earning a living. Explain to them that economists generally rate a country's economic status as developing or developed by how much of its workforce is engaged in agriculture, industry, and services. Generally, countries with a high percentage of the workforce in agriculture (whether it be subsistence, commercial, or another type) are placed in the category of "developing."

Have the class begin by using the three production criteria maps of the world to determine whether a country is classified as developed or developing.

Before starting the exercise on the computer, ask the following questions to elicit knowledge, beliefs, or ideas your students may already have about countries around the world:

- Name several countries that have a high percentage of their workforce participating in agriculture.
- Name a country that has a high percentage of its workforce participating in services.
- Which countries do we generally think of as highly industrialized (have high percentages of the workforce participating in industry)?
- Can they think of other factors that might be helpful in determining whether a country is developed or developing?

Student activity

 Before completing this lesson with students, we recommend that you complete it as well. Doing so will allow you to modify the activity to accommodate the specific needs of your students.

After the initial discussion, have the students work on the computer component of the lesson individually or in groups of no more than two or three. Distribute the GIS Investigation sheets to the students. Explain to them that they will learn and use GIS skills to enable them to gather data for the assessment worksheet.

In addition to instructions, the handout includes questions to help students focus on key concepts.

 Teacher Tip: In order for students to complete the assessment, they must save their project. Be sure to have a suggested naming convention and location on a computer or computer network for your students.

Things to look for while students are working on this activity:

- Are all students in each group participating in the activity, taking turns using the computer and writing information on the chart?
- Are the students using a variety of tools to obtain the information they need?

Conclusion   After the students complete the lesson and the assessment, discuss their findings. If you have time, you can have each group share findings on an overhead projector and explain how they came to their conclusions. Students can also take turns presenting the thematic maps they have created, either in printed format or on a computer projection device from the front of the room. Conclude the lesson by asking the students to explain which factors they feel are most important in deciding if a country is developed or developing, and to provide support for their evaluation.

Assessment   *Middle school and high school: Highlights skills appropriate to grades 5 through 12*

The middle- and high-school assessments ask students to choose one country they believe is developed and another they believe is developing. Students will use the thematic mapping skills they have learned in this lesson to map additional indicators and draw conclusions about the two countries they have chosen. They will individually write an essay that answers specific questions. The middle- and high-school assessments differ in the types of questions asked.

Extensions
- Ask students to look in local newspaper employment advertisements to classify the jobs in their own community as agriculture, industry, or service. How would they classify the types of employment in their own community?

- Research employment classifications of two different counties, and compare and contrast the percentages of jobs in each category.

- Research the GDP for the top 10 trading partners of the United States, compute the GDP per capita, and map it.

- Have students choose a country and analyze all of the standard of living and economic indicator data, and present their findings orally to the rest of the class.

- Check out the Resources by Module section of the Teacher Resource CD for print and media resources on the topics of economic indicators, developing countries, and economic production of countries or visit *www.esri.com/mappingourworld* for Internet links.

NAME _____   DATE _____

The Wealth of Nations
A GIS investigation

ACQUIRE

ASK   EXPLORE

ACT   ANALYZE

Answer all questions on the student answer sheet handout

Economists generally classify a country's economic status as developing or developed by determining the percentage of its workforce engaged in each of three sectors of the economy—agriculture, industry, and services. In general, a country with a high percentage of its workforce in agriculture is considered to be "developing," while a country with a high percentage of its workforce in services and industry is considered to be "developed."

In this GIS Investigation, you will use world maps of the workforce in the three employment sectors to explore patterns of development around the world. You will also examine two other economic indicators—energy use and GDP per capita—and compare the maps of employment sectors to the maps of GDP and energy use. You will evaluate whether or not the employment criteria are good indicators of a country's economic status.

Step 1   Start ArcView

a   Double-click the ArcView icon on your computer's desktop.

b   If the Welcome to ArcView dialog appears (pictured below), click **Open an Existing Project** and click OK. If it doesn't appear, proceed to step 2.

Step 2   **Open the global6.apr file**

a   In this exercise, a project file has been created for you. To open the file, go to the **File** menu and choose **Open Project**.

b   Navigate to the exercise data directory (**C:\esri\mapworld\mod6**) and choose **global6.apr** from the list.

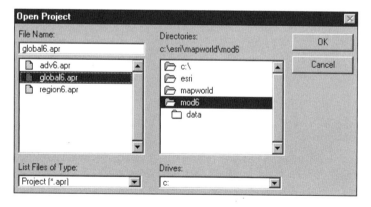

c   Click OK.

When the project opens, you see three views, each with a world map for one of the employment sectors.

d   Maximize the ArcView window. If you prefer larger views, stretch each view to make it bigger. Make sure you can see all three views at the same time.

Step 3   Evaluate the legends and patterns of the maps

    a   Look at each map's legend.

          (1)   *What do the darkest colors represent?*

          (2)   *What do the lightest colors represent?*

    b   Study the % Workforce Involved in Agriculture view.

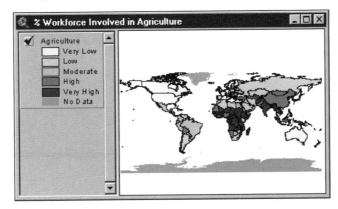

          (1)   *What does the description "Very High" refer to in this map?*

          (2)   *Where are the countries with a high percentage of agricultural workers generally located?*

          (3)   *Where are the countries with a low percentage of agricultural workers generally located?*

    c   Study the % Workforce Involved in Service map.

          (1)   *Where are the countries with a high percentage of service workers generally located?*

          (2)   *Where are the countries with a low percentage of service workers generally located?*

    d   Study the % of Workforce Involved in Industrial map.

       What patterns do you see on the map?

    e   Using the workforce information in all three maps, in what part or parts of the world do you find the greatest number of developing countries?

Step 4   Analyze data on Bolivia and other countries

    a   Click the title bar for the % of Workforce Involved in Industrial view. The view's title bar turns blue, indicating this is the active View window.

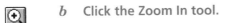   *b*   Click the Zoom In tool.

c   Click and drag a box around South America.

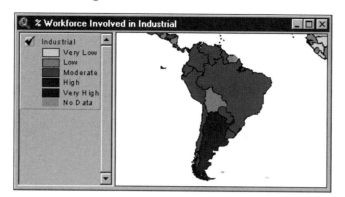

d   Zoom in to South America on the other views.

e   Click the title bar for the % Workforce Involved in Agriculture view to make it active.

f   Click the Find tool.

g   Type **Bolivia** in the dialog box. Click OK.

Bolivia is highlighted yellow on the map. Now you will use ArcView to gather and record data about Bolivia.

h   Click the Identify tool, then click Bolivia on the map.

An Identify Results window appears with data on Bolivia.

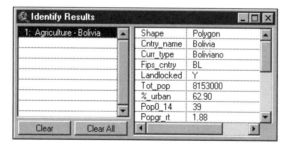

i   Scroll down the right side of the window and complete the following questions:

    (1)   *What percentage of workers in Bolivia are involved in agriculture?*

    (2)   *What percentage of workers in Bolivia are involved in the industrial field?*

    (3)   *What percentage of workers in Bolivia are involved in service?*

    (4)   *Assume that a developing country has a high percentage of its workforce in agriculture and lower percentages of its workforce in industrial- and service-related occupations and analyze the data on Bolivia. Based on these criteria, would you classify Bolivia as a developed or developing country?*

j   Click the Clear Selected Features button to deselect Bolivia. Close the Identify Results window.

The yellow highlight color disappears from the map.

k   On your answer sheet, you see a table that has some information completed for Bolivia. Check the classification (developed or developing) you gave Bolivia in the previous question.

In order to complete the table on your answer sheet, you will need to find each country on the maps, interpret the legends for each map, write your answers in the columns under Agriculture, Industry, and Service, and decide whether each country is developing or developed.

l   If you know how to find each country and complete the table, do so now. You can skip to step 4s when you're finished. If you need instruction on how to find the countries and interpret the legends, go to the next step.

m   In the % Workforce Involved in Agriculture view, click the Find tool.

n   Type **India** in the text box. The view centers on India and it turns yellow in the view.

o   Click the Clear Selected Features button so you can see the legend for India. Write the correct percentage of workers in agriculture in the table on your answer sheet.

p   Click the blue title bar for the % Workforce Involved in Industrial view. Click the Find tool and find India.

q   After clicking the Clear Selected Features button, interpret the legend and record your results in the table.

r   Click the blue title bar for the % Workforce Involved in Service view. Click the Find tool and find India.

s   After clicking the Clear Selected Features button, interpret the legend and record your results in the table.

t   Based on the workforce criteria previously used to determine if a country is developed or developing, write your answer for India in the table.

u   Repeat this process (step 4m–step 4t) for the other countries listed in the table.

v   Zoom to the full extent of each view. You now can see the entire world in all three views.

Step 5   **Create a new view and add data**

Now you will add economic and energy data and determine whether this data supports your initial conclusions about which countries are developed and which are developing.

Gross Domestic Product (GDP) is the total value of all goods, services, and products produced in a given country. Typically, developing countries have a low GDP. The total amount of energy consumed by a given country is also an indicator of development. If a country has a low level of energy consumption, it tends to be a developing country. Developed countries are high in both GDP and energy use.

a   Click the Project window title bar.

b   Make sure the Views icon is highlighted, then click the New button.

A new empty view entitled View1 is displayed.

    c   Click the Add Theme button.

d   Navigate to the exercise data directory (**C:\esri\mapworld\mod6\data**) and choose **economics.shp** from the list. Click OK.

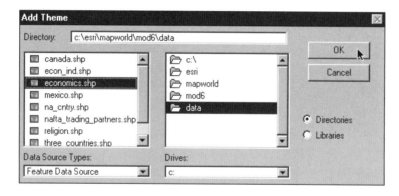

e   Add Economics.shp again so there are two copies of it in the table of contents.

f   Click the box next to one of the Economics.shp themes in the table of contents to turn it on.

The theme is displayed with only one color. ArcView randomly assigns the color, so the color on your screen may not match the color on your neighbor's screen. ArcView uses a basic legend like this one whenever you add a new theme to a view, unless someone already created and saved a special legend for the data.

Step 6   Thematically map GDP per capita and energy use

To see the pattern of GDP per capita, you need to change the legend. GDP per capita is the gross domestic product per person in a given year (1998). It is calculated by dividing the GDP by the total population of a country.

a   Double-click one Economics.shp theme in the table of contents. The Legend Editor is displayed.

b   In the Legend Type drop-down menu, select Graduated Color.

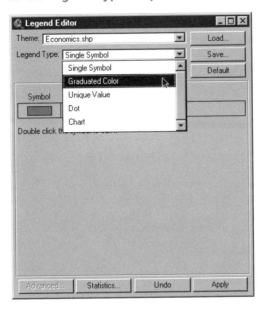

c   For Classification Field, scroll down and select **Gdp_pc8**.

The Legend Editor updates and you see a list of five colored symbols, values, and labels.

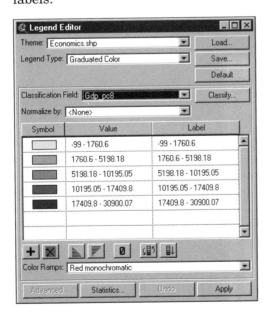

The classification field (Gdp_pc8) represents the total population for each country in the world. By default, ArcView has divided the GDP values into five groups.

Some countries have no data available for GDP. In this case, the Gdp_pc8 field has a value of –99.

Because the value of –99 represents no data and not the number –99, this data needs to be taken out of the range of GDP figures. You will fix this by assigning values of –99 to a No Data class.

d Click the Null Value button.

e Type **–99** in the Null Value field. Click the box to the left of Display No Data Class. Click OK.

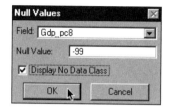

f In the Legend Editor, double-click the blank space in the symbol column next to the value –99.

g Click the Color Palette button in the Fill Palette window.

h Click the medium-gray color box. Close the Color Palette.

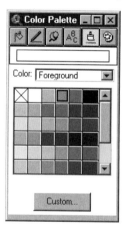

The grey symbol for the No Data class is updated in the Legend Editor. Now you will change the labels for the legend

i Click the top label (488.4 - 2498.64) once and press Delete. Type **Very Low**.

j Click the next label down (5380.19 - 10195.05) once and press Delete. Type **Low**.

k Change the rest of the labels so they read **Moderate**, **High**, and **Very High**. When you are finished, click Apply and close the Legend Editor. The legend for GDP_pc8 is updated in the table of contents.

l   Click the Theme Properties button. Change the theme name to **GDP per capita** by typing it into the dialog. Click OK.

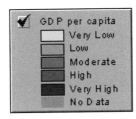

Now you will map energy use. Energy use is a measure of the total energy consumption from all energy sources as measured by a trillion BTUs (British thermal units). One BTU is the amount of energy needed to heat one pound of water one degree Fahrenheit.

m   Double-click the other Economics.shp theme to open its legend editor. Repeat step 6b–step 6l. Map the classification field, **Ener_use9**, use the orange monochromatic color ramp, and change its theme name to **Energy Use**.

n   Turn on Energy Use.

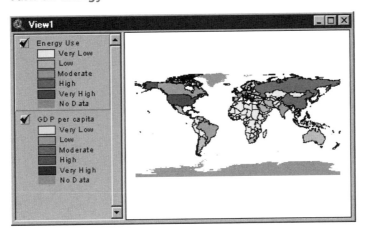

Step 7   **Analyze GDP per capita and energy use data**

Now you will take this new data on GDP per capita and energy use into consideration, and you will reevaluate how you classified countries as developed or developing.

a   Make GDP per capita the active theme and turn it on. Turn off Energy use.

b   Click the Find tool and find **Bolivia**.

    (1)   What level is the GDP per capita for Bolivia?

    (2)   Based on this new information and the workforce data, should Bolivia be classified as a developing or developed country? Why?

c   Turn off GDP per capita and turn on Energy use. Make Energy use the active theme.

    (1)   What level is the Energy use for Bolivia?

    (2)   Based on this new information and previous data, should Bolivia be classified as a developing or developed country?

    (3)   Why does energy use increase when a country develops?

Now you will explore the data as it pertains to other countries and you will see if your earlier classification of each country has changed.

d   Explore GDP per capita and Energy use for each of the countries listed on your answer sheet. You may consult your previous classification of these countries by looking at the step 4K table.

?   (1)   Complete the table on your answer sheet.

?   (2)   Name one country from above that you earlier classified as developed and that has GDP and Energy Use data that indicates it's developing.

?   (3)   For the country you named in question 2, explain your current classification.

Step 8   Save the project and exit ArcView

In this exercise, you used workforce data to determine whether countries should be classified as developed or developing. You added new data, thematically mapped the data, and reevaluated your previous classifications.

 a   Click the Clear Selected Features button and then the Zoom to Full Extent button so you can see the entire world in the view.

b   Ask your teacher for instructions on how to rename the project and where to save it. You must save this project because you will need to use it for the assessment.

?   c   Write the new name you gave the project and where you saved it in the space on your answer sheet.

d   From the File menu, click Exit. When asked if you want to save changes to global6.apr, choose No.

NAME _____   DATE _____

Student answer sheet
Module 6
Human Geography III: Economic Geography

Global perspective: The Wealth of Nations

Step 3   Evaluate the legends and patterns of the maps
a-1   What do the darkest colors represent?

_____

a-2   What do the lightest colors represent?

_____

b-1   What does the description "Very High" refer to in this map?

_____

b-2   Where are the countries with a high percentage of agricultural workers generally located?

_____

b-3   Where are the countries with a low percentage of agricultural workers generally located?

_____

c-1   Where are the countries with a high percentage of service workers generally located?

_____

c-2   Where are the countries with a low percentage of service workers generally located?

_____

d   What patterns do you see on the map?

_____

e   Using the workforce information in all three maps, in what part or parts of the world do you find the greatest number of developing countries?

_____

_____

Step 4   Analyze data on Bolivia and other countries
i-1   What percentage of workers in Bolivia are involved in agriculture? _____

i-2   What percentage of workers in Bolivia are involved in the industrial field? _____

i-3   What percentage of workers in Bolivia are involved in service? _____

i-4   Would you classify Bolivia as a developed or developing country? Explain.

_____

_____

_____

_____

k   Complete the table below.

COUNTRY	AGRICULTURE	INDUSTRY	SERVICE	DEVELOPING	DEVELOPED
Bolivia	High	Low	High		
India					
New Zealand					
South Korea					
Portugal					
Uruguay					

Step 7   Analyze GDP per capita and energy use data

b-1   What level is the GDP per capita for Bolivia? _____

b-2   Based on this new information and the workforce data, should Bolivia be classified as a developing or developed country?

_____

Why?

_____

_____

c-1   What level is the Energy use for Bolivia? _____

c-2   Based on this new information and previous data, should Bolivia be classified as a developing or developed country?

_____

c-3   Why does energy use increase when a country develops?

_____

d-1   Complete the table below.

COUNTRY	GDP PER CAPITA	ENERGY USE	DEVELOPED OR DEVELOPING	IS THIS A CHANGE FROM YOUR EARLIER CLASSIFICATION?
Bolivia				
India				
New Zealand				
South Korea				
Portugal				
Uruguay				

d-2   Name one country from above that you earlier classified as developed and that has GDP and energy use data that indicates it's developing.

_____

d-3   For the country you named in question 2, explain your current classification.

_____

_____

_____

_____

_____

_____

_____

_____

_____

_____

Step 8   Save the project and exit ArcView

c   Write the new name you gave the project and where you saved it in the space below.

_____   _____
        (Name of project.   **(Navigation path to where project is saved.**
   For example: kzeglobal6.apr)   **For example: C:\student\kze\mapworld)**

NAME _____   DATE _____

The Wealth of Nations
Middle school assessment

1   Use the three economic activity maps in the global6.apr and select a country you believe is a developed country and one that you believe is a developing country. Write the names of those countries on the sheet provided.

2   Go back to your saved version of the global6.apr and thematically map the developing country indicators in the Economics.shp theme. (Hint: You can add Economics.shp and copy it in the new view.) Refer to the attached assessment sheet for guidelines on which data to map (page 342).

3   Record the information you observe from the new maps on the attached sheet. You may use other developing country indicators that are not listed on the sheet. Write those in the blank lines at the end of the table.

4   Look at the data in the two columns and decide if your initial prediction is correct.

5   Write an essay that includes or answers the following information and questions:

  • Name and description of the three economic production models.

  • Does the data you found about your selected countries support what economists say about economic production as an indicator of developed or developing status?

  • Your own definitions, based on your research, of a developed country and a developing country.

The Wealth of Nations

Assessment rubric

Middle school

STANDARD	EXEMPLARY	MASTERY	INTRODUCTORY	DOES NOT MEET REQUIREMENTS
The student understands how to use maps and databases to analyze spatial distributions and patterns.	Uses GIS to analyze economic data by creating at least seven thematic maps that compare and contrast the different economic indicators for two countries. Uses additional data from outside sources.	Uses GIS to analyze economic data by creating seven thematic maps that compare and contrast the different economic indicators for two countries.	Uses GIS to analyze economic data by creating five or six thematic maps that compare and contrast the economic indicators for two countries.	Uses GIS to create four or fewer thematic maps based on economic data for one or two countries.
The student understands ways to classify economic activity.	Clearly describes the three economic production models and provides an example of each. Creates detailed and accurate original definitions of developed and developing countries. Provides ample evidence for the definition.	Clearly describes the three economic production models and then creates an accurate and original definition of developed and developing countries.	Describes the three economic production models, and attempts to create original definition of developed and developing countries.	Has difficulty describing the three economic production models, and does not attempt to create original definitions for developed and developing countries.
The student understands the spatial organization of human activities and physical systems and is able to make informed decisions.	The student understands how varying points of view about geographic context influence plans for change.	Compares own findings with the predefined economic status of the selected countries. Identifies any inconsistencies in the findings, and either accepts or rejects his/her hypothesis.	Compares and contrasts own findings with the predefined economic status of the selected countries.	Lists own ideas on the economic status of the selected country (or countries), but does not draw any comparisons with predefined economic status.

This is a four-point rubric based on the National Standards for Geographic Education. The "Mastery" level meets the target objective for grades 5–8.

NAME _____   DATE _____

The Wealth of Nations
High school assessment

1 Use the three economic activity maps in the global6.apr and select a country you believe is a developed country and one that you believe is a developing country. Write the names of those countries on the sheet provided.

2 Go back to your saved version of the global6.apr and thematically map the developing country indicators in the Economics.shp theme. (Hint: You can add Economics.shp and copy it in the new view.) Refer to the attached assessment sheet for guidelines on which data to map (page 342).

3 Record the information you observe from the new maps on the attached sheet. You may use other developing country indicators that are not listed on the sheet. Write those in the blank lines at the end of the table.

4 Look at the data in the two columns and decide if your initial prediction is correct.

5 Write an essay that includes or answers the following information and questions:

• Name and description of the three economic production models. Give a brief explanation of the relationship between those models. For example: Will a country usually have high percentages or low percentages in all three production areas?

• Does the data you found about your selected countries support what economists say about economic production as an indicator of developed or developing status?

• Your own definitions, based on your research, of a developed country and a developing country.

• Prediction of changes in your countries' economic status by the year 2020, and supporting data.

Assessment rubric

High school

The Wealth of Nations

STANDARD	EXEMPLARY	MASTERY	INTRODUCTORY	DOES NOT MEET REQUIREMENTS
The student understands how to use geographic representations and tools to analyze and explain geographic problems.	Uses GIS to analyze economic data by creating at least seven thematic maps that compare and contrast the different economic indicators for two countries. Uses additional data from outside sources.	Uses GIS to analyze economic data by creating seven thematic maps that compare and contrast the different economic indicators for two countries.	Uses GIS to analyze economic data by creating five or six thematic maps that compare and contrast the economic indicators for two countries.	Uses GIS to create four or fewer thematic maps based on economic data for one or two countries.
The student understands the classification, characteristics, and spatial distribution of economic systems.	Clearly describes the three economic production models, explains the relationship between these models, and then creates original detailed definitions of developed and developing countries. Provides ample evidence and examples for each.	Clearly describes the three economic production models, explains the relationship between these models, and then creates original detailed definitions of developed and developing countries. Provides some evidence and examples for each.	Describes the three economic production models, and attempts to create original definition of developed and developing countries.	Has difficulty describing the three economic production models, and does not attempt to create original definitions for developed and developing countries.
The student understands how to use geographic knowledge, skills, and perspectives to analyze problems and make decisions.	Compares own findings with the predefined economic status of the selected countries. Identifies any inconsistencies in the findings, and either accepts or rejects his/her hypothesis. Refines original hypothesis based on the findings. Predicts future changes for the economic status of the countries giving specific examples.	Compares own findings with the predefined economic status of the selected countries. Identifies any inconsistencies in the findings, and either accepts or rejects his/her hypothesis. Predicts any future changes for the economic status of the countries.	Compares and contrasts own findings with the predefined economic status of the selected countries. Attempts to predict future changes in economic status, but does not provide enough evidence.	Lists own ideas on the economic status of selected country (or countries), but does not draw any comparisons with predefined economic status.

This is a four-point rubric based on the National Standards for Geographic Education. The "Mastery" level meets the target objective for grades 9–12.

DEVELOPING COUNTRY INDICATORS	DEVELOPED COUNTRY _____	DEVELOPING COUNTRY _____
High birth rates		
Moderately high death rates		
High infant mortality rate		
Low life expectancy at birth		
Large percentage of population under 15		
Literacy rates low		
High urban populations		
Low gross domestic product		
Low consumption of energy		
Higher percentage per person communication facilities		
Economic production criteria—write whether they are predominantly agriculture-, industry-, or service-based.		
Other criteria: _____		
Other criteria: _____		
Other criteria: _____		

Share and Share Alike

A regional case study of North America and NAFTA

Lesson overview

Students will explore trade in North America focusing on the three trading partners in the North American Free Trade Agreement (NAFTA)—Canada, Mexico, and the United States. They will study export data for the past 10 years from each of the NAFTA countries, then use this information to identify trading trends before and after NAFTA, and to assess its effectiveness. Finally, students will make an ArcView layout containing a map and charts that support their opinions.

Estimated time   Two to three 45-minute class periods

Materials   ✔ Student handouts from this lesson to be copied:
- NAFTA objectives handout (page 347)
- GIS Investigation sheets (pages 349 to 364)
- Student answer sheet (pages 365 to 369)
- Assessment(s) (pages 370 to 373)

Standards and objectives

National geography standards

GEOGRAPHY STANDARD	MIDDLE SCHOOL	HIGH SCHOOL
1 How to use maps and other geographic representations, tools, and technologies to acquire, process, and report information from a spatial perspective	The student understands how to make and use maps, globes, charts, models, and databases to analyze spatial distributions and patterns.	The student understands how to use geographic representations and tools to analyze, explain, and solve geographic problems.
11 The patterns and networks of economic interdependence on Earth's surface	The student understands the basis for global interdependence.	The student understands the increasing economic interdependence of the world's countries.
13 How the forces of cooperation and conflict among people influence the division and control of Earth's surface	The student understands how cooperation and conflict among people contribute to economic and social divisions of Earth's surface.	The student understands why and how cooperation and conflict are involved in shaping the distribution of social, political, and economic spaces on Earth at different scales.

Objectives

The student is able to:
- Explain the concept of imports and exports related to trade balances of a country.
- Evaluate the effectiveness of NAFTA.
- Create a layout that displays the results of the student's research.

GIS skills and tools

  Identify a feature on the map or chart

  Open a theme's attribute table

  Select and highlight a feature in the map or table and change the focus of the chart

  Clear the selected features in the map or table

  Modify the colors in the chart

  Modify the properties of the chart

  Add a chart to a layout

  Select and move or resize an object in a layout

  Add text to the layout

- Link a table to an attribute table
- Create a new chart

For more on geographic inquiry and these steps, see Geographic Inquiry: Thinking Geographically (pages xxi to xxiii).

Teacher notes

Lesson introduction

This lesson is designed to correspond with a unit on NAFTA. Before you start this lesson, we suggest that you review the history of NAFTA. You may want to visit the following Web sites:

- *www.customs.ustreas.gov/nafta*
  Provides links to NAFTA customs and other information from each of the three countries.
- *www.dfait-maeci.gc.ca/nafta-alena/menu-e.asp*
  Presents clear and concise information on the history of NAFTA and continuing NAFTA issues.
- *www.sice.oas.org/indexe.asp*
  Contains excellent information on NAFTA as well as on foreign trade in general.

Note: Due to the dynamic nature of the Internet, the URLs listed above may have changed. If the URLs do not work, refer to this book's Web site for an updated link: www.esri.com/mappingourworld.

Additional resources on NAFTA are listed in the GIS Resources section on the Teacher Resource CD included with this book.

Your students should have a working knowledge of basic economics in addition to background information on NAFTA. Make sure they are familiar with the terms trade, imports, exports, trade balance, and tariffs. Basic graph-reading skills will also be essential.

Begin the lesson with a discussion of the North American Free Trade Agreement (NAFTA) as spelled out in the NAFTA objectives handout included with this lesson. Ask your students to explain the objectives of NAFTA in their own words. This exercise could be done as a class activity on the board or in groups with several sets of goals being developed. Note: You may want to save the results so students can refer to them when they evaluate NAFTA in step 9.

When you are satisfied that your students have a handle on NAFTA, discuss the flow of goods into and out of a country and the concept of trade balance. Introduce the following formula for trade balance:

EXPORTS – IMPORTS = TRADE BALANCE (trade surplus or deficit)

When exports are more than imports, you have a trade surplus.

When imports are more than exports, you have a trade deficit.

Ask the students to predict the trade balance between each of the NAFTA countries.

Student activity   **Before completing this lesson with students, we recommend that you work through it yourself. Doing so will allow you to modify the activity to accommodate the specific needs of your students.**

Student work on the computer component of the lesson follows the initial discussion. Ideally each student should be at an individual computer, but the lesson can be modified to accommodate a variety of instructional settings.

On the first day, distribute the GIS Investigation sheets to the students. Explain that in this activity they will create charts and maps to explore and report on the patterns of trade over the past 10 years between the NAFTA countries. In addition, they will use the charts to identify whether NAFTA achieved its goals as outlined in the objectives discussed in class.

The worksheets will provide students with detailed instructions for their investigations. As they navigate through the lesson, they will be asked questions that will help keep them focused on key concepts. Some questions will have specific answers while others will require creative thought.

Things to look for while the students are working on this activity:

- Are students using a variety of tools?
- Are they answering the questions as they work through the procedure?
- Do they need help with the lesson's vocabulary?

 Teacher Tip: Step 8 of this GIS Investigation contains instructions for students to stop and save their work. This is a good spot to stop the class for the day and to pick up the investigation the next day. Be sure your students know how to rename their project and where to save it. We recommend that you spend two class periods doing the GIS Investigation and allow one period for work on the assessment.

Conclusion   At the completion of the lesson, have the students present to the class their trading history layouts and answers to their initial predictions. Ask them to describe the trade balances between each of the countries and also whether they think NAFTA has succeeded in achieving its goals. Finally, the students should evaluate whether or not NAFTA should remain in place. Tally up the results and report to the class what their combined opinion suggests.

Assessment   *Middle school: Highlights skills appropriate to grades 5 through 8*

Students will create and present layouts that illustrate the history of trade between Mexico and the United States, using maps they have made, charts, text, and graphics. Presenters will describe the trade balance between the countries, and indicate whether they think NAFTA has achieved its goals. Presentations will conclude with students offering opinions about whether NAFTA should remain in place, using information from their layout to support those opinions.

High school: Highlights skills appropriate to grades 9 through 12

Students will create and present layouts that illustrate the history of trade between Mexico and the United States, using maps they have made, charts, text, and graphics. Presenters will describe the trade balance between the countries, and indicate whether they think NAFTA has achieved its goals. Presentations will conclude with students offering opinions about whether NAFTA should remain in place, using information from their layout to support those opinions. Finally, students will write a paragraph describing how they would change the NAFTA agreement to improve or enhance future trading for all three countries.

Extensions
- Have the students research the history of NAFTA.
- Have students acquire similar data for the European Union, OPEC, and the Organization of American States, and compare and contrast the outcomes of NAFTA and the European Union.
- Have students map the movements of commodities from the NAFTA countries in which they were produced, to countries where they are bought and used.
- Go to Statistics Canada, U.S. Census Bureau, and INEGI sites to obtain total international export information and map that information.
- NAFTA affects other aspects of each nation's economy, such as employment rates. Have students research employment trends since the advent of NAFTA. Have the NAFTA countries experienced increased employment or unemployment since 1994?
- Ask the students to do research on the employment/unemployment figures in the United States, Canada, and Mexico since NAFTA was initiated in 1994, and to report if there was a substantial change in employment/unemployment figures since January 1, 1994.
- Check out the Resources by Module section of the Teacher Resource CD for print and media resources on the topics of economics, trade, and NAFTA or visit *www.esri.com / mappingourworld* for Internet links.

NAFTA objectives handout

The North American Free Trade Agreement (NAFTA) came into effect on January 1, 1994. This agreement created the world's largest free trade area. Among the agreement's main objectives are the liberalization of trade between Canada, Mexico, and the United States, stimulation of economic growth in all three countries, and equal access to each other's markets.

NAFTA Articles 101 and 102

Article 101: Establishment of the Free Trade Area

The Parties to this Agreement, consistent with Article XXIV of the General Agreement on Tariffs and Trade, hereby establish a free trade area.

Article 102: Objectives

1   The objectives of this Agreement, as elaborated more specifically through its principles and rules, including national treatment, most-favored-nation treatment and transparency are to:

 (a) eliminate barriers to trade in, and facilitate the cross border movement of, goods and services between the territories of the Parties;

 (b) promote conditions of fair competition in the free trade area;

 (c) increase substantially investment opportunities in the territories of the Parties;

 (d) provide adequate and effective protection and enforcement of intellectual property rights in each Party's territory;

 (e) create effective procedures for the implementation and application of this Agreement, for its joint administration and the resolution of disputes; and

 (f) establish a framework for further trilateral, regional and multilateral cooperation to expand and enhance the benefits of this Agreement.

2   The Parties shall interpret and apply the provisions of this Agreement in the light of its objectives set out in paragraph 1 and in accordance with applicable rules of international law.

From the North American Free Trade Agreement between the Government of Canada, the Government of the United Mexican States, and the Government of the United States of America, published January 1, 1994, as written in the U.S. Customs Department Web site (www.customs.gov/impoexpo/nafta_newf.htm).

NAME _____   DATE _____

Share and Share Alike
A GIS investigation

ACQUIRE

ASK

EXPLORE

ACT

ANALYZE

Answer all questions on the student answer sheet handout

On January 1, 1994, the North American Free Trade Agreement (NAFTA) was enacted to enhance trade and increase access to the total trade market available to businesses in North America. Tariffs and quotas were eliminated to increase the competitiveness of goods produced by all North Americans. Now, almost a decade after NAFTA's inception, your task will be to evaluate whether these objectives have been accomplished for all three countries involved. In this lesson you will create and analyze graphs of exports and trade balances for the NAFTA trading partners. As you are doing these exercises, ask yourself the following questions:

- Have there been any changes in exports or trade balance since the inception of NAFTA?
- Is there a larger market available for business in North America?
- Is there greater competition forcing businesses to provide the best products and best prices for their goods?
- Has NAFTA been equally beneficial for all countries involved?
- Does the trade balance of a country tell the whole story of how effective NAFTA is?

Ultimately, you will have to decide if you think NAFTA is effective!

Step 1 Start ArcView

a Double-click the ArcView icon on your computer's desktop.

b If the Welcome to ArcView dialog appears (pictured below), click **Open an Existing Project** and click OK. If it doesn't appear, proceed to step 2.

Step 2 Open the region6.apr file

a In this exercise, a project file has been created for you. To open the file, go to the **File** menu and choose **Open Project**.

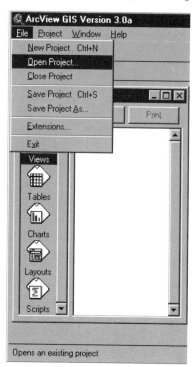

b   Navigate to the exercise data directory (**C:\esri\mapworld\mod6\data**) and choose **region6.apr** from the list.

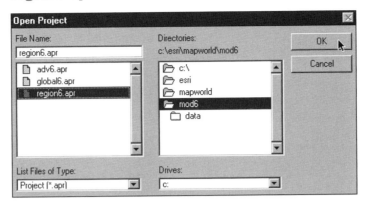

c   Click OK.

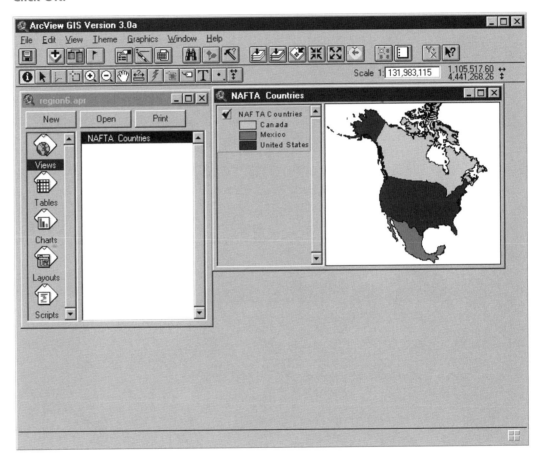

When the project opens you see a map of North America showing the three countries participating in NAFTA.

Step 3   Examine the map and attribute table

a   Make sure the NAFTA Countries theme is active.

b   Click the Identify tool, then click any country on the map.

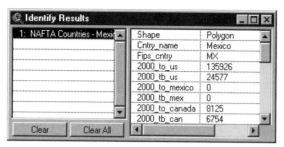

c   Look at the attribute names in the list on the right side of the Identify Results window. Scroll down the list.

> *Note: Field names with "tb" in them (e.g., 2000_tb_us) stand for trade balance.*

(1)   Which years does the theme contain data for?

(2)   How many attributes are there for each year?

d   Close the Identify Results window.

e   Click the Open Theme Table button.

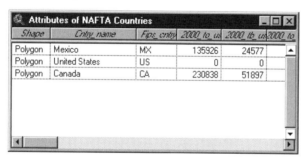

(1)   What is the name of the table?

(2)   How many rows are there for each country on the map?

f   Click any row in the table.

(1)   What happened on the table?

(2)   What happened on the map?

The first trade attribute in the table is **2000_to_us**. This field represents the value of goods and services a country exported **to** the **United States** in the year 2000. The numbers in the table are expressed in millions of U.S. dollars. That means you have to multiply the number in the table by 1,000,000 to get the actual dollar value.

g   Look at the table and answer the following question.

?   *What was the value of goods and services exported from Canada to the United States in 2000?*

h   Scroll the table from left to right. Notice that the table contains many fields for different countries and years, making it difficult to understand.

In the next step you will open another table with the same data that is easier to understand. Then you will link (connect) it to the theme table so that both tables are tied to the map.

   i   Click the Select None button.

Step 4   Link another table to the theme table

a   Maximize your ArcView window, if you haven't already done so.

b   Click the title bar (blue bar) of the Project window (region6.apr) and click the Tables icon.

c   Double-click the table entitled "NAFTA Trading Statistics 1991 - 2000."

d   Click the title bar of the table and drag it down below the other table window, which is partially showing.

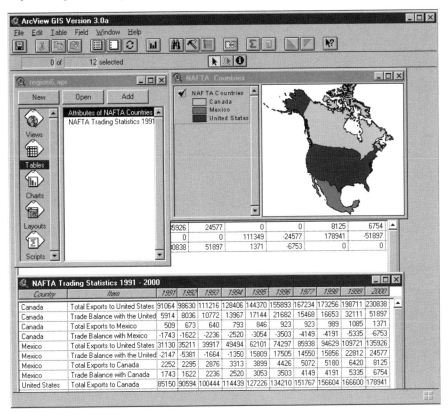

e **Examine the NAFTA Trading Statistics 1991 - 2000 table and answer the questions below.**

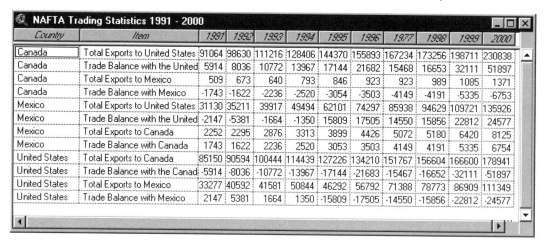

Country	Item	1991	1992	1993	1994	1995	1996	1977	1998	1999	2000
Canada	Total Exports to United States	91064	98630	111216	128406	144370	155893	167234	173256	198711	230838
Canada	Trade Balance with the United	5914	8036	10772	13967	17144	21682	15468	16653	32111	51897
Canada	Total Exports to Mexico	509	673	640	793	846	923	923	989	1085	1371
Canada	Trade Balance with Mexico	-1743	-1622	-2236	-2520	-3054	-3503	-4149	-4191	-5335	-6753
Mexico	Total Exports to United States	31130	35211	39917	49494	62101	74297	85938	94629	109721	135926
Mexico	Trade Balance with the United	-2147	-5381	-1664	-1350	15809	17505	14550	15856	22812	24577
Mexico	Total Exports to Canada	2252	2295	2876	3313	3899	4426	5072	5180	6420	8125
Mexico	Trade Balance with Canada	1743	1622	2236	2520	3053	3503	4149	4191	5335	6754
United States	Total Exports to Canada	85150	90594	100444	114439	127226	134210	151767	156604	166600	178941
United States	Trade Balance with the Canad	-5914	-8036	-10772	-13967	-17144	-21683	-15467	-16652	-32111	-51897
United States	Total Exports to Mexico	33277	40592	41581	50844	46292	56792	71388	78773	86909	111349
United States	Trade Balance with Mexico	2147	5381	1664	1350	-15809	-17505	-14550	-15856	-22812	-24577

(1) *How many rows are there for each country?*

(2) *What information is collected under the field title "item" for Canada?*

(3) *Describe in general terms the information collected in the field titled "item."*

(4) *How many years of data are represented in the table?*

In the next steps, you will link the two tables by using common information to both tables (country names).

f **Click the field name "Country." The column heading turns dark to show that it is selected.**

Country	Item	1991	1992
Canada	Total Exports to United States	91064	98630
Canada	Trade Balance with the United	5914	8036
Canada	Total Exports to Mexico	509	673
Canada	Trade Balance with Mexico	-1743	-1622
Mexico	Total Exports to United States	31130	35211

g **Click in the white space of the other table (the area below the rows) to bring the window forward.**

h **Click the field name "Cntry_name" in this table.**

Attributes of NAFTA Countries

Shape	Cntry_name	Fips_cntry	2000_to_u
Polygon	Mexico	MX	135926
Polygon	United States	US	0
Polygon	Canada	CA	230838

Now you have a field selected in each table containing exactly the same data—country names. ArcView will use the country names to link the two tables together.

i **Click the Table menu, then click Link.**

j **Click the United States row in the Attributes of NAFTA Countries table.**

What happens in the two tables and the map?

k **Click the title bar of the NAFTA Countries view.**

l Click the Select Feature tool and then click Canada on the map.

What happens in the two tables and the map?

m Click a row for Mexico in the NAFTA Trading Statistics 1991 - 2000 table.

    (1) *What happens in the two tables and the map?*

    (2) *What have you observed about the way the NAFTA Trading Statistics table is tied to the NAFTA Countries attribute table and map theme?*

n Click the title bar of the Attributes of NAFTA Countries table, then click the Select None button.

Step 5 Examine export charts

a Click the title bar of the Project window. Click the Charts icon.

 Note: In ArcView, a graph is referred to as a "chart."

b Double-click Exports to Canada. A Chart window opens.

The Chart window actually contains three small charts: one for each country.

    (1) *Which country exported more goods and services to Canada—Mexico or the United States?*

    (2) *Why is the chart empty in the space for Canada?*

c Drag the Chart window to the right, completely off the other windows if there is room.

d Click Mexico on the map. Mexico turns yellow. (If it doesn't turn yellow, click the Select Feature button and try again.)

What happened to the chart?

Because the chart is created from the data in the attribute table, it is tied to the map through the table. When you select a country on the map, only the data for that associated row in the table is displayed in the chart.

e Answer the questions below using the chart and tables.

    (1) *How many years of data are represented on the chart?*

    (2) *What year does the first bar on the left represent?*

    (3) *Compare the numbers on the y-axis with those in the NAFTA Trading Statistics table. Are the numbers on the chart in thousands, millions, or billions of dollars?*

    (4) *How do you know?*

    (5) *Looking at the chart, how would you describe the trend of Mexican exports to Canada over the 10-year period?*

    (6) *What was the approximate value of Mexican exports to Canada in 1991? In 2000?*

    (7) *Approximately how many times greater is the 2000 export figure than the 1991 export figure?*

f Close the Exports to Canada chart.

g Go to the Project window and open the chart entitled Exports to United States.

    (1) *How would you describe the trend of Mexican exports to the United States over the 10-year period?*

    (2) *Approximately how many times greater is the 2000 export figure than the 1991 export figure?*

MODULE 6 • HUMAN GEOGRAPHY III: ECONOMIC GEOGRAPHY

Step 6   Examine a trade balance chart

a   Go to the Project window and open the chart entitled Trade Balance with United States.

b   Drag the chart to a position below the first chart so you can see both charts at the same time.

Remember: Trade balance compares how much a country exports to a trading partner with how much it imports from the same partner.

c   Use the Trade Balance with United States chart to answer the questions below.

    (1)   *Did Mexico have a trade surplus or deficit with the United States for 1992?*

    (2)   *What was the approximate value of the trade balance for 1992? (Remember, the y-axis is in millions of dollars.)*

    (3)   *What was the first year that Mexico exported more to the United States than it imported from the United States?*

    (4)   *Describe the trend of Mexico's trade balance with the United States over the 10-year period.*

d   Click Canada on the map. Look at the Trade Balance chart again.

    (1)   *Did Canada have a deficit trade balance with the United States anytime during the 10-year period?*

    (2)   *In 1998, was Canada's trade balance with the United States greater, smaller, or about the same as Mexico's?*

    (3)   *In 2000, was Canada's trade balance with the United States greater, smaller, or about the same as Mexico's?*

    (4)   *Referring to the NAFTA Trading Statistics table, what was the exact value of Canada's trade balance with the United States in 2000?*

e   Click the Clear Selected Features button.

Step 7   Create a trade balance chart

In this step, you will create the chart that is missing from the region6 project: the Trade Balance with Mexico chart. First you will close the open windows to make your work easier.

a   Click the File menu and click Close All.

    Caution: If your File menu does not have a Close All option, click in one of the charts and make sure it has a blue title bar. Then go to the File menu again.

b   Make sure the Charts icon is selected in the Project window.

c   Click the New button.

d   Make sure Attributes of NAFTA Countries is selected in the Pick a Table box. Click OK.

The Chart Properties dialog is displayed.

e   Delete Chart 1 in the Name box. Type **Trade Balance with Mexico**.

Next you will choose the fields that contain the data you want to chart. The first field you select will become the first bar on the left of the chart, so you must add the field for 1991 first.

f   Scroll down to the end of the Fields list.

g   Click **1991_tb_mex**. (Hint: Make sure you do *not* click on the field named **1991_to_mexico**.)

h   Click the Add button.

The field is added to the Groups list.

i   Scroll up a little in the Fields list and find **1992_tb_mex**, then click the Add button. Repeat this procedure until the Mexico trade balance fields for all 10 years are in the Groups list.

> *Hint: if you make a mistake, for example if you add the wrong field or skip a year, click the field that is wrong in the Groups list and then click the Delete button. You can reorder a field in the Groups list by clicking and dragging it up or down.*

When you are sure your Groups list is correct, go on to the next step. (Do not click Finish.)

j   Click the Label series using box. Choose Cntry_name.

k   Click OK. The new chart displays.

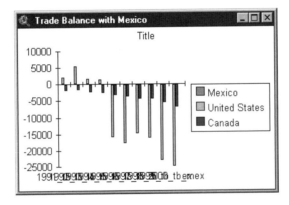

Step 8   **Change the look of your chart**

The next steps will show you how to change the look of your chart. You will begin by changing the way the data in the chart is organized. Right now, the legend represents countries and the bars are grouped along the x-axis by year. You will switch it around so the legend contains years and the bars are grouped by country.

a   Click the Series From Records/Fields button.

b   Click the Chart menu, then click Hide Title.

c   Click the Chart menu again, then click Hide Legend.

d   Click the Chart Color tool. Move the Palette window to the right, off the chart.

e Click the Color button in the Palette window. Scroll to the bottom and click the bright green box in the lower left corner.

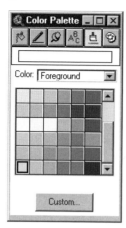

f Click each bar in the chart to change them all to the same color.

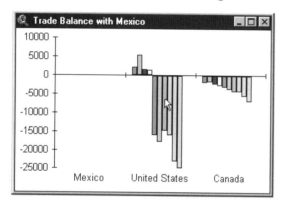

:⚡: Note: Try to avoid clicking on the x-axis line, or it will turn green. If this happens, you will need to choose black in the Color Palette and click the x-axis again to return it to black.

g Close the Color Palette window.

 h Click the Chart Element Properties tool.

:⚡: Note: The tool you want is next to the Chart Color tool that you just used. Be sure not to click the similar-looking Chart Properties button in the top row.

i Click on one of the country names at the bottom of the chart.

The Chart Axis Properties dialog for the x-axis opens.

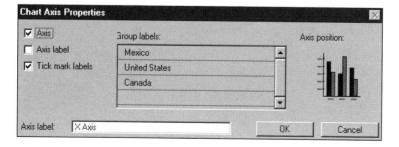

j   Click the box next to Axis label to turn that option on.

k   Click in the Axis label box and type the following text:
All charts cover the period 1991 - 2000, left to right

l   Click OK.

m   Look at the numbers along the y-axis in your chart.

?   *What are the minimum and maximum y-axis values?*

You will make a few changes to the y-axis so your chart matches the other two trade balance charts.

n   Click once on the numbers in the y-axis to open the Chart Axis Properties dialog for the y-axis.

o   For Scale min, type **–60000**. For Scale max, type **60000**. For Major unit, type **20000**.

p   Click the box for Major Grid. Make sure your dialog looks like the one below, then click OK.

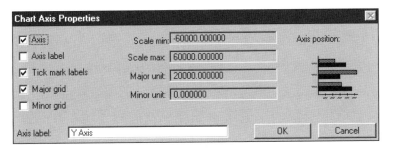

?   *What changes did this make on your chart?*

q   Close the chart window.

?   *r*   Ask your teacher if you should stop here and save this ArcView project. Follow your teacher's instructions on how to rename the project and where to save it. If you do not need to save the project, proceed to step 9. Write the new name you gave the project and where you saved it on the answer sheet.

Step 9   Evaluate the effectiveness of NAFTA

In this lesson, you are looking at the level of trade among the NAFTA trading partners—Canada, Mexico, and the United States. You are also looking at changes in the level of trade among the three countries over time.

Remember: It is nearly impossible for any country to produce everything it needs to support its people. Countries must import (purchase) goods from other countries. One country's imports are the other country's exports. In this project, you have no charts for imports because you can figure that information out by looking at the export charts.

For example, to find the United States' imports from Canada, you switch the question in your mind and look for Canada's exports to the United States. You would find this information on the chart entitled Exports to United States.

To decide whether or not NAFTA is meeting its goals, you will need to further explore the data in the charts and tables in the ArcView project. In this step, you will gather the information that you need to decide if you think NAFTA is effective. You may want to refer to the NAFTA objectives handout or the list of NAFTA goals you developed with your teacher at the beginning of the lesson.

Remember these formulas as you examine the charts and answer the questions that follow:

Exports – Imports = Trade Balance

Trade Balance:
Exports > Imports = Trade Surplus
Imports > Exports = Trade Deficit

a Click the Views icon in the Project window. Open the NAFTA Countries view.

b Make sure that no country is selected. (Hint: If a country is selected, click the Clear Selected Features button.)

c Click the Charts icon in the Project window.

d Hold down the Shift key and click each of the Exports chart titles.

e Click the Open button.

f Arrange the view and chart windows so they are not overlapping.

g Use the charts to find the value of exports between each set of countries for the year 2000. Write the information in the table on your answer sheet.

h Add the export values together for each pair of countries (for example, United States to Mexico plus Mexico to United States) and write that number in the Total Volume Between Partners column.

i Rank the trading partners by the overall volume of trade between the two countries. Use 1 for the partners trading the most and 3 for the partners trading the least.

   (1) *Do you think that NAFTA had a positive (+), negative (–), or neutral (n) effect on trade volume between each set of partner countries?*

   (2) *Do you think that any one of these three countries benefited more than the other two by NAFTA? If so, which country? Explain your answer.*

j Close the three chart windows.

k Go to the Project window and open the three Trade Balance charts. Arrange them so you can compare them.

Remember that most countries prefer to export a higher volume and dollar value of goods than they import, but either a large trade surplus or large trade deficit may negatively affect a country's overall economic health.

l Look at the Trade Balance with Canada chart.

   (1) *What country has a healthier trade balance with Canada—Mexico or the United States?*

   (2) *On what chart do you find a set of bars that looks like a mirror image of those for the U.S. trade balance with Canada?*

m　Compare the trade balance charts.

?　　(1)　*What country had the most dramatic change for the better after NAFTA came into being? (Remember, NAFTA went into effect in 1994.)*

?　　(2)　*Estimate the U.S. trade deficit with Canada and Mexico for 1999 and 2000. Express the values in billions of dollars, using round numbers that you estimate from the charts. Write the answers in the space provided.*

?　　(3)　*Did the U.S. combined trade balance get better or worse between 1999 and 2000? By how much?*

Step 10　**Prepare charts for the layout**

In step 11 you will be creating a layout that presents the information about the U.S. trade balance and exports to Canada. The layout will include the NAFTA countries map and two charts. Later, in the assessment, you will use the techniques you learn to create a layout showing the results of your evaluation of NAFTA.

Before you start your layout, you need to make sure to open the charts you want to include. You also need to make sure they look the way you want them to appear on the layout.

    a　Close the Trade Balance with Mexico and Trade Balance with United States chart windows. (The Trade Balance with Canada chart should still be open.)

    b　In the Project window, double-click Exports to Canada.

    c　Arrange the view and the two chart windows so they don't overlap.

    d　Click the title bar of the View window. Click the Select Feature tool if it isn't selected.

    e　Click the United States on the map.

The charts change to focus on U.S. data.

    f　Click the title bar of the Trade Balance with Canada chart.

    g　From the Chart menu, select Show title.

    h　Click the Chart Element Properties tool, then click the word "Title" in the chart.

    i　Type **Trade Balance with Canada** in the Chart Title Properties dialog. Click OK.

?　　*What happens to your chart?*

    j　Click the title bar of the Exports to Canada chart window.

    k　Repeat steps 10h–k to add the title to the export chart. Type **Exports to Canada** for the title.

MODULE 6 • HUMAN GEOGRAPHY III: ECONOMIC GEOGRAPHY

Step 11   Create and print a layout

a   Click the Layouts icon in the Project window.

b   Open the layout entitled US Trade with CA.

A window opens with a layout that looks like a piece of paper. The map and a title have already been added.

c   If your screen is large enough, stretch the Layout window until it fills most of your ArcView window. (It is OK for the Layout window to cover the other windows.)

d   Click and hold on the View Frame tool to display the drop-down tool menu. Choose the Chart Frame tool (third from the bottom).

You will add the two charts in the blank area of the layout to the left of the map.

e   Click at the upper left side of the layout, below the title. Drag a box for the first chart.

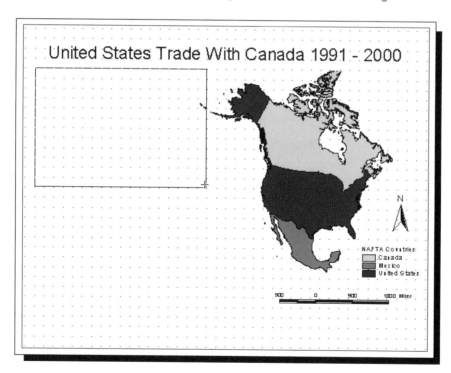

f **Select Exports to Canada in the Chart Frame Properties dialog and click OK.**

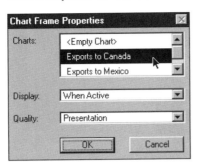

You will be able to move or change the size of the chart frame in a later step if it isn't exactly the way you want it to be.

g **Drag another box in the area below the first chart. This time scroll the list and choose Trade Balance with Canada. Click OK.**

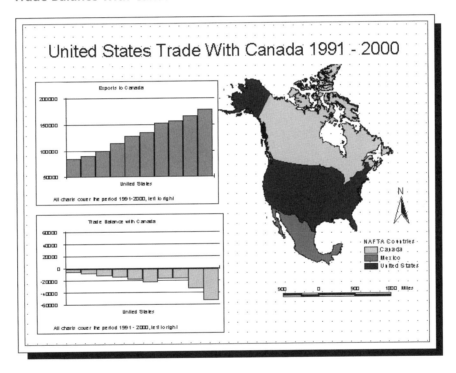

h **If you wish to adjust the position or size of your charts, click the Pointer tool. First click anywhere on the chart you want to work with to display the black handles (squares). Click anywhere within the handles and drag the chart to move it. Click on a handle and drag it to change the size of the chart.**

When you are satisfied with the look of your layout, go on to the next step.

i **Click the Text tool. Now you will add your name and date to the map.**

j   Click in the lower right of the layout (below the scale bar) and type your name and the date in the Text Properties box.

For example:

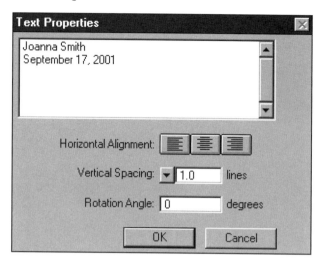

k   Click the Pointer tool and, if necessary, use it to reposition your name and date.

l   From the File menu, select Print.

Step 12   Exit ArcView

In this exercise, you used ArcView tables, charts, and a map to explore trade data for the three North American countries. After analyzing this data, you created a layout to present some of the information.

a   Ask your teacher for instructions on where to save this ArcView project and on how to rename the project. Save it according to your teacher's instructions.

b   If you are not going to save the project, exit ArcView by choosing Exit from the File menu. When it asks if you want to save changes to region6.apr, click No.

NAME _____   DATE _____

Student answer sheet
Module 6
Human Geography III: Economic Geography

Regional case study: Share and Share Alike

Step 3   Examine the map and attribute table

c-1   Which years does the theme contain data for? _____   to   _____

c-2   How many attributes are there for each year? _____

e-1   What is the name of the table?   _____

e-2   How many rows are there for each country on the map? _____

f-1   What happened on the table?   _____

f-2   What happened on the map?   _____

g   What was the value of goods and services exported from Canada to the United States in 2000?

    $ _____

Step 4   Link another table to the theme table

e-1   How many rows are there for each country?   _____

e-2   What information is collected under the field title "item" for Canada?

    _____

    _____

    _____

e-3   Describe in general terms the information collected in the field titled "item."

    _____

    _____

e-4   How many years of data are represented in the table? _____

j   What happens in the two tables and the map?

    _____

    _____

l   What happens in the two tables and the map?

    _____

    _____

m-1   What happens in the two tables and the map?

_____

_____

m-2   What have you observed about the way the NAFTA Trading Statistics table is tied to the NAFTA Countries attribute table and map theme?

_____

_____

Step 5   Examine export charts

b-1   Which country exported more goods and services to Canada—Mexico or the United States?

_____

b-2   Why is the chart empty in the space for Canada?

_____

_____

d   What happened to the chart?

_____

_____

e-1   How many years of data are represented on the chart? _____

e-2   What year does the first bar on the left represent? _____

e-3   Compare the numbers on the y-axis with those in the NAFTA Trading Statistics table. Are the numbers on the chart in thousands, millions, or billions of dollars?

_____

e-4   How do you know?

_____

e-5   Looking at the chart, how would you describe the trend of Mexican exports to Canada over the 10-year period?

_____

e-6   What was the approximate value of Mexican exports to Canada in 1991?

_____

In 2000? _____

e-7   Approximately how many times greater is the 2000 export figure than the 1991 export figure?

_____

g-1   How would you describe the trend of Mexican exports to the United States over the 10-year period?

_____

g-2   Approximately how many times greater is the 2000 export figure than the 1991 export figure?

_____

Step 6   Examine a trade balance chart

c-1   Did Mexico have a trade surplus or deficit with the United States for 1992? _____

c-2   What was the approximate value of the trade balance for 1992? (Remember, the y-axis is in millions of dollars.)

_____

c-3   What was the first year that Mexico exported more to the United States than it imported from the United States?

_____

c-4   Describe the trend of Mexico's trade balance with the United States over the 10-year period.

_____

_____

d-1   Did Canada have a deficit trade balance with the United States anytime during the 10-year period?

_____

d-2   In 1998, was Canada's trade balance with the United States greater, smaller, or about the same as Mexico's?

_____

d-3   In 2000, was Canada's trade balance with the United States greater, smaller, or about the same as Mexico's?

_____

d-4   Referring to the NAFTA Trading Statistics table, what was the exact value of Canada's trade balance with the United States in 2000?

_____

Step 8   Change the look of your chart

m   What are the minimum and maximum y-axis values?

Minimum: _____

Maximum: _____

p   What changes did this make on your chart?

_____

_____

r   Write the new name you gave the project and where you saved it.

_____   _____
(Name of project.   **(Navigation path to where project is saved.**
For example: kzeregion6.apr)   **For example: C:\student\kze\mapworld)**

Step 9   Evaluate the effectiveness of NAFTA

g   Use the charts to find the value of exports between each set of countries for the year 2000. Write the information in the table.

DIRECTION OF EXPORT FLOW	VALUE OF EXPORTS (MILLION $)	TOTAL VOLUME BETWEEN PARTNERS (MILLION $)
United States to Mexico		
Mexico to United States		
United States to Canada		
Canada to United States		
Canada to Mexico		
Mexico to Canada		

h   Add the export values together for each pair of countries (for example, United States to Mexico plus Mexico to United States) and write that number in the Total Volume Between Partners column.

i   Rank the trading partners by the overall volume of trade between the two countries. Use 1 for the partners trading the most and 3 for the partners trading the least.

United States–Mexico: _____

United States–Canada: _____

Canada–Mexico: _____

i-1   Do you think that NAFTA had a positive (+), negative (–), or neutral (n) effect on trade volume between each set of partner countries?

United States–Mexico: _____

United States–Canada: _____

Canada–Mexico: _____

i-2   Do you think that any one of these three countries benefited more than the other two by NAFTA? If so, which country? Explain your answer.

_____

_____

_____

l-1   What country has a healthier trade balance with Canada—Mexico or the Untied States?

_____

l-2   On what chart do you find a set of bars that looks like a mirror image of those for the U.S. trade balance with Canada?

_____

m-1   What country had the most dramatic change for the better after NAFTA came into being? (Remember, NAFTA went into effect in 1994.)

_____

m-2 Estimate the U.S. trade deficit with Canada and Mexico for 1999 and 2000. Express the values in billions of dollars, using round numbers that you estimate from the charts.

	1999 ($ BILLION)	2000 ($ BILLION)
U.S. trade balance with Mexico		
U.S. trade balance with Canada		
Combined deficit (total)		

m-3 Did the U.S. combined trade balance get better or worse between 1999 and 2000? _____

By how much? _____

Step 10 Prepare charts for the layout

i What happens on your chart?

_____

NAME _____   DATE _____

Share and Share Alike
Middle school assessment

1 Using the 10 years of export and trade balance data you have for the NAFTA trading partners, Canada, Mexico, and the United States, make a determination if you think that NAFTA has been effective, marginally effective, or ineffective and whether you think NAFTA should continue or not.

Remember that there are several factors that you can use in drawing your conclusions. You have looked at several of those factors during this lesson. For instance, you can compare the exports of each country as well as look at each country's trade balances. Refer back to the NAFTA Objectives handout to remind yourself of the original intent of NAFTA.

2 To display your findings, create a layout that graphically displays your conclusion. Use the layout in the region6.apr ArcView project entitled NAFTA Assessment that has already been started for you. You may rearrange and resize the existing layout components and add your own. Your final layout should include the following components:

(a) Title

(b) Map

(c) Map legend

(d) Orientation (north arrow or compass rose)

(e) At least two charts

(f) Text labels or descriptions

(g) Author (your name)

(h) Today's date

3 Present your layout to the class as evidence for your conclusion on NAFTA's effectiveness.

Assessment rubric

Share and Share Alike

Middle school

STANDARD	EXEMPLARY	MASTERY	INTRODUCTORY	DOES NOT MEET REQUIREMENTS
The student knows and understands how to make and use maps, globes, graphs, charts, models, and databases to analyze spatial distributions and patterns.	Creates a layout using GIS with eight or more components including several maps and charts that illustrate the effectiveness of NAFTA.	Creates a layout using GIS with eight components including a map and minimum of two charts.	Creates a layout using GIS with most of the eight specified components including a map and at least one chart.	Creates a layout using GIS with some of the eight specified components including a map.
The student knows and understands the basis for global interdependence.	Layout shows a clear understanding of the concepts of balance of trade for NAFTA trading partners through multiple charts and maps.	Layout includes charts and maps showing an understanding of the concept of balance of trade for NAFTA trading partners.	Layout includes charts and maps showing some understanding of the concept of balance of trade for NAFTA trading partners.	Layout includes chart and or map showing trade partners, but does not illustrate an understanding of balance of trade.
The student knows and understands how cooperation and conflict among people contribute to economic and social divisions of Earth's surface.	Presents a logical argument to his/her conclusions on the effectiveness of NAFTA using a variety of resources to support his/her findings including, but not limited to, data and maps.	Presents a logical argument to his/her conclusions on the effectiveness of NAFTA using data and maps to support his/her ideas.	Describes conclusions on the effectiveness of NAFTA and provides ample argument or evidence.	Identifies NAFTA participants and trading partners, but does not present any personal conclusions as to its effectiveness.

This is a four-point rubric based on the National Standards for Geographic Education. The "Mastery" level meets the target objective for grades 5–8.

NAME _____ DATE _____

Share and Share Alike
High school assessment

1 Using the 10 years of export and trade balance data you have for the NAFTA trading partners, Canada, Mexico, and the United States, make a determination if you think that NAFTA has been effective, marginally effective, or ineffective and whether you think NAFTA should continue or not.

Remember that there are several factors that you can use in drawing your conclusions. You have looked at several of those factors during this lesson. For instance, you can compare the exports of each country as well as look at each country's trade balances. Refer back to the NAFTA objectives handout to remind yourself of the original intent of NAFTA.

2 To display your findings, create a layout that graphically displays your conclusion. Use the layout in the region6.apr ArcView project entitled NAFTA Assessment that has already been started for you. You may rearrange and resize the existing layout components and add your own. Your final layout should include the following components:

  (a) Title

  (b) Map

  (c) Map legend

  (d) Orientation (north arrow or compass rose)

  (e) At least two charts

  (f) Text labels or descriptions

  (g) Author (your name)

  (h) Today's date

3 Present your layout to the class as evidence for your conclusion on NAFTA's effectiveness.

4 Compose a paragraph describing how you would change the NAFTA agreement to improve or enhance future trading results for all three countries. Include specific reasons why you believe your changes to NAFTA would be effective.

Assessment rubric

High school

Share and Share Alike

STANDARD	EXEMPLARY	MASTERY	INTRODUCTORY	DOES NOT MEET REQUIREMENTS
The student knows and understands how to use geographic representations and tools to analyze, explain, and solve geographic problems.	Creates a layout using GIS with eight or more components including several maps and charts that illustrate the effectiveness of NAFTA.	Creates a layout using GIS with eight components including a map and minimum of two charts that illustrate the effectiveness of NAFTA.	Creates a layout using GIS with most of the eight specified components including a map and at least one chart that illustrate the effectiveness of NAFTA.	Creates a layout using GIS with some of the eight specified components including a map that illustrates the effectiveness of NAFTA.
The student knows and understands the increasing economic interdependence of the world's countries.	Composes a brief essay that defines specific improvements to the NAFTA agreement that will benefit future trading for all three countries. Provides examples to illustrate these ideas.	Composes a paragraph that defines specific improvements to the NAFTA agreement that will benefit future trading for all three countries.	Composes a paragraph that lists ideas for improvement to the NAFTA agreement, but the improvements may not benefit all parties.	Lists ideas for improvement to the NAFTA agreement, but does not define how these improvements will enhance future trading.
The student knows and understands why and how cooperation and conflict are involved in shaping the distribution of social, political, and economic spaces on Earth at different scales.	Presents a logical argument to his/her conclusions on the effectiveness of NAFTA at a local, regional, and global level using a variety of resources to support his/her findings including, but not limited to, data and maps.	Presents a logical argument to his/her conclusions on the effectiveness of NAFTA at a local, regional, and global level using data and maps to support his/her ideas.	Describes conclusions on the effectiveness of NAFTA at a local, regional, and global level, but cannot provide ample argument or evidence.	Identifies NAFTA participants and trading partners, but does not present any personal conclusions as to its effectiveness.

This is a four-point rubric based on the National Standards for Geographic Education. The "Mastery" level meets the target objective for grades 9–12.

Live, Work, and Play
An advanced investigation

Lesson overview

Students will use the Internet to acquire the most recent data on employment in California. By exploring this data through a GIS, they will construct a thematic map for California counties that displays population, number of jobs in the finance field, and average wage data, then perform a complex query and create a map that illustrates their results. Students will be learning how to examine data to choose a place to live that satisfies a set of criteria.

Estimated time   Two to three 45-minute class periods

Materials   ✔ Student handouts from this lesson to be copied:
- Live, Work, and Play worksheet (page 378)
- GIS Investigation sheets (pages 379 to 388)

Standards and objectives

National geography standards

	GEOGRAPHY STANDARD	MIDDLE SCHOOL	HIGH SCHOOL
1	How to use maps and other geographic representations, tools, and technologies to acquire, process, and report information from a spatial perspective	The student understands how to make and use maps, globes, graphs, charts, models, and databases to analyze spatial distributions and patterns.	The student understands how to use geographic representations and tools to analyze, explain, and solve geographic problems.
12	The processes, patterns, and functions of human settlement	The student understands the causes and consequences of urbanization.	The student understands the evolving forms of present-day urban areas.
18	How to apply geography to interpret the present and plan for the future	The student understands how varying points of view on geographic context influence plans for change.	The student understands how to use geographic knowledge, skills, and perspectives to analyze problems and make decisions.

Objectives

The student is able to:
- Use the Internet to locate data for recent economic activity.
- Use a GIS to map data acquired from the Internet.
- Thematically map various data to select an appropriate place to live.

GIS skills and tools
- Locate data on the Internet and prepare it in a comma-delimited text file
- Add tables to a project
- Join tables
- Perform a complex query
- Select a site according to different location criteria
- Create a layout
- Print a presentation-style map

For more on geographic inquiry and these steps, see Geographic Inquiry: Thinking Geographically (pages xxi to xxiii).

Teacher notes

Lesson introduction

Open a class discussion with the following prompts:
- Business owners frequently use GIS to select a suitable site for a new business location.
- Individuals can also use GIS to decide where they would like to live, work, and spend their leisure time by looking for areas that have various characteristics, such as affordable housing, employment opportunities, and recreational areas.
- What criteria are important to students when they think about where they would like to live after they graduate from high school and enter adulthood?

Explain to your students that they will use the Internet to find data on recent economic activity. They will use a GIS to map economic data they find and will analyze the data to locate the best place to live, work, and play. This GIS Investigation will require students to work independently, with little guidance.

Student activity

 Before completing this lesson with students, we recommend that you complete it as well. Doing so will allow you to modify the activity to accommodate the specific needs of your students.

Distribute this lesson's GIS Investigation sheets and the Live, Work, and Play Worksheet. Students should fill out the worksheet, identifying the criteria they consider the most important in choosing a place to live, then follow the guidelines to retrieve county economic data from the Internet. They will create a text file of that data, import that data into ArcView, and create thematic maps that display concentrations of jobs and wages. Once the data is in ArcView, students will analyze it to find areas with desirable characteristics (as listed on the worksheet).

 Teacher Tip: Students will be creating text files that need to be saved. Be prepared to advise them on where to save these files.

Conclusion

Engage students in a discussion of the patterns they discovered as they mapped different economic criteria. Were they able to find a place to live that satisfied *all* of the criteria they had? Could they add more criteria? Could they use this same method to download data for other states and broaden the search area (e.g., including Washington and Oregon)? How might site analysis be useful for businesses?

Assessment

The last step of the GIS Investigation asks students to perform their own complex query on where to live, and create a layout of their results. In addition to the map, you may wish to ask students to submit a written report that summarizes the process they went through in selecting an optimal living site and explains their query of the data. A written report could answer any or all of the following questions:

- What steps were involved in the query process each student went through to make their site selection?
- Were they surprised at the results they got when they mapped the various criteria? Why or why not?
- Was there an alternate site that might have been as acceptable as the first site chosen? Why or why not?

 Note: Due to the independent nature of this lesson, there is no supplied assessment rubric. You are free to design assessment rubrics that meet the needs of your specific adaptation.

Extensions

- Ask students to broaden their search area to other states. Download economic data for other states from the Bureau of Economic Analysis Web site and perform the same analysis. Students should note any changes they see.
- Have students query more fields in the CA counties table to narrow their search. They can explore, for example, median rent, median house values, and the percentage of residents who are married.
- Have students download employment data for an industry that interests them and perform a query based on their individual priorities.
- Check out the Resources by Module section of the Teacher Resource CD for books and media resources on the topics of employment data and demographics or visit *www.esri.com/mappingourworld* for Internet links.

NAME _____   DATE _____

Live, Work, and Play worksheet

Imagine: You have just graduated from college with a degree in finance. You want to move to California, but you don't know which county would be best for you. You need to find a good job in the finance industry (banking, investments, insurance, or real estate). You know that salary is important (after all, you would like to start paying off your student loans as soon as possible), but you also know there are many other things to consider when deciding where to live and work. For example, you might want to consider the cost of housing, the length of your commute, and what you'll do in your free time (do you want to be near the beach, the mountains, a city with good nightlife?).

To start your investigation of good, better, and best places to live, complete the table below. For each category, designate it as a Very Important, Average, or Not Important criterion for determining where you want to live.

THEMES AVAILABLE	VERY IMPORTANT	AVERAGE	NOT IMPORTANT
# of people ages 18 - 29			
# of finance jobs available			
Average wage			

NAME _____   DATE _____

Live, Work, and Play
An advanced investigation

> *Note: Due to the dynamic nature of the Internet, the URLs listed in this lesson may have changed, and the graphics shown below may be out of date. If the URLs do not work, refer to this book's Web site for an updated link: www.esri.com/mappingourworld.*

ACQUIRE

ASK

EXPLORE

ACT

ANALYZE

Answer all questions on the student answer sheet handout

Step 1   **Locate economic data on the World Wide Web**

The World Wide Web is a great source to find geographic data. In this step, you will go to the Bureau of Economic Analysis (BEA) Web site to find economic data.

a   Open your Web browser and go to the BEA economic accounts page at www.bea.doc.gov.

b   Under Regional, click **State and local area data**.

c   Under the Description column, click **Local area personal income**.

d   Scroll down the page to the section titled **Detailed county annual tables**.

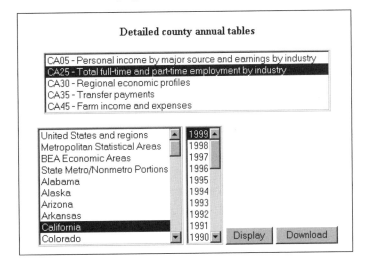

e   Click **CA25 - Total full-time and part-time employment by industry**, **California**, and **1999**.

f   Click the Display button to see your report.   [Display]

g   Scroll down the industry classifications and click **700**. A table appears showing the number of people employed in finance jobs per county or metropolitan area (MSA) in California.

h   Scroll down and click the Download button. The data table is reformatted as a plain text list and displayed in your Web browser.   [Download]

Step 2   **Download data and prepare it for ArcView**

Before you can see a map of this data in ArcView, you need to save it to your computer and modify it so that ArcView can understand it. The modifications are done in a text editor program.

a   From your Web browser, click Edit, click File, and Save As.

b   Save the file as text (.txt) and rename it **abcfinance.txt**, where abc are your initials. Save it in a folder your teacher specifies.

c   Open a text editor such as WordPad or Notepad. Open the abcfinance.txt file you just saved. The data appears in your new document.

For example:

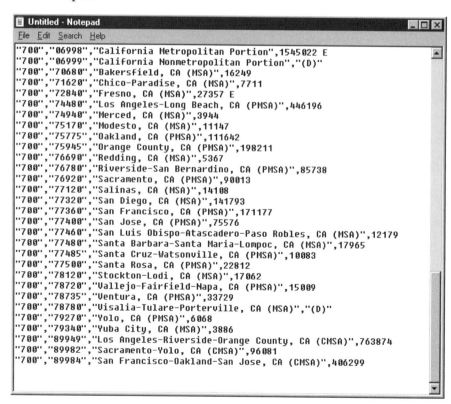

d   Scroll to the top of the document and make the following changes:

- Replace "Private employment: Finance, insurance, and real estate" with **"Finance"**. (Keep the quotation marks.)
- Change "1999" in the first line of the text file to **Fin_1999**. (Omit the quotation marks.)
- Replace all "(D)" with **–99**. (Omit the quotation marks.)

Consult the graphic below to make sure you made all the necessary changes.

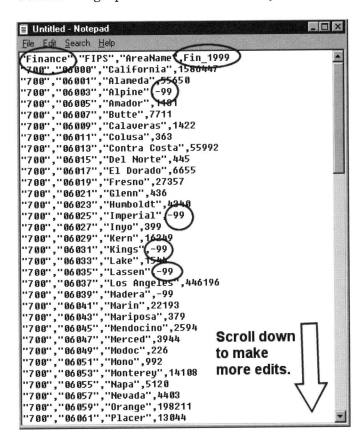

The data you saved includes statistics for counties and metropolitan areas, but you need only the county statistics.

e Delete all the rows after Yuba County; the last line in your file now should read "700","06115","Yuba",12771. See example below:

```
"700","06095","Solano",188051
"700","06097","Sonoma",677605
"700","06099","Stanislaus",247651
"700","06101","Sutter",44368
"700","06103","Tehama",21901
"700","06105","Trinity",2608
"700","06107","Tulare","(D)"
"700","06109","Tuolumne",21362
"700","06111","Ventura",939560
"700","06113","Yolo",163828
"700","06115","Yuba",12771
```

f Save the file.

g Minimize the text editor program and your Web browser. You will download additional data in step 5.

Now that your data is in a format that ArcView can read, you will bring the data into ArcView and map it.

Step 3   Start ArcView and add your data

a   Start ArcView.

b   Open the **adv6.apr** project from your exercise directory (**C:\esri\mapworld\mod6 \adv6.apr**).

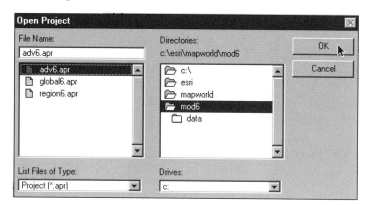

When the project opens, you see a map in the view with two themes turned on (Pop. 18 - 29 yrs. old and CA counties).

c   From the Project Window, click the Tables icon.

d   Click the Add button to open the Add Table dialog box.

e   In the lower left corner of the Add Table dialog, change the List Files of Type to **Delimited Text [*.txt]**. Navigate to the location of **abcfinance.txt**, select it, and click OK.

f   Check the table to make sure the Fin_1999 field displays numbers.

You will join this table to the attribute table of the CA Counties theme.

g   Open the Attributes of CA Counties theme table.

h   Arrange the tables so that the attribute table is on the left and the abcfinance table is on the right.

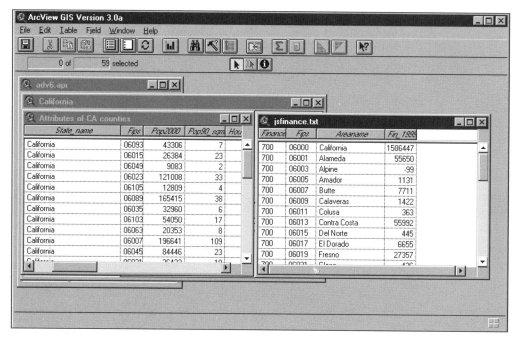

i   Click the column heading Fips in the abcfinance.txt table. Click the column heading Fips in the Attributes of CA counties table.

 j   Click the join button to join the two tables. The abcfinance window has closed.

  k   Scroll to the right of the Attribute table and notice the fields, Finance, Areaname, and Fin_1999. Now the information in the abcfinance table is joined to the Attributes of CA counties table. Close the table.

In the next step, you will classify its legend so it thematically maps the finance employment data you just downloaded.

Step 4   Thematically map finance employment data

  a   From the Edit menu, click Copy. From the Edit menu, click Paste. Now you have two copies of CA counties theme.

  b   Open the Legend Editor for the copied theme.

  c   In the Legend Type drop-down menu, select Graduated Color. For Classification Field, scroll down and select Fin_1999.

The classification field (Fin_1999) represents the number of finance jobs for each county in California in 1999. By default, ArcView has divided the number of finance jobs into five groups.

  d   Click the Classify button and change Type to Quantile.

The "Quantile" method will group the counties into five categories based on the number of finance jobs available in that county (i.e., counties with the fewest jobs, counties with the second fewest jobs, and so forth, up to counties with the most jobs). The Legend Editor updates and you see a list of five colored symbols, values, and labels.

Some counties have no data available. In such cases, the Fin_1999 field has a value of –99. Because the value of –99 represents no data and not the number –99, those counties need to be taken out of the categories you just created. You will do this by assigning values of –99 to a No Data class.

 e   Click the Null Value button.

  f   Type **–99** in the Null Value field. Click the box to the left of Display No Data Class. Click OK.

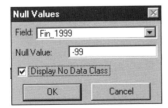

  g   In the Legend Editor, assign a medium-gray color to symbolize the No Data class (–99).

h   Change the labels for the five classes to the following phrases:

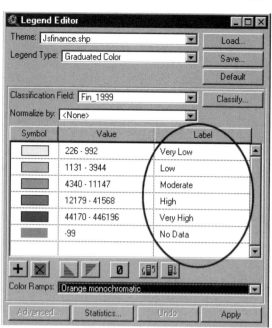

i   Change the color ramp to orange monochromatic. Click Apply and close the Legend Editor.

j   Turn on the CA counties theme with the graduated color legend. Use the Theme Properties button to change its theme name to **# of Finance Jobs**.

k   Ask your teacher where you should save your work and how you should name the project. We recommend that you save your project as **abcadv6.apr** where abc are your initials.

l   Save your project.

m   Look at the Finance jobs data. Where are the counties with a high number of finance jobs? You will refer back to this data after you've downloaded average wage data for the California counties.

Step 5   Download average wage data for California counties

Now, you will go back to the BEA Web site, search for data on average wages for California counties, download the data, and import it to ArcView.

a   Maximize your Web browser and go to the BEA economic accounts page at *www.bea.doc.gov*.

b   Under Regional, click **State and local area data**.

c   Under the Description column, click **Local area personal income**.

d Scroll down the page to the section titled **CA34 - wage and salary summary estimates**.

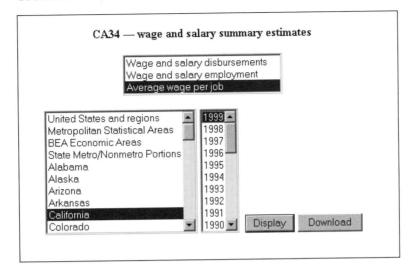

e Click **Average wage per job**, **California**, and **1999**. Click the Display button to see your report.

f Scroll down and click the Download button.

g From the File menu, click Save As. Save the file as a text file (.txt) and rename it abcwage.txt, where abc are your initials. Save it in a folder your teacher specifies.

h Maximize your text editor program and open abcwage.txt.

Step 6   Prepare average wage data for ArcView

a Replace the double quotes and the year "1999" in the first line of the text file with **Avg_wage**. Replace any "(D)" with **–99** (Note: There may be no data fields to replace with –99 and you must omit the quotes). Replace the long title "Average Wage per Job" with the word **"Wage"**. (Keep the quotes here.)

b Delete everything after the line with the Yuba County data.

c Ask your teacher where to save this document. Save it as a text file and name it **abcwage.txt** where abc are your initials. Exit the text editor program.

Step 7   Add average wage data to ArcView

a Make the CA counties theme active.

b Open the Attributes of CA counties table.

c Repeat steps 3g–3j above to add the new abcwage.txt table to your project and to join it to the Attributes of CA counties table.

d Copy the CA counties theme and rename it CA Wages. Close any open tables in the view.

Step 8   Thematically map California average wage data

a Double-click CA Wages to open the Legend Editor.

b Map the theme using Legend Type = graduated color, Classification Field = Ave_wage, and Classify using Natural Breaks.

c   Edit the Labels to reflect the following descriptions:

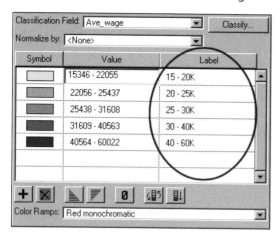

d   Click Apply and close the Legend Editor. Turn on CA Wages.

e   Save your project.

Step 9   **Query the data**

With population data, number of finance jobs data, and average wage data, you will perform a complex query to narrow down the counties where you could live in California. For the purpose of this exercise, you will perform an example query with detailed instructions. In step 10, you will perform your own query, using the preferences you made on your worksheet.

a   Look at the different themes by turning them on and off. Try to visually identify a couple of counties that appear to meet the criteria you identified earlier.

Now you will perform a query. For this example query, assume that the number of people aged 18–29 years old is an important criterion, followed by number of finance jobs, and then average wage. Because this is countywide data, it's important to take that into consideration when you develop the query.

b   Turn off all themes except CA counties. Make it active.

 c   Click the Query button.

d   For Fields, scroll down and double-click [Age_18_29].

e   Click the greater than or equal to symbol.

The 18- to 29-year-old population ranges from 203 in one county to 1,988,668 in another county. Because we don't want to necessarily eliminate counties with a low overall population, we are going to query for counties with an 18- to 29-year-old population size of 10,000 or more.

f   For Values, double-click 11225 (it's the closest number in the set to 10,000). Click New Set.

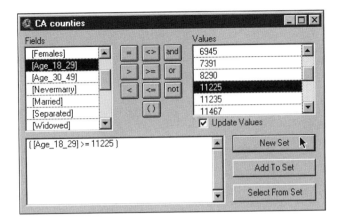

g   Close the Query Builder so you can see all the counties in California with an 18- to 29-years-old population of 11,225 or more.

h   Click the Open Theme table button. Take note of how many counties are selected (37). Close the table.

Now you will search for counties with many finance jobs as well as many people aged 18–29. This might narrow the list of places you would consider living, because some of the counties selected in your first query might not have many finance jobs. Because we don't want to necessarily eliminate counties with a low overall population, we are going to query for counties with 10,000 or more finance jobs.

i   Click the Query button.

j   Double-click Fin_1999, click greater than or equal to, double-click 10083, and click Select From Set. Close the Query builder.

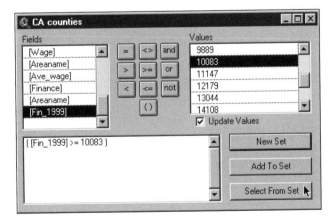

The number of selected counties has decreased in the view. There are now 22 selected counties that have a population aged 18–29 years old of 11,225 or greater and more than 10,083 finance jobs available. Now you will add average wage data to your search. These numbers reflect annual salary in dollars. In this example, you will query for counties with an average salary of $36,972 or more.

k   Click the Query button.

l   Double-click Ave_wage, click greater than or equal to, double-click 36972, and click Select From Set.

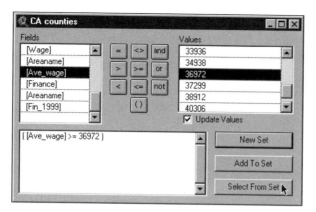

The number of selected counties has decreased significantly.

 m   Open the theme table and click the Promote button.

The eight counties that meet the criteria you queried are highlighted in yellow at the top of the table.

n   Close the Attributes of CA counties table.

o   Look at the map and ask yourself the following questions: Are the selected counties the ones you identified in step 9a? Do you know anything else about those counties that would make them an attractive place to live?

Step 10   Perform your own query and map your observations

a   Use the instructions in step 9 and the worksheet you completed at the beginning of the GIS Investigation to create your own query of CA counties. Be sure to record the criteria you used for your query. You will need to include that information in your layout.

b   After you've performed your query, prepare your view to look the way you want it to look in a layout.

c   Create a layout and title it **Live, Work and Play - Your Name**. For example, *Live, Work and Play, Joanna Smith* would be an appropriate title for a map.

Remember to include the following items in your layout:
- Title
- Legend
- List of criteria used for query
- The Attributes of CA counties table open, displaying the results of your query
- Map scale
- North arrow
- Author of map
- Date of map creation

> *Note: If you experience difficulty creating the map layout, use ArcView online Help.*

d   Print your map.

e   Save your project according to your teacher's instructions.

Human/Environment Interaction

Physical processes influence patterns of human activity, just as human activities have an effect on the environment.

Water World: A global perspective
Students will investigate and explore changes that might occur to the surface of the earth if the major ice sheets of Antarctica melted. They will begin their exploration at the South Pole by studying images and information relating to the physical geography of Antarctica. Proceeding according to the steps of the geographic inquiry method, they will consider the consequences of projected changes on human structures, both physical and political. The final assessment will call for students to create an action plan for a major city of the world that would be flooded in the event of catastrophic polar meltdown.

In the Eye of the Storm: A regional case study of Latin America and the impact of Hurricane Mitch
Students will study the destructive force of Hurricane Mitch, the deadliest storm of the twentieth century, which devastated Central America. Students will analyze information about the storm itself, compare the region before, during, and after the storm, and reflect on the impact it had on the society it ravaged. The students will end their study by developing a hurricane relief/rebuilding plan for the region, and then compare their theories with the plans that were actually made.

Data Disaster: An advanced investigation
Students will act as data detectives in this scenario-based investigation. Due to a recent storm, the main computer was flooded at the International Wildlife Conservancy (a fictitious organization). They lost all their metadata for their ecozones. The data and maps have survived, but they are unsure what each zone represents. They know that the zones represent areas that are critically challenged and are in danger. The students must analyze the data that is available to them and create a new metadata document for the data.

Water World
A global perspective

Lesson overview

Students will investigate and explore changes that might occur to the surface of the earth if the major ice sheets of Antarctica melted. They will begin their exploration at the South Pole by studying images and information relating to the physical geography of Antarctica. Proceeding according to the steps of the geographic inquiry method, they will consider the consequences of projected changes on human structures, both physical and political. The final assessment will call for students to create an action plan for a major city of the world that would be flooded in the event of catastrophic polar meltdown.

Estimated time   Two to three 45-minute class periods

Materials   ✔ Student handouts from this lesson to be copied:
- GIS Investigation sheets (pages 397 to 404)
- Student answer sheets (pages 405 to 408)
- Assessment(s) (pages 410 to 414)

Standards and objectives   *National geography standards*

	GEOGRAPHY STANDARD	MIDDLE SCHOOL	HIGH SCHOOL
1	How to use maps and other geographic representations, tools, and technologies to acquire, process, and report information from a spatial perspective	The student understands the relative advantages and disadvantages of using maps, globes, aerial and other photographs, satellite-produced images, and models to solve geographic problems.	The student understands how to use geographic representations and tools to analyze, explain, and solve geographic problems.
4	The physical and human characteristics of places	The student understands how different physical processes shape places.	The student understands the changing physical and human characteristics of places.
11	The patterns and networks of economic interdependence on Earth's surface	The student understands the basis for global interdependence and the reasons for the spatial patterns of economic activities.	The student understands the increasing economic interdependence of the world's countries.
18	How to apply geography to interpret the present and plan for the future	The student understands how the interaction of physical and human systems may shape present and future conditions on Earth.	The student understands how to use geographic knowledge, skills, and perspectives to analyze problems and make decisions.

Standards and objectives (continued)

Objectives

The student is able to:

- Manipulate map projections and orientations using GIS technology.
- Compare locations on a map to photos and satellite imagery using GIS.
- Analyze the impact on major human systems (such as transportation networks) if various parts of the Antarctic ice sheet melted causing a significant rise in sea level.
- Predict how such a catastrophe might change the nature of cities and societies around the world, and propose ways to minimize danger and hardship.

GIS skills and tools

 Add a theme

 View a hot link to a photograph, image, or text

      Zoom in and out of the view

 Use the Identify tool to learn more about a selected record

 Change the latitude and longitude coordinates for the center of the view

Find features on the map

- Change the map projection
- Open a second view

ASK geographic questions

ACQUIRE geographic resources

EXPLORE geographic data

ANALYZE geographic information

ACT on geographic knowledge

For more on geographic inquiry and these steps, see Geographic Inquiry: Thinking Geographically (pages xxi to xxiii).

Teacher notes

Lesson introduction

Begin the lesson with a discussion of Antarctica. Use these questions as a guide.

- What is the climate of Antarctica like?
- What does the place look like?
- Are there any human settlements there?

After a brief discussion introducing the subject, share with your students some of the work that scientists have been doing in the region. You may want to have them explore some of the Internet sites associated with this lesson on the book's Web site. These resources provide information on the latest research into snow and ice melt in Antarctica and its impact on mean sea level. In the GIS investigation, students will look at visual representations of the rising sea level and analyze current data about cities, transportation networks, and other important human structures. They will be challenged in the closing assessment to save a major city from the rising floodwaters by using the data they gather in the course of their investigation.

Student activity

 Before completing this lesson with students, we recommend that you complete it as well. Doing so will allow you to modify the activity to accommodate the specific needs of your students.

After the initial discussion, have your students work on the computer component of the lesson. Ideally, each student should be at an individual computer, but the lesson can be modified to accommodate a variety of instructional settings.

Distribute the GIS Investigation sheets to the students. This investigation has two main parts: observing Antarctica, and analyzing the earth at various sea levels. The investigation sheets will provide students with detailed instructions for their investigations. As they work through the steps, they will explore how the changes in sea level could affect important human structures. They will begin to make hypotheses that they will attempt to prove in their final assessment.

The investigation sheets include questions to help students focus on key concepts. Some questions will have specific answers; others require creative thought.

Things to look for while the students are working on this activity:

- Are the students using a variety of tools?
- Are the students answering the questions as they work through the procedure?
- Are the students beginning to ask their own questions of the data they are observing?

Conclusion

Before beginning the assessment, briefly discuss the initial observations your students have made. Review their findings on the areas most affected by the changing sea level. How similar or different were these observations?

Assessment

Middle school: Highlights skills appropriate to grades 5 through 8

The middle school assessment will provide students a list of cities that will be greatly affected by the 50-meter rise in sea level. Students will work in teams of three. They will select one city from the list (or you can assign each group a city) and create an action plan for relocating the city and its resources. They will focus on basic modes of transportation such as major roads and railways. Each group may also address how the change in sea level will affect industry and commerce.

Within each group, students should assign themselves to the following roles:

- Team leader—organizes the group and assists in creating the final product
- Cartographer—focuses on manipulating the GIS and printing of maps (if necessary) for the final product
- Data expert—focuses on research and determines which data is best to use

Each student group will work together as a team to create the final product. Team presentations can be made at your discretion. Some suggestions include oral presentation, a formal written report with printed maps, or a "science fair"– type poster presentation.

High School: Highlights skills appropriate to grades 9 through 12

The high school assessment will provide students a list of 10 major cities that will be greatly affected by the 50-meter rise in sea level. Students will work in teams of three. They will select one city from the list and create an action plan to relocate the city, shift the roles that city plays in the national and international scheme of things to another city, adapt the city to its new environment, or choose another option. They must take into account transportation, utilities, economics, and politics in their solutions.

Within each group, students should assign themselves to the following roles:

- Team leader—organizes the group and assists in creating the final product
- Cartographer—focuses on manipulating the GIS and printing of maps (if necessary) for the final product
- Data expert—focuses on research and determines which data is best to use

Each student group will work together as a team to create the final product. Team presentations are at your discretion. Some suggestions include oral presentation, a formal written report with printed maps, or a "science fair"–type poster presentation.

Extensions

- Have students research other details for the affected cities and countries by obtaining additional data from outside sources such as the Internet.
- Have students create action plans for an entire country by having small groups research specific details. For example, one group could focus on political boundary issues, another on transportation issues, another on export and trade. Each group would need to present its findings to the whole class (or country) and then the group must decide as a whole team how the different plans will work together.
- Identify a city along the Mississippi River that frequently floods. Research and analyze its flood disaster plans. How could those plans be improved?
- Check out the Resources by Module section of the Teacher Resource CD for books and media resources on the topic of Antarctica or visit *www.esri.com/ mappingourworld* for Internet links.

NAME _____   DATE _____

Water World
A GIS investigation

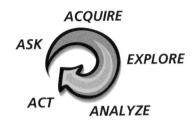

Answer all questions on the student answer sheet handout

Imagine that the year is 2100. Scientists have determined that the rapidly warming climate of the earth will cause the ice sheets of Antarctica to break apart and melt at a much faster rate than was predicted a hundred years earlier. You and your GIS investigation team are presented with the challenge of studying the impact this change will have on the planet.

Part 1: A South Pole point of view

Step 1   Start ArcView

a   Double-click the ArcView icon on your computer's desktop.

b   If the Welcome to ArcView dialog appears (pictured below), click **Open an Existing Project** and click OK. If it doesn't appear, proceed to step 2.

Step 2   Open the global7.apr file

a   In this exercise, a project file has been created for you. To open the file, go to the **File** menu and choose **Open Project**.

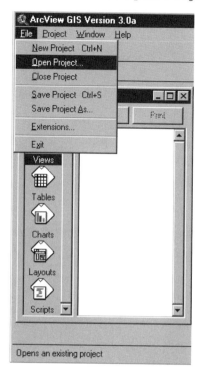

b   Navigate to the exercise data directory (**C:\esri\mapworld\mod7**) and choose **global7.apr** from the list.

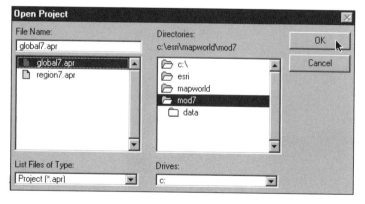

c   Click OK.

When the project opens, you see a world map in the View window with the following themes displayed: Oceans, Continents, and Latitude & Longitude. A check mark next to the theme name indicates that it is displayed.

Step 3   Look at Antarctica

a   Use the legend to locate the continent of Antarctica.

Take a look at where the melted water is coming from. Scientists believe that the first area to melt will be the Western Ice Shelf of Antarctica. The western part of Antarctica is considerably smaller than the eastern portion. It lies on the west of the Transantarctic Mountain Range, which is basically all the land to the west of the prime meridian. The prime meridian is the line that runs north–south in the view.

The map is now viewed in a geographic map projection. Because this is a flat projection of the spherical earth, some parts of the map are skewed (out of shape). This affects either size or shape of features (such as landforms), distance, or all three.

?   *Do you think this map gives you a realistic representation of Antarctica? Explain.*

b   Click the View menu and select Change Map Projection.

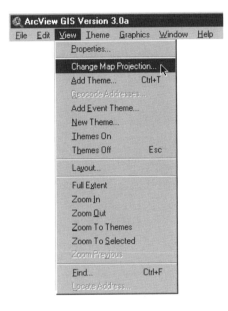

The Projection dialog appears.

c   Select Albers. Click OK.

The map in the view looks conical.

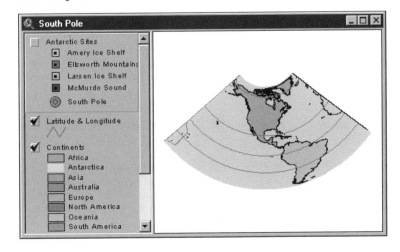

?   *Does this projection give you a better view of the region around the South Pole? Why or why not?*

d   Repeat the steps to change the map projection until you find one that provides the best perspective on Antarctica. Answer the following questions.

?   *(1)   Do any of these projections work well for viewing Antarctica?*

?   *(2)   Which projection do you think works best? Why?*

Step 4   View the South Pole

As you reviewed the various projections, you may have thought that none of them would give you a good perspective of the South Pole, or you may have wanted to flip the map upside down or change its center. Of the map projections you explored, the recommended projection to view either pole is the Orthographic projection. This is a picture of the earth as though you were looking at it from space. In order to see the South Pole, you will change the center of the map.

a   Change the map projection to Orthographic.

 b   Click the Change Center button. A dialog box appears.

c   Type in the coordinates of the South Pole.

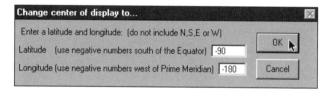

d   Click OK. The view changes and is focused on the South Pole.

Step 5   Picture Antarctica

a   Click the box next to Antarctic Sites to display the theme points.

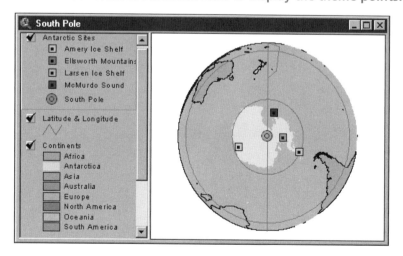

Each point on the map is hot linked to an image. Like an Internet link, if you click on a hot link in ArcView, you are taken to additional information.

b   Make sure Antarctic Sites is the active theme by clicking on the theme name.

c   Click the Hot Link tool. When you move the cursor over the view area, it turns into a lightning bolt.

d   Move the tip of the lightning bolt over the South Pole and click. A photograph of the South Pole appears.

e   If the photograph appears cut off, click its title bar and move it to where you can see the whole thing.

> Geographic South Pole

f   Click the photograph to find out more information about the picture. Another window appears with text in it. Read the information.

g   Close the text window and the photograph window. Do *not* close the View window.

h   Find out more about what Antarctica looks like by clicking on the other points in the view. Remember: You can find out more information about the images by clicking on the images themselves. You can also stretch or maximize the image windows to see the images at a larger size.

i   When you're finished looking at the hot links, close the View window.

j   Ask your teacher if you should stop here and save this ArcView project. Follow your teacher's instructions on how to rename the project and where to save it. Exit ArcView by choosing Exit from the File menu.

If you do not need to save the project and exit ArcView, proceed to part 2.

?   *On the answer sheet, write the new name you gave the project and where you saved it.*

Part 2: Just add water

Antarctica has two major ice sheets: the western and the eastern. The western sheet is smaller than the eastern, and covers Antarctica from the Transantarctic Mountains, which run from the South Pole westward. The eastern sheet is on the opposite side of the mountain range and includes the majority of the continent. Both of these enormous sheets of ice are moving from the continental center toward the ocean. For example, as the western ice sheet moves into the ocean it forms the Ross and Ronne Ice Shelves that float on top of the ocean. It is at this point where the ice begins to break apart and melt.

Now you will examine what might happen to the water levels of the oceans if part of these ice sheets were to melt.

Step 6   Open the Water World view

a   Make sure the correct ArcView project is open. If you renamed this project in part 2, open it using the name you recorded. Otherwise, continue with step 6b.

b   Click the Project Window title bar.

> global7.apr

c Click Water World and click Open.

The Water World view displays on your screen. You see a world map with the Country Outlines from the year 2000. You also see a theme named "20,000 Years Ago." This theme shows an elevation map of the earth as scientists believe it looked 20,000 years ago. At that time, sea level was 400 feet lower than it is today.

d Stretch the View window by dragging the lower right corner outward with your mouse.

e What significant differences do you see between today's country outlines and the elevation map of 20,000 years ago? List at least three.

Step 7 Analyze global sea levels if Antarctic ice sheets melted

If the western ice sheet melted, scientists predict that the oceans would rise approximately 5 meters. If the eastern ice sheet melted, sea level would rise approximately 50 meters. If all the ice at the South Pole melted, including all the ice shelves and glaciers, sea level would increase by 73 meters.

One by one, you will turn on the themes Today, Plus 5 meters, Plus 50 meters, and Antarctic Total Thaw and make observations in the table on your answer sheet. Remember: ArcView draws the themes starting with the bottom of the table of contents and moves upward. Therefore, a theme that's turned on at the top of the table of contents will "draw over" a theme below it.

a Turn the themes on and off and compare each change in sea level. Record your general observations of each theme in the table on the answer sheet.

b Turn off all themes except Plus 50 meters.

Step 8   View changes in water levels

   a   Click the Add Theme button.

b   Navigate to the module 7 world data directory (**C:\esri\mapworld\mod7\data\world**). Click **lakes.shp** and hold the Shift key down while you scroll and click **rivers.shp**. Click OK.

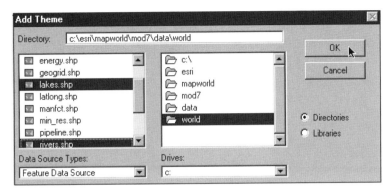

Both themes appear in the table of contents.

c   Click the box next to Lakes.shp and Rivers.shp to turn them on.

   d   Use the Zoom In tool to drag a box around South America.

(1)   What kinds of changes do you see in the rivers and lakes? Provide a specific example.

(2)   With a sea level increase of 50 meters, what kinds of consequences do you foresee for the major river ecosystems of South America? Provide a specific example.

(3)   There are several locations around the globe that are on the interior of landmasses and are below sea level. One of them is in South America. Hypothesize how these low-lying areas were formed.

   e   Click the Zoom to Full Extent button to see the entire world in the view.

f   Turn off Rivers.shp and Lakes.shp.

Step 9   View changes in political boundaries

The oceans of the world form the coastlines of many nations. In this step, you will focus on coastal boundaries and how the 50-meter rise would affect political boundaries.

   a   Click the Add Theme button.

b   Navigate to the module 7 world data directory (**C:\esri\mapworld\mod7\data\world**) and double-click **cities.shp**.

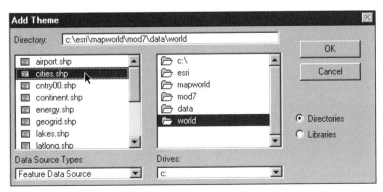

c   Click the Zoom In tool. Zoom to focus on the Middle East.

     d   Turn on Cities.shp.

The black dots that represent city locations appear in the view. Note how some of them are now in the ocean or on the edge of the ocean.

     e   Turn on Country Outlines so you can view current country boundaries.

?   *f*   Take note of significant changes in the amount of land now remaining in the countries of that region. Predict possible consequences to the societies that live in those areas (political disputes, trade and economic issues, transportation problems, and so on) and record them in the table on your answer sheet. Use the ArcView tools and buttons you've learned in this lesson to help you complete the table.

     g   Click the Zoom to Full Extent button. Repeat the process of zooming and identifying potential consequences of the rising sea level for the other major regions of the world.

?   *Record your results in the table on the answer sheet.*

     h   Click the Zoom to Full Extent button.

Step 10   Look for additional data to explore

You have now only begun to scratch the surface of understanding a world with a significant rise in sea level.

?   *Based on your previous observations, list other possible themes or data you would like to analyze to study the impact of this phenomenon. In your final assessment, you will have the opportunity to explore many other data sets; this list will help to guide you in further explorations.*

Step 11   Exit ArcView

In this exercise, you used ArcView to investigate the continent of Antarctica. You also explored and analyzed the potential effect of thawing the Antarctic Ice Sheets on the global environment.

     a   Ask your teacher for instructions on where to save this ArcView project and on how to rename the project.

     b   If you are not going to save the project, exit ArcView by choosing Exit from the File menu. When it asks if you want to save changes to global7.apr, click No.

NAME _____   DATE _____

Student answer sheet
Module 7
Human/Environment Interaction

Global perspective: Water World

Part 1: A South Pole point of view

Step 3   Look at Antarctica

a   Do you think this map gives you a realistic representation of Antarctica? Explain your answer.

_____

_____

_____

_____

_____

c   Does this projection give you a better view of the region around the South Pole? Why or why not?

_____

d-1   Do any of these projections work well for viewing Antarctica?   _____

d-2   Which projection do you think works best? Why?

_____

_____

_____

_____

_____

Step 5   Picture Antarctica

j   Write the new name you gave the project and where you saved it.

_____   _____
(Name of project.   **(Navigation path to where project is saved.**
For example: kzeregion5.apr)   **For example: C:\student\kze\mapworld)**

Part 2: Just add water

Step 6 Open the Water World view

e   What significant differences do you see between today's country outlines and the elevation map of 20,000 years ago? List at least three.

_____

_____

_____

_____

_____

Step 7 Analyze global sea levels if Antarctic ice sheets melted

a   Turn the themes on and off and compare each change in sea level. Record your general observations of each theme in the table below.

SEA LEVEL	OBSERVATIONS
Today	
Plus 5 meters	
Plus 50 meters	
Total Thaw (plus 73 meters)	

Step 8   View changes in water levels

d-1   What kinds of changes do you see in the rivers and lakes? Provide a specific example.

_____

_____

_____

_____

_____

d-2   With a sea level increase of 50 meters, what kinds of consequences do you foresee for the major river ecosystems of South America? Provide a specific example.

_____

_____

_____

_____

_____

d-3   There are several locations around the globe that are on the interior of landmasses and are below sea level. One of them is in South America. Hypothesize how these low-lying areas were formed.

_____

_____

_____

_____

_____

_____

_____

_____

_____

_____

Step 9   View changes in political boundaries

g   Record your results in the table below.

REGION	COUNTRIES/AREAS AFFECTED	POSSIBLE CONSEQUENCES
Middle East		
Asia		
Europe		
Africa		
Oceania		
North America		
Latin America		

Step 10   Look for additional data to explore

Based on your previous observations, list other possible themes or data you would like to analyze to study the effect of this phenomenon. In your final assessment, you will have the opportunity to explore many other data sets; this list will help to guide you in further explorations.

_____

_____

_____

_____

_____

_____

_____

_____

_____

NAME _____ DATE _____

Water World
Middle school assessment

You and your teammates have been selected to be part of an elite team of GIS experts who will determine the fate of one of the following six world cities. Over the next 50 years, the rise in sea level up to 50 meters will affect these cities.

List of world cities affected by 50-meter rise in sea level:

San Francisco, USA   Miami, USA

London, England   Calcutta, India

Tokyo, Japan   Houston, USA

As part of your task, select a city and develop an action plan for relocating the city and its resources. The plan must take into account the following factors:

Major roads   Ocean ports and shipping lanes

Railroads   Utilities

Airports   Relocation of people

Your Project Director (teacher) will provide you with a detailed listing of the available data you can use. You will add this data to the Water World view in the global7.apr you used in the GIS Investigation.

Your action plan must include each of the following:

- Time line—It describes the various phases of your plan. For example, one five-year phase might include relocating people, while another includes relocating specific businesses.

- Map—It displays proposed changes and could be a series of maps generated in ArcView or on paper.

- Data—It will support your suggested changes. This data will come from the GIS Investigation and can be displayed through maps, charts, or tables.

- Written report—This is written explanation of your plan and can include the time line.

Your Project Director (teacher) will provide you with instructions on how you will present your action plan to the class.

Water World

Assessment rubric

Middle school

STANDARD	EXEMPLARY	MASTERY	INTRODUCTORY	DOES NOT MEET REQUIREMENTS
The student knows and understands the relative advantages and disadvantages of using maps, globes, aerial and other photographs, satellite-produced images, and models to solve geographic problems.	Uses GIS to create a map in an appropriate projection (Ortho-graphic) to display information on the region of the South Pole. Uses GIS to create a digital map illustrating the effects of a 50-meter rise in sea level for the selected city.	Uses GIS to create a map in an appropriate projection (Ortho-graphic) to display information on the region of the South Pole. Uses a GIS to create a paper map illustrating the effects of a 50-meter rise in sea level for the selected city.	Uses GIS to create a map in an appropriate projection (Ortho-graphic) to display information on the region of the South Pole. Creates a map of the selected city, but it does not adequately illustrate the effects of the 50-meter rise in sea level.	Does not select an appropriate projection for view data on the South Pole, and cannot make distinctions between various projections. Does not include a map of the selected city.
The student knows and understands how different physical processes shape places.	Creates a clear and concise hypothesis on the impact of a 50-meter rise in sea level on a city. Provides data and maps that support ideas.	Creates a clear and concise hypothesis on the impact of a 50-meter rise in sea level on a city. Provides a map that supports ideas, but does not highlight supporting data.	Identifies some factors that will affect a city with a significant rise in sea level, but does not create a formal hypothesis. Provides some data or a basic map.	Identifies some places that will be affected by a rise in sea level, but does not identify what changes could take place.
The student knows and understands the basis for global interdependence and the reasons for the spatial patterns of economic activities.	Analyzes the effect on trade and transportation routes for a flooded city. Develops an action plan that relocates the city and provides details on plan implementation. The plan uses data and maps to support ideas and includes a time line.	Analyzes the impact on trade and transportation routes for a flooded city. Develops an action plan on how to deal with this change that includes a time line.	Attempts to analyze appropriate data and creates an outline for a plan, but does not provide detail on implementation. Time line is incomplete.	Does not select appropriate data sets (such as transportation) to analyze the issue of spatial patterns of economic activities, and therefore has difficulty creating a plan. The plan lacks a time line.
The student knows and understands how the interaction of physical and human systems may shape present and future conditions on Earth.	Develops a coherent argument that supports the relocation plan. Uses a variety of data to support the findings and creates additional data and maps that illustrate changes to the human infrastructure.	Develops a coherent argument that supports the relocation plan. Uses a variety of data to support the findings.	Creates a relocation plan for the selected city, but does not offer a variety of data to support ideas.	Relocates a selected city, but does not formalize a plan on how the change will take place. Uses little or no data to support ideas.

This is a four-point rubric based on the National Standards for Geographic Education. The "Mastery" level meets the target objective for grades 5–8.

NAME _____   DATE _____

Water World
High school assessment

You and your teammates have been selected to be part of an elite team of GIS experts who will determine the fate of one of the following 10 world cities. Over the next 50 years, the rise in sea level up to 50 meters will affect these cities.

List of World Cities affected by 50-meter rise in sea level:

San Francisco, USA	Houston, USA
London, England	Odessa, Ukraine
Tokyo, Japan	Rome, Italy
Miami, USA	Sydney, Australia
Calcutta, India	Buenos Aires, Argentina

As part of your task, select a city and develop an action plan that relocates the city, relocates the roles of the city to another city, adapts the city to its new environment, or chooses another option. The plan must take into account the following factors:

Major roads	Utilities
Railroads	Relocation of people
Airports	Economics and trade relations
Ocean ports and shipping lanes	Agriculture and manufacturing

Your Project Director (teacher) will provide you with a detailed listing of the available data you can use. You will add this data to the Water World view in the global7.apr you used in the GIS Investigation.

Your action plan must include each of the following:

- Time line—It describes the various phases of your plan. For example, one five-year phase might include relocating people, while another includes relocating specific businesses.
- Map—It displays proposed changes and could be a series of maps generated in ArcView or on paper.
- Data—It will support your suggested changes. This data will come from the GIS Investigation and can be displayed through maps, charts, or tables.
- Written report—This is written explanation of your plan and can include the time line.

Your Project Director (teacher) will provide you with instructions on how you will present your action plan to the class.

Water World

Assessment rubric

High school

STANDARD	EXEMPLARY	MASTERY	INTRODUCTORY	DOES NOT MEET REQUIREMENTS
The student knows and understands how to use geographic representations and tools to analyze, explain, and solve geographic problems.	Uses GIS to create a map in an appropriate projection (Ortho-graphic) to display information on the region of the South Pole. Uses GIS to create a digital map illustrating the effects of a 50-meter rise in sea level for the selected city.	Uses GIS to create a map in an appropriate projection (Ortho-graphic) to display information on the region of the South Pole. Uses a GIS to create a paper map illustrating the effects of a 50-meter rise in sea level for the selected city.	Uses GIS to create a map in an appropriate projection (Ortho-graphic) to display information on the region of the South Pole. Creates a map of the selected city, but it does not adequately illustrate the effects of the 50-meter rise in sea level.	Has difficulty projecting maps using GIS, and does not show an understanding of variations in projections. Does not include a map of the selected city.
The student knows and understands the changing physical and human characteristics of places.	Clearly defines how the characteristics of a city could change, should a 50-meter rise in sea level occur. Includes data from the GIS Investigation to create an original map to illustrate the changes.	Clearly defines how the characteristics of a city could change, should a 50-meter rise in sea level occur. Includes data from the GIS Investigation.	Defines how some characteristics of a city change, should a 50-meter rise in sea level occur. Provides little evidence from the data.	Attempts to define how some characteristics of a city change with a 50-meter rise in sea level. Does not provide any evidence from the data.
The student knows and understands the increasing economic interdependence of the world's countries.	Provides specific examples of how the loss of the selected city would affect the global marketplace. Provides an economic action plan to prevent this loss from occurring.	Identifies how the global economic infrastructure will be altered with a 50-meter rise in sea level, using the perspective of the selected city.	Identifies how the selected city will be affected economically, but does not make a connection to the global marketplace.	Attempts to identify how the selected city will be affected economically, but does not make a connection to the global marketplace.
The student knows and understands how to use geographic knowledge, skills, and perspectives to analyze problems and make decisions.	Develops a detailed action plan to sustain the selected city in the event of a 50-meter rise in sea level. Creates original maps with a GIS that uses a variety of data sources to support the ideas in the plan.	Develops a detailed action plan to sustain the selected city in the event of a 50-meter rise in sea level. Uses a variety of data sources to support the ideas in their plan.	Develops a plan to sustain the selected city in the event of a 50-meter rise in sea level. Provides little or no data to support ideas.	Creates an outline of a plan to sustain the selected city in the event of a 50-meter rise in sea level. Provides little or no data to support ideas.

This is a four-point rubric based on the National Standards for Geographic Education. The "Mastery" level meets the target objective for grades 9–12.

Student assessment handouts data sheet

The files below can be found in the World section of your module 7 data directory
(C:\esri\mapworld\mod7\data\world).

SHAPEFILE NAME	DESCRIPTION OF DATA
wroads.shp	Lines that represent main and smaller roads and major and minor railroads for the world.
wrldcity.shp	Points that represent cities that are major population centers of the world.
uscities.shp	Points that represent a detailed listing of U.S. cities.
airport.shp	Points representing the world's airports.
energy.shp	Points that represent major power plants for the world and includes type of energy source (atomic, thermal, and so on).
pipeline.shp	Lines that represent major oil and gas pipelines throughout the world.
manfct.shp	Points that list major manufacturing places around the world.
min_res.shp	Points of mineral resource mining sites around the world.
rivers.shp	Lines that represent the major rivers of the world.
lakes.shp	Polygons that represent the major lakes of the world.

In the Eye of the Storm
A regional case study of Latin America and the impact of Hurricane Mitch

Lesson overview

Students will study Hurricane Mitch, the deadliest storm of the twentieth century, and the havoc it wreaked on several Central American countries. They will analyze information about the storm itself, compare the region before, during, and after the storm, and consider the consequences of such a disaster for the society it ravaged. The lesson will conclude with the development of a hurricane relief and rebuilding plan for the region, and comparison of student ideas with the plans that were actually made.

Estimated time   Three 45-minute class periods

Materials   ✓ Student handouts from this lesson to be photocopied:
- GIS Investigation sheets (pages 419 to 429)
- Central America Prior to Hurricane Mitch handout (pages 432 to 433)
- Student answer sheets (pages 431 to 436)
- Assessment(s) (pages 437 to 442)

Standards and objectives

National geography standards

GEOGRAPHY STANDARD	MIDDLE SCHOOL	HIGH SCHOOL
1 How to use maps and other geographic representations, tools, and technologies to acquire, process, and report information from a spatial perspective	The student understands the characteristics, functions, and applications of maps, globes, aerial and other photographs, satellite-produced images, and models to solve geographic problems.	The student understands how to use technologies to represent and interpret Earth's physical and human systems.
7 The physical processes that shape the patterns of Earth's surface	The student understands how to predict the consequences of physical processes on Earth's surface.	The student understands the spatial variation in the consequences of physical processes across Earth's surface.
15 How physical systems affect human systems	The student understands how natural hazards affect human activities.	The student understands strategies of response to constraints and stresses placed on human systems by the physical environment, and how humans perceive and react to natural hazards.
18 How to apply geography to interpret the present and plan for the future	The student understands how to apply the geographic point of view to solve social and environmental problems by making geographically informed decisions.	The student understands how to use geographic knowledge, skills, and perspectives to analyze problems and make decisions.

Standards and objectives (continued)

Objectives

The student is able to:

- Analyze human and physical characteristics of the region of Central America.
- Follow the development and impact of Hurricane Mitch on Central American countries.
- Compare satellite imagery of Hurricane Mitch to ground tracking observations.
- Develop a disaster relief plan for the areas most heavily affected by Hurricane Mitch.

GIS skills and tools

  Add a theme

       Zoom in or out of an area

  Pan the view

  Identify the attributes of a feature

  Open an attribute table

  Sort a table field in ascending order

  Sort a table field in descending order

  Promote a selected record to the top of an attribute table

  Measure the size of a feature

- Open a second view
- Hide or show a theme's legend in the table of contents
- Use Auto-Label
- Remove labels
- Change a theme's symbol

For more on geographic inquiry and these steps, see Geographic Inquiry: Thinking Geographically (pages xxi to xxiii).

Teacher notes

Lesson introduction

First, discuss the weather hazards of your own region. Use the following questions as a guide:

- What weather hazards are specific to our hometown or region?
- When do they typically occur? (Year-round, or in a particular season?)
- What are some characteristics of these phenomena?
- How do you prepare for one? (You may want to review school procedures, or have the students share plans they have from home.)

After the discussion, tell your class that in this GIS Investigation, they will be studying the impact of Hurricane Mitch on a large part of Central America. They will explore characteristics of the storm, including how it developed, and how it changed the physical and human characteristics of the region. In the final assessment, they will be part of a special team developing an action plan for dealing with the devastation the storm caused.

Student activity

 Before completing this lesson with students, we recommend that you complete it yourself. Doing so will allow you to modify the activity to suit the specific needs of your students.

After the initial discussion, have the students work on the computer component of the lesson. Ideally, each student should be at an individual computer, but the lesson can be modified to accommodate a variety of instructional settings.

Distribute the GIS Investigation sheets to the students and the "Central America Prior to Hurricane Mitch" handout. This project has two parts: a general investigation of the region, then tracking of the storm and consideration of its impact. The GIS Investigation will provide them with detailed instructions for their ArcView exploration.

 Teacher Tip: This GIS Investigation is divided into two parts, each appropriate for a 45-minute class period. At the end of part 1, students are instructed to ask you how to rename the project and where to save it. Because this data will be valuable to the students during the assessment, be sure they record the new name and location on their answer sheet.

In part 2, the students will add image data on Hurricane Mitch and track its path over Central America. Because these image files are large, you may not want students to save individual copies of the project on a school server or computer hard drive. Your decision will depend on a combination of technical considerations and whether you want students to have access to part 2 of the project while they complete the assessment.

In addition to instructions, the handout includes questions to help students focus on key concepts. Some questions will have specific answers, while others require creative thought.

Things to look for while the students are working on this activity:

- Are the students using a variety of tools?
- Are the students answering the questions as they work through the procedure?
- Are the students beginning to ask their own questions of the data they are observing?

Conclusion

Once students have completed the GIS Investigation, have them share their findings, either in small groups or as a class. They should have a basic understanding of the region and the effect of the storm in terms of rainfall amounts, wind speeds, and so forth. Ask your students which parts of the region suffered the most damage and why. To find out more about the storm and its impact on Central America, refer to the Teacher Resource CD for print and media resources on Hurricane Mitch and the companion Web site (www.esri.com/mappingourworld) for Internet links. The USGS has an enormous amount of data on the impact of Hurricane Mitch (www.usgs.gov). Encourage students to research this topic while completing their assessment. Allow class time for each team to meet and plan how they will complete the assessment.

Assessment

The middle school and high school assessments vary in the specific tasks the students undertake, but the basic design of the assessments is the same.

Assign students to small groups of three to four students per team. Each team will have the following (some roles may be combined):

Within each group, students should assign themselves to the following roles:

- Team leader—organizes the group and takes the lead in creating the final product/presentation.
- Cartographer—creates the maps using GIS.
- Data expert—gathers data sources and determines which data is best to use.
- Multimedia specialist—responsible for creating the multimedia presentation of your choice.

Assign each team to a particular job focus: flood hazards or volcano and landslide hazards. The choices are detailed in the assessment handouts. There is a different assessment handout created for each team type. Also assign each team a particular country to focus on. That way, they will have more depth to their research. The goal for each team is to assess the possible damage and then to develop a disaster relief plan based on their findings. You may modify this assessment into a longer research project or shorten it to a two-night homework assignment.

Extensions

- Have students create a disaster relief plan for your city using a local natural hazard as the springboard. Research disasters in your city or region's past to find out how they have changed the local area.
- Watch excerpts from films such as *The Perfect Storm* or *Twister* and discuss how these fictional storms relate to those in real life.
- Conduct a book study on *Isaac's Storm,* by Erik Larson. Compare the Galveston storm of 1900 to Hurricane Mitch.
- Check out the Resources by Module section of the Teacher Resource CD for print and media resources on the topics of Hurricane Mitch, Central America, and tropical storms or visit *www.esri.com/mappingourworld* for Internet links.

NAME _____   DATE _____

In the Eye of the Storm
A GIS investigation

ACQUIRE

ASK

EXPLORE

ACT

ANALYZE

Answer all questions on the student answer sheet handout

Part 1: The calm before the storm

October 21, 1998

A tropical storm is brewing in the Atlantic Ocean. It began as a tropical wave a few weeks earlier, off the coast of western Africa. Today it is causing some rain and thunderstorms over the Caribbean. Later, the barometric pressure of the system will continue to drop and it will soon be identified as a tropical depression—the beginning of a hurricane. By the time Hurricane Mitch left the Central America region, 9,086 people were dead and 9,191 were declared missing.

Central America consists of the small chain of countries that link the North and South American continents. Explore these countries using ArcView and gather data to complete your chart titled Central America Prior to Hurricane Mitch. The data you are collecting will help you to gain an understanding of the complexity of this region's delicate infrastructure. The steps that follow will guide you in obtaining the information using GIS.

> **Step 1**   **Start ArcView**
>
> *a*   Double-click the ArcView icon on your computer's desktop.
>
>
>
> *b*   If the Welcome to ArcView dialog appears (pictured below), click **Open an Existing Project** and click OK. If it doesn't appear, proceed to step 2.
>
>

Step 2   Open the region7.apr file

a   In this exercise, a project file has been created for you. To open the file, go to the **File** menu and choose **Open Project**.

b   Navigate to the exercise data directory (**C:\esri\mapworld\mod7**) and choose **region7.apr** from the list.

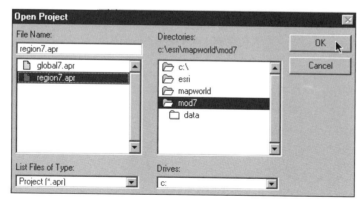

c   Click OK.

The project opens and you see a world map in the View window. A check mark next to the following themes tells you they are displayed: Ocean, Continents, and Central America.

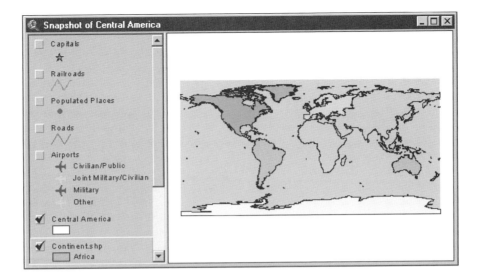

Step 3   Focus on the capital cities of Central America

a   Click on the Central America theme name to make it active. A raised box appears around the theme name.

b   Click the Zoom to Active Theme button. The view immediately centers around the countries of Central America.

Before you look at the path and effect of Hurricane Mitch, you will collect data about Central America prior to Mitch. Your teacher has given you a data sheet to record this information. The information for the country of Belize has been completed for you as an example of the data you will need to find. First, you will record the country capitals.

c   Turn on the Capitals theme by clicking the box next to the theme name. Make the theme active.

You can find out the names of the capitals by using the Identify tool or by using Auto-Label. Auto-Label is a quick way to get information about a group of features.

d   From the Theme menu, click Auto-Label. The Auto-Label dialog displays. Notice that Name is already chosen as the field to use for labeling.

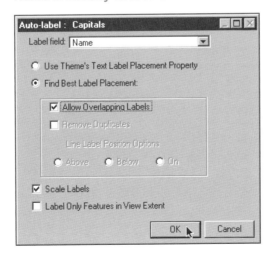

e   Click the box to Allow Overlapping Labels and click OK. The name of each capital city displays in the view. San Salvador appears in a different color because it is an overlapping label.

?   *f*   Use this information to record the capital cities for each of the countries in Central America. Record them in the Populated Places column.

> *Note: If you don't know the name of a country, use the Identify tool.*

g   When you have finished recording the names of each capital, go to the Theme menu and click Remove Labels. The text labels disappear.

h   Close the Identify Results window if it is open.

Step 4   **Focus on Central America prior to Hurricane Mitch**

There are four themes in the table of contents that are not turned on. Each of these themes provides important data about the transportation network in Central America. This table is a summary of the themes.

Populated Places	Points showing major cities and populated areas of Central America
Roads	Lines representing Central American roads and trails
Railroads	Lines representing Central American railroads
Airports	Points showing airport location and type (civilian, military/civilian, military, other)

a   Turn these themes on one at a time to obtain important data for your handout.

> *Remember: Check the order of the themes in the table of contents. Themes that are at the top of the table of contents will cover up themes that are listed lower. Turn on each theme individually to see it clearly. When using the Identify tool, make sure the appropriate theme is active.*

?   *b*   Record this data in the Populated Places and Transportation Network columns on your answer sheet.

Now you will add more data needed to complete your handout.

c   Click the Add Theme button.

d   Navigate to the module 7 data directory (**C:\esri\mapworld\mod7\data\camerica**).

e   Hold down the Shift key and click the following four themes:
- agruse.shp
- coastal.shp
- landfrm.shp
- precip.shp

f   Click OK. All of the themes display in the table of contents.

The table below summarizes the data in each theme:

agruse.shp	Polygons showing agricultural use of land in Central America
coastal.shp	Points representing coastal features of Central America
landfrm.shp	Points representing mountains, mountain chains, and volcanoes in Central America
precip.shp	Polygons showing annual precipitation in millimeters (mm) for Central America

g   Turn on Precip.shp.

The precipitation data appears in the map, but you cannot determine the average precipitation for each individual country. You will move the Central America theme and change its legend so you can view each country's average precipitation data.

h   Click and drag the Central America theme to the top of the table of contents. It covers up Precip.shp.

i   Double-click Central America to open its Legend Editor.

j   Double-click the yellow symbol box. The Fill Palette displays.

k   Click the first box in the top row.

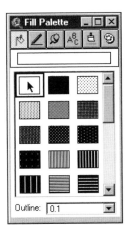

l   Click Apply. Close the Fill Palette and the Legend Editor.

The view now displays Central American countries as outlines.

?   *m*   Analyze the precipitation for each country and record the precipitation pattern in the Average Precipitation column on your answer sheet.

n   Turn off Precip.shp.

o   Turn on each of the remaining themes individually. Use the available data to complete the remaining columns in your table.

> *Remember: Check the order of the themes in the table of contents. Themes that are represented by lines and points (streets, rivers, cities) will be covered up by themes represented by polygons (countries, states). You may need to drag line or point themes above the polygon themes in order to see them. When using the Identify tool, make sure the appropriate theme is active.*

p After you complete the data table, close the Identify Results window and close the Snapshot of Central America view.

   (1) Which country has the most area devoted to agriculture?

   (2) Which country has the most area covered by mountains?

   (3) Which country has the most extensive transportation network?

Step 5 Save the project and exit ArcView

In the first part of the GIS Investigation, you analyzed and recorded information about Central America from before Hurricane Mitch.

a Ask your teacher for instructions on where to save this ArcView project and on how to rename the project. If you are not going to save the project, proceed to step 5b now. Otherwise, record the project's new name and where you saved it on your answer sheet.

b From the File menu, click Exit.

c Click No when asked if you want to save changes.

Part 2: The storm

October 24–26, 1998

In a span of less than two days, Tropical Storm Mitch develops into a category 5 hurricane with winds in excess of 155 mph. Category 5 is the deadliest rating on the Saffir-Simpson Hurricane Potential Damage Scale. Barometric pressure drops to 905 millibars, the lowest pressure ever observed in the Atlantic basin.

Step 1 Open the project

a Double-click the ArcView icon on your computer's desktop.

b Refer to your answer sheet and determine where region7.apr is saved and how you renamed it.

c Navigate to the location of the saved project and open it. If you didn't rename or save the project in part 1, navigate to the module 7 data directory (**C:\esri\mapworld\mod7**) and select **region7.apr**.

Step 2 Track Hurricane Mitch

a Click the title bar of the Project window. Note: The title bar will show the new name of the project if you changed it in part 1.

b   **Click Hurricane Mitch to highlight it. Click Open.**

The Hurricane Mitch view displays with the following themes turned on: Latitude & Longitude, Central America, Continents, and Oceans.

   c   **Turn on Pre-Hurricane Mitch and Mitch2.shp.**

Both of these themes, and similarly named ones to follow, will show the location of the center or eye of the storm. At locations where it was declared a hurricane, the legend reflects the category number in the point. You will now explore the data that these themes represent.

   d   **Make Pre-Hurricane Mitch the active theme.**

   e   **Click the Identify tool. Click the most southeastern dark square that is a tropical storm. An Identify Results window displays with information about Tropical Storm Mitch.**

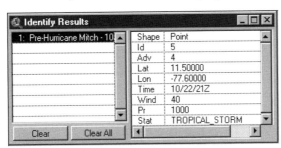

   f   **Answer the following questions:**

       (1)   *At what time was Tropical Storm Mitch at this location?*

       (2)   *What does the "Z" mean in the time?*

       (3)   *What was Mitch's wind speed at this location?*

   g   **Make Mitch2.shp the active theme. Click the location of Hurricane Mitch, category 1 in the view. The Identify Results window updates.**

   h   **Answer the following questions:**

       (1)   *What are the latitude and longitude coordinates for Hurricane Mitch at this location?*

       (2)   *At what time was Hurricane Mitch at this location?*

       (3)   *What was Mitch's wind speed at this location?*

   i   **Click the last mapped location for Hurricane Mitch-5 before it struck land. The Identify Results window updates.**

   j   **Answer the following questions.**

       (1)   *At what time was Hurricane Mitch at this location?*

       (2)   *What was Mitch's wind speed at this location?*

Now you will determine how long it took Mitch to develop from a tropical storm to a category 5 hurricane.

   k   **Close the Identify Results window. Make sure Mitch2.shp is the active theme.**

   l   **Click the Open Theme Table button. The Attributes of Mitch2.shp table opens and part of the View window is covered.**

m   Scroll right in the table until you see a field named Stat. Click the name Stat. It turns a darker shade of gray.

n   Click the Sort Descending button.

o   Click the first entry (Tropical Storm). Scroll down and locate the entry representing the first time Hurricane-5 hit land. Hold the Shift key down and click the Hurricane-5 entry. Both entries are highlighted yellow in the table and in the view.

p   Click the Promote button. Both entries appear at the top of the table where they are easier to compare.

q   Use the information in the attribute table to determine how long it took for Tropical Storm Mitch to become a category 5 hurricane. On the answer sheet, write down the times for each event in the appropriate spot and determine the difference. The time is written in this format: Month/Day/Hour (of 24).

r   Examine the attribute table further and identify the maximum wind speed. Record this on your answer sheet.

s   Find the point where Mitch made landfall. Answer the following questions:

     (1)   *What was the wind speed at this location?*

     (2)   *Did the speed of the wind increase or decrease when the hurricane hit land?*

     (3)   *Why do you think the wind speed changed?*

t   Close the Attributes of Mitch2.shp table.

Step 3   Measure the size of the storm

The National Oceanic and Atmospheric Administration (NOAA), in partnership with the National Aeronautics and Space Administration (NASA), used special storm-tracking satellites to take several high-resolution photographs of Mitch from space. You will view these images and measure the massive size of this storm.

a   Click the Add Theme button and navigate to the module 7 data directory (**C:\esri\mapworld\mod7\data\camerica**).

b   Select Image Source Data from the Data Source Types drop-down menu. Click **mitch2sat.tif** and click OK.

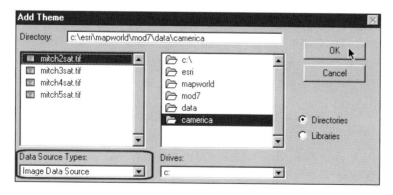

     Note: Files that end in .tif and .jpg are image files and will always be added as image data sources.

c   Turn on Mitch2sat.tif.

The satellite image now sits on top of your basemap in the Hurricane Mitch view. Unfortunately, you cannot see the track of the storm now. This can be fixed easily by rearranging the themes in the table of contents.

d   Click and drag the theme Mitch2.shp so it's at the top of the table of contents.

Now you can see the highlighted points of Tropical Storm and Hurricane-5. Hurricane-5 is almost directly over the eye of the storm. The eye is the center of the cloud mass that looks like a donut hole.

 You will use the Measure tool to measure several parts of the storm.

e   Turn off Mitch2.shp so you have a better view of the eye of Hurricane Mitch.

 f   Use the Zoom In tool and Pan tool so the satellite image fills the view. Do not zoom in too close or it will be difficult to view the image.

 g   Click the Measure tool. Your cursor turns into a right-angle ruler with cross hairs.

h   Click the left edge of the eye once, hold the cursor down, and move it directly across the diameter of the eye. Double-click when your cursor is at the right edge of the eye.

> *Note: If you accidentally clicked the wrong spot, you can double-click to end the line and simply start over.*

A segment length appears on the bottom left of the View window.

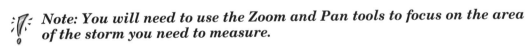
Segment Length: 24.94 mi   Length: 24.94 mi

    What is the diameter of the eye of Hurricane Mitch?

 i   Use the Measure tool to measure the total diameter of the storm at its widest point and the distance of the eye to the coastline of Honduras. Record your answers for Mitch2sat.shp in the table on the answer sheet.

> *Note: You will need to use the Zoom and Pan tools to focus on the area of the storm you need to measure.*

Now that you have recorded data for the Mitch2sat.shp, you will follow the same procedures for adding new themes and measuring the storm at other locations.

 j   Click the Add Theme button. Navigate to the module 7 data directory (**C:\esri \mapworld\mod7\data\camerica**).

k   Add the following Image data sources: **Mitch3sat.tif**, **Mitch4sat.tif**, and **Mitch5sat.tif**.

l   Hold the Shift key down and click the following Feature data sources: **Mitch3.shp**, **Mitch4.shp**, and **Mitch5.shp**.

With six new themes added, it's important to organize your table of contents so you can view the themes easily.

m   Change the theme order so that Mitch3.shp is directly above Mitch3sat.tif. Do this for all the themes.

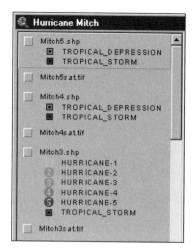

n   Turn each set of themes on and off individually. Use the Zoom and Pan tools to see different parts of each theme. Use the Measure tool to collect data on the size of Hurricane Mitch at these different locations.

> *Note: If you don't remember how to use the Measure tool, refer to steps 3e–3i.*

o   Record your measurements in the table on the answer sheet.

Step 4   Analyze rainfall from Hurricane Mitch

Once Hurricane Mitch made landfall, the winds weakened to a point where it was downgraded to a tropical storm. Nonetheless, Mitch still had not shown its worst side. In the days that followed, Mitch dumped over 30 inches of rain in the region. You will take a closer look at the precipitation that fell in the region on October 29 and 30, 1998.

a   Turn off all themes except Mitch3sat.tif, Central America, Continents, and Ocean.

b   Click the Add Theme button. Add the following themes to your view: **rain3.shp**, **rain4.shp**, and **rain5.shp**.

c   Turn on Rain3.shp. The rain pattern appears overlaid on top of Mitch3sat.tif. Answer the following questions:

   (1)   What pattern do you notice in the amount of rainfall within the storm?

   (2)   Is this a pattern you expected to find? Why or why not?

d Rearrange the order of the themes so that the three rain themes are above their corresponding satellite images.

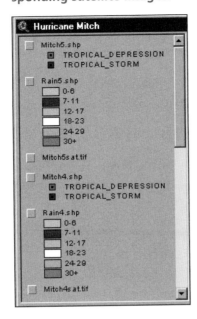

e While turning themes on and off, look at each set of rain and satellite themes. Answer the following questions:

(1) At the Mitch4 location, what was the highest range of rainfall measured?

(2) Which country received the majority of this heavy rain?

(3) Describe the difference between the rainfall patterns on October 30 and October 31, 1998.

(4) What kind of damage do you expect to find with this type of storm? What aspects of the region will be most affected? Elaborate on your answer using your table, Central America Prior to Hurricane Mitch, as a resource.

Step 5 **Save the project and exit ArcView**

In this GIS Investigation, you used ArcView to analyze a large region of Central America and to track Hurricane Mitch as it approached and made landfall. You more than likely have many questions as to the extent of the damage Mitch caused. The assessment will have you take on the role of emergency management personnel. Your job will be to identify those areas where danger is still high as a result of the storm, and to develop emergency action plans for affected Central American countries. You will use the data from this investigation to help you in the assessment.

a Ask your teacher for instructions on where to save this ArcView project and on how to rename the project. If you already renamed the project in part 1, save it under that name.

b Record the new name of the project and its location on your answer sheet.

c Choose Exit from the File menu. When it asks if you want to save changes to region7.apr, click No.

NAME _____   DATE _____

Student answer sheet

Module 7
Human/Environment Interaction

Regional case study: In the Eye of the Storm

Part 1: The calm before the storm

Step 3   Focus on the capital cities of Central America

f   Record the capital cities for each of the countries in Central America in the Populated Places column in the table on the next two pages.

Step 4   Focus on Central America prior to Hurricane Mitch

b   Record this data in the Populated Places and Transportation Network columns in the table on the next two pages.

m   Analyze the precipitation for each country and record the precipitation pattern data in the Average Precipitation column in the table on the next two pages.

p-1   Which country has the most area devoted to agriculture? _____

p-2   Which country has the most area covered by mountains? _____

p-3   Which country has the most extensive transportation network? _____

Step 5   Save the project and exit ArcView

a   Write the new name you gave the project and where you saved it.

_____   _____
(Name of project.   **(Navigation path to where project is saved.**
For example: kzeregion5.apr)   **For example: C:\student\kze\mapworld)**

PHOTOCOPY

Central America Prior to Hurricane Mitch

COUNTRY	POPULATED PLACES	TRANSPORTATION NETWORK	MAJOR EXPORTS
Belize	Capital: Belmopan Distribution: Throughout the country, but concentrated around the capital	Roads: Sparse road network Railways: none Airports: 1 civilian	Bananas
Guatemala	Capital: Distribution:	Roads: Railways: Airports:	Coffee
Honduras	Capital: Distribution:	Roads: Railways: Airports:	Coffee
El Salvador	Capital: Distribution:	Roads: Railways: Airports:	Coffee
Nicaragua	Capital: Distribution:	Roads: Railways: Airports:	Coffee
Costa Rica	Capital: Distribution:	Roads: Railways: Airports:	Garments
Panama	Capital: Distribution:	Roads: Railways: Airports:	Bananas

MODULE 7 • HUMAN/ENVIRONMENT INTERACTION

Central America Prior to Hurricane Mitch (continued)

COUNTRY	LAND USE	AVERAGE PRECIPITATION	PHYSICAL LANDMARKS
Belize	Primarily forest with some irrigated land and little cropland	Primarily 2,000–2,800 mm	Maya Mountains
Guatemala			
Honduras			
El Salvador			
Nicaragua			
Costa Rica			
Panama			

Part 2: The storm

Step 2 Track Hurricane Mitch

f-1   At what time was Tropical Storm Mitch at this location? _____

f-2   What does the "Z" mean in the time?

_____

f-3   What was Mitch's wind speed at this location? _____

h-1   What are the latitude and longitude coordinates for Hurricane Mitch at this location?

_____

h-2   At what time was Hurricane Mitch at this location? _____

h-3   What was Mitch's wind speed at this location? _____

j-1   At what time was Hurricane Mitch at this location? _____

j-2   What was Mitch's wind speed at this location? _____

q   Write down the times for each event in the appropriate spot and determine the difference. The time is written in this format: Month/Day/Hour (of 24).

Hurricane - 5: _____

Tropical_Storm: _____

Time Difference: _____

r   Examine the attribute table further and identify the maximum wind speed. _____

s-1   What was the wind speed at this location? _____

s-2   Did the speed of the wind increase or decrease when the hurricane hit land? _____

s-3   Why do you think the wind speed changed?

_____

_____

_____

Step 3 Measure the size of the storm

h   What is the diameter of the eye of Hurricane Mitch? _____

i, o

	DIAMETER OF EYE	DIAMETER OF STORM	DISTANCE OF EYE TO COASTLINE OF HONDURAS	HOW HAS THE STORM CHANGED FROM THE PREVIOUS IMAGE?
Mitch2sat.tif				
Mitch3sat.tif				
Mitch4sat.tif				
Mitch5sat.tif				

Step 4   Analyze rainfall from Hurricane Mitch

c-1   What pattern do you notice in the amount of rainfall within the storm?

_____

c-2   Is this a pattern you expected to find? Why or why not?

_____

e-1   At the Mitch4 location, what was the highest range of rainfall measured?

_____

e-2   Which country received the majority of this heavy rain?

_____

e-3   Describe the difference between the rainfall patterns on October 30 and October 31, 1998.

_____

_____

_____

e-4   What kind of damage do you expect to find with this type of storm? What aspects of the region will be most affected? Elaborate on your answer using your table, Central America Prior to Hurricane Mitch as a resource.

_____

_____

_____

Step 5   Save the project and exit ArcView

b   Write the new name you gave the project and where you saved it.

_____   _____
(Name of project.   **(Navigation path to where project is saved.**
For example: kzeregion7.apr)   **For example: C:\student\kze\mapworld)**

NAME _____   DATE _____

Volcano and Landslide Hazard Team
Middle school assessment

Your assigned country is: _____

The focus of this team is potential hazards of landslides and debris flows from volcanoes in the region devastated by the intense rainfall of Hurricane Mitch. Begin by studying the volcanoes in your assigned country and the amounts of rainfall they received from the storm. Compare this rainfall to the typical precipitation received in that area. Use the ArcView project and add data from the module 7 data folder (cautily.shp) to help you with your presentation. This additional data provides information on utility lines in Central America.

In your report/presentation, you need to do the following:

- Predict which towns are in the most danger from flooding.
- Create an emergency action plan for these towns, including:
  - An evacuation plan in case of major landslides/debris flows
  - Ways to reroute power
  - A plan to provide medical and humanitarian aid to the affected areas
  - Alternative routes to transportation networks that will be affected
- Identify agricultural areas that will be damaged (if any).

NAME _____   DATE _____

Flood Hazard Team
Middle school assessment

Your assigned country is: _____

The focus of this team is damage from rising floodwaters caused by Hurricane Mitch. It will be helpful to look at the rainfall themes from Hurricane Mitch, typical precipitation patterns for the region, and data from the module 7 data folder (cadrain.shp and cautily.shp). This additional data provides information on drainage patterns of creeks and utility lines in Central America.

In your report/presentation, you need to do the following:
*   Predict which towns are in the most danger from flooding.
*   Create an emergency action plan for these towns, including:
    *   An evacuation plan in case of major landslides/debris flows
    *   Ways to reroute power
    *   A plan to provide medical and humanitarian aid to the affected areas
    *   Alternative routes to transportation networks that will be affected
*   Identify agricultural areas that will be damaged (if any).

In the Eye of the Storm

Assessment rubric

Middle school

STANDARD	EXEMPLARY	MASTERY	INTRODUCTORY	DOES NOT MEET REQUIREMENTS
The student knows and understands the relative advantages and disadvantages of using maps, globes, aerial and other photographs, satellite-produced images, and models to solve geographic problems.	Uses GIS to gather a variety of data about Hurricane Mitch including new data from outside sources. Analyzes this information to determine cities at greatest risk for particular hazards.	Uses GIS to gather a variety of data about Hurricane Mitch and analyzes this information to determine cities at greatest risk for particular hazards.	Uses GIS to gather data about Hurricane Mitch and attempts to determine cities at risk for particular hazards. Because their selection of data is limited, some predictions may not be accurate.	Only uses data provided from the original GIS project and does not make accurate predictions.
The student knows and understands how to predict the consequences of physical processes on Earth's surface.	Accurately predicts the impact of Hurricane Mitch on the physical environment of their assigned Central American country. Creates a map of the affected areas using a variety of data.	Accurately predicts the impact of Hurricane Mitch on the physical environment of their assigned Central American country. Provides ample data to support their predictions.	Attempts to predict the impact of Hurricane Mitch on the physical environment of their assigned Central American country. Provides some data to support their predictions.	Is not able to identify major effects of Hurricane Mitch on the physical environment of their assigned Central American country.
The student knows and understands how natural hazards affect human activities.	Accurately predicts the impact of Hurricane Mitch on human activities in their assigned Central American country. Creates a map of the affected areas using a variety of data.	Accurately predicts the impact of Hurricane Mitch on human activities in their assigned Central American country. Provides ample data to support their predictions.	Attempts to predict the impact of Hurricane Mitch on human activities in their assigned Central American country. Provides some data to support their predictions.	Is not able to identify the effects of Hurricane Mitch on human activities in their assigned Central American country.
The student knows and understands how to apply the geographic point of view to solve social and environmental problems by making geographically informed decisions.	Creates an emergency action plan for their country that takes into account infrastructure and environmental changes. Uses a variety of data to support the ideas in their plan. Creates a map with evacuation routes and other important factors.	Creates an emergency action plan for their country that takes into account infrastructure and environmental changes. Uses a variety of data to support the ideas in their plan, including an evacuation route.	Creates an emergency action plan, but does not take into account both infrastructure and environmental changes—focuses only on one. Uses data to support the ideas in their plan, but is missing some important factors.	Creates an outline for an emergency action plan, but does not take into account both infrastructure and environmental changes—focuses only on one. Provides little or no data to support their plan.

This is a four-point rubric based on the National Standards for Geographic Education. The "Mastery" level meets the target objective for grades 5–8.

NAME _____   DATE _____

Volcano and Landslide Hazard Team
High school assessment

Your assigned country is: _____

The focus of this team is on the potential hazards of landslides and debris flows from volcanoes in the region devastated by the intense rainfall of Hurricane Mitch. Begin by studying the volcanoes in your assigned country and the amounts of rainfall they received from the storm. Compare this rainfall to the typical precipitation received in that area. Use the ArcView project and add data from the module 7 data folder (cautily.shp) to help you with your presentation. This additional data provides information on utility lines in Central America.

In your report/presentation, you need to do the following:

- Predict which towns are in the most danger from debris flow and landslide threat.
- Create an emergency action plan for these towns, including:
  - An evacuation plan in case of major landslides/debris flows
  - Ways to reroute power
  - A plan to provide medical and humanitarian aid to the affected areas
  - Alternative routes to transportation networks that will be affected
- Identify agricultural areas that will be damaged (if any) and discuss how this would hinder the economy of the country.

NAME _____   DATE _____

Flood Hazard Team
High school assessment

Your assigned country is: _____

The focus of this team is damage from rising floodwaters caused by Hurricane Mitch. It will be helpful to look at the rainfall themes from Hurricane Mitch, typical precipitation patterns for the region, and data from the module 7 data folder (cadrain.shp and cautily.shp). This additional data provides information on drainage patterns of creeks and utility lines in Central America.

In your report/presentation, you need to do the following:
- Predict which towns are in the most danger from flooding.
- Create an emergency action plan for these towns, including:
  - An evacuation plan in case of major landslides/debris flows
  - Ways to reroute power
  - A plan to provide medical and humanitarian aid to the affected areas
  - Alternative routes to transportation networks that will be affected
- Identify agricultural areas that will be damaged (if any) and discuss how this would hinder the economy of the country.

In the Eye of the Storm

Assessment rubric

High school

STANDARD	EXEMPLARY	MASTERY	INTRODUCTORY	DOES NOT MEET REQUIREMENTS
The student knows and understands how to use technologies to represent and interpret Earth's physical and human systems.	Uses GIS to analyze a variety of data from satellite imagery to social infrastructure themes to determine what areas are at greatest risk from Hurricane Mitch. In addition, the students import data from outside sources.	Uses GIS to analyze a variety of data from satellite imagery to social infrastructure themes to determine what areas are at greatest risk from Hurricane Mitch.	Uses GIS to analyze data to determine what areas are at greatest risk from Hurricane Mitch. Because their selection of data is limited, some predictions may not be accurate.	Only uses data provided from the original GIS project and does not make accurate predictions.
The student knows and understands the spatial variation in the consequences of physical processes across Earth's surface.	Analyzes and makes accurate predictions on the impact of Hurricane Mitch on the physical environment of their assigned country. Uses GIS to create a map that details how the storm affects the region.	Analyzes and makes accurate predictions on the impact of Hurricane Mitch on the physical environment of their assigned country. Provides details on how the storm affects the region.	Reviews data and attempts to make predictions about the storm impact on the physical environment of their assigned country.	Makes inaccurate predictions on the storm impact on the physical environment of their assigned country.
The student knows and understands strategies to respond to constraints placed on human systems by the physical environment and how humans perceive and react to natural hazards.	Analyzes and makes accurate predictions on the impact of Hurricane Mitch on human systems of their assigned country. Uses GIS to create a map that details how the storm affects the region.	Analyzes and makes accurate predictions on the impact of Hurricane Mitch on human systems of their assigned country. Provides details on how the storm affects the region.	Reviews data and attempts to make predictions about the storm impact on human systems of their assigned country.	Makes inaccurate predictions on the storm impact on human systems of their assigned country.
The student knows and understands how to use geographic knowledge, skills, and perspectives to analyze problems and make decisions.	Creates an emergency action plan for their country that takes into account infrastructure and environmental changes. Uses a variety of data to support the ideas in their plan. Creates a map with evacuation routes and other important factors.	Creates an emergency action plan for their country that takes into account infrastructure and environmental changes. Uses a variety of data to support the ideas in their plan, including an evacuation route.	Creates an emergency action plan, but does not take into account both infrastructure and environment changes—focuses only on one. Uses data to support the ideas in their plan, but is missing some important factors.	Creates an outline for an emergency action plan, but does not take into account both infrastructure and environmental changes—focuses only on one. Provides little or no data to support their plan.

This is a four-point rubric based on the National Standards for Geographic Education. The "Mastery" level meets the target objective for grades 9–12.

Data Disaster
An advanced investigation

Lesson overview

Students will act as data detectives in this scenario-based investigation. Due to a recent storm, the main computer was flooded at the International Wildlife Conservancy (a fictitious organization). All documentation for their ecozone maps was lost. The data and maps have survived, but staff members are unsure what each field in the database represents. What they do know is that the maps show areas in critical situations, and even in some cases in outright danger. The students must analyze the data that is available to them and create a new metadata document for the data.

Estimated time   Two 45-minute class periods

Materials   ✔ Student handouts from this lesson to be copied:
  - GIS Investigation sheets (pages 447 to 452)
  - Student answer sheet (pages 453 to 456)

Standards and objectives

National geography standards

GEOGRAPHY STANDARD	MIDDLE SCHOOL	HIGH SCHOOL
1 How to use maps and other geographic representations, tools, and technologies to acquire, process, and report information from a spatial perspective	The student understands how to make and use maps, globes, graphs, charts, models, and databases to analyze spatial distributions and patterns.	The student understands how to use maps and other graphic representations to depict geographic problems.
5 That people create regions to interpret Earth's complexity	The student understands the connections among regions.	The student understands how multiple criteria can be used to define a region.
8 The characteristics and spatial distribution of ecosystems on Earth's surface	The student understands the local and global patterns of ecosystems.	The student understands the distribution and characteristics of ecosystems.

Objectives

The student is able to:
  - Analyze and interpret various kinds of ecoregion data.
  - Understand the importance of metadata in using spatial data.
  - Create a useful metadata document for a data set.

GIS skills and tools
- Change map symbolization with the Legend Editor
- Create variations of a particular theme using different sets of attributes from the theme table
- Determine the best use of color for accurately depicting spatial information
- Build a new project
- Understand how to read and write a simple metadata document

For more on geographic inquiry and these steps, see Geographic Inquiry: Thinking Geographically (pages xxi to xxiii).

Teacher notes

Lesson introduction

Begin the lesson with a discussion of what metadata is. Metadata is data about data—it is information that describes the content and quality of GIS data. Metadata includes information about who created the data, when it was created, what resources were used, how it was made, and also defines various fields in an attribute table. Without metadata, tables can end up as meaningless sets of numbers and words.

Student activity

 Before completing this lesson with students, we recommend that you complete it as well. Doing so will allow you to modify the activity to accommodate the specific needs of your students. The lesson is designed for students working individually at the computer, but it can be modified to accommodate a variety of instructional settings.

After the initial discussion, distribute and explain the Data Disaster GIS Investigation. In the GIS Investigation, students will act as data detectives. They will ultimately create a metadata file for an ecoregions theme. They will also build this ArcView project from the ground up, beginning with a new view.

 Teacher Tip: Make sure you advise students on how to name and where to save their ArcView project. Be sure to have instructions ready for the students if you would like them to save their projects.

Steps 5–7 can be done as part of the assessment. They include instructions on how to look at a sample metadata document and how to write a simple one.

Things to look for while the students are working on this activity:

- Are students selecting appropriate colors for the data they are trying to map?
- Are they documenting their hypotheses about the data as they create the various maps?
- When creating the metadata document, are they referring back to the sample metadata for guidance?

Conclusion

Each student will have completed the holes in the data definition tables provided in the handout. The GIS Investigation reviews sample metadata documentation. You should be present to answer any questions students might have. Show them how to access the metadata for the data included with this book. (Scroll through a module data directory and click on a file that ends in .htm. These are metadata documents.) Visit *www.esri.com/mappingourworld* to access the Federal Geographic Data Committee (FGDC) Web site.

Assessment

Middle school: Highlights skills appropriate to grades 5 through 8

In the middle school assessment, students will create a metadata set based on their findings in the GIS Investigation. Items that they should include are:

- Identification—identifies where the data came from, who created it, and what the data represents.
- Attribute information—defines each of the fields in the attribute table.
- Metadata reference—identifies who created the metadata and contact information.

High school: Highlights skills appropriate to grades 9 through 12

In the high school assessment, students will create a metadata set based on their findings in the GIS Investigation. Items that they should include are:

- Identification—identifies where the data came from, who created it, and what the data represents.
- Attribute information—defines each of the fields in the attribute table.
- Spatial data organization—identifies the type of data and the scale.
- Spatial reference information—identifies coordinate system and coverage information.
- Metadata reference—identifies who created the metadata and contact information.

Extensions

- Have students create their own attribute tables and corresponding metadata using school information or by conducting a local survey, and so forth.
- Students can create their own data disasters and try to stump the class in trying to define other mystery data sets.
- View metadata documents for other data included in this book. Have students determine basic information about the data.
- Check out the Resources by Module section of the Teacher Resource CD for print and media resources on the topic of metadata or visit *www.esri.com/ mappingourworld* for Internet links.

NAME _____ DATE _____

Data Disaster
An advanced investigation

ACQUIRE

ASK

EXPLORE

ACT

ANALYZE

Answer all questions on the student answer sheet handout

International Wildlife Conservatory (a fictitious organization) headquarters have been flooded. All of the metadata for their ecosystems data has been destroyed. This metadata contained detailed information describing the ecosystems database. It included descriptions of the fields of the database, the names of types of data, and definitions for the data.

The IWC has created a list of endangered spaces known as Earth 200. According to the IWC, Earth 200 represents a list of the world's most unique and diverse environments. If any of these areas were lost or destroyed, the impact the planet's ecosystem would be felt at a global level.

The data set that you must investigate contains important information about these regions and identifies which are most threatened. The problem is that without the metadata, no one knows which areas are under the greatest threat. Your job is to review and map the data, rebuild the metadata, and use it to determine which regions are at a critical stage.

Step 1 **Start ArcView**

a Double-click the ArcView icon on your computer's desktop.

b If the Welcome to ArcView dialog appears, click **Create a new project with a new View** and click OK. Otherwise, proceed to step 1c.

In this exercise, you will create your own project from scratch.

c From the File menu, choose New Project. Click the New button. A new blank view opens.

 d Click the Add theme button.

e Navigate to the module 7 data directory (**C:\esri\mapworld\mod7\data\world**) and add the following themes:

world30.shp—ocean background with 30-degree grid

continent.shp—continents of the world

f Turn on both themes. Stretch your View window to make it larger.

ArcView randomly assigns a symbol color to each added theme. You will change these colors using the Legend Editor and then change the theme name.

g Double-click the world30.shp theme to open its Legend Editor.

h Change the symbol color to a shade of blue.

i   Click the Theme Properties button.

j   Change the theme name to **Oceans**.

k   Make continent.shp the active theme and move it to the top of the table of contents.

l   Change the symbol color for continent.shp to one of your choice.

 m   Click the Theme Properties button.

n   Change the name of the theme to **Continents**.

Step 2   Add the mystery data

 a   Click the Add Theme button and add **wwf_eco.shp** to the view.

The theme may load with a complex legend with many colors.

b   If it loads with many colors, proceed to step 2g. Otherwise, follow the steps below to load the correct legend file.

c   Double-click the wwf_eco.shp theme name to open the Legend Editor.

d   Click the Load button.

e   Navigate to the module 7 data directory (**C:\esri\mapworld\mod7\data\world**) and select **wwf_eco.avl**. The Load Legend dialog appears.

f   Make sure the All box is checked and click OK. Click Apply.

g   Turn on wwf_eco.shp. It now displays a detailed legend.

h   Widen the table of contents so you can read the legends and enlarge the view so you can see the map more easily.

 What do you think this map is showing?

Step 3   Evaluate the attribute table and map two themes

The theme legend that you just mapped is created from the attribute table that is linked to the wwf_eco.shp file. Take a closer look at the information in the table to begin to solve the mystery of the data.

a   Make wwf_eco.shp the active theme.

 b   Click the Open Theme Table button.

The Attributes of wwf_eco.shp table displays. There are six fields in the table. Write each field name and the type of data it contains (text, numeric, or string) in the table on the answer sheet.

c   Look at the data in the table and compare it to the Wwf_eco.shp theme legend.

Which field does this theme map?

In order to evaluate all of the data, you will map each field, name each theme to match the field being mapped, and compare it to the others.

d   Click the View1 title bar to make the view active.

 e   Click the Theme Properties button. Rename the Wwf_eco.shp theme to **Mht_name**.

Now you will remember that the data mapped in this theme matches the field Mht_name in the attribute table. In order to map the other fields, you will make copies of this theme and change the legend appropriately.

f   Turn off Mht_ name theme and make sure it is still active. From the Edit menu, select Copy Themes.

g   Click the Edit menu again and select Paste. A duplicate of the Mht_name theme is now added to the top of the table of contents.

h   Rename the pasted theme **Ecoregion**.

i   Open the Legend Editor for Ecoregion.

j   Select Ecoregion. Open the Values Field drop-down menu and notice that all the fields from the attribute table can be mapped.

k   Keep the default color scheme or change it by using the Color Schemes drop-down menu.

l   Click Apply. Close the Legend Editor.

m   Look at the map and observe the similarities and differences between the Ecoregion and Mht_name themes. Do the following:

   (1)   *Record the similarities and differences between these two themes.*

   (2)   *Write a hypothesis on what you think this data represents.*

The legend for Ecoregion is very long. You can hide it in the table of contents until you want to look at it closely.

n   Make sure the Ecoregion theme is active. From the Theme menu, click Hide/Show Legend.

o   Ask your teacher how to rename and where to save this project. Save the project.

p   Record the new name and location on your answer sheet.

Step 4   **Map the data and analyze it**

In order to get a good look at all the data, you will need to map each of the three remaining fields the same way you mapped the Ecoregion theme.

a   Refer to step 3e–step 3k for instructions on how to copy the theme Mht_name, rename it, and change its legend.

b   Map the remaining fields: BDI, Threat, and Final. You will need to copy and paste three times.

   Hint: Adapt the legend type and the colors for each theme so they represent the data effectively. Remember: You are dealing with numeric fields. What do you think "0" represents? Be sure to use an appropriate color scheme.

c   Save the project.

   (1)   *After mapping out all the data fields, what additional conclusions can you make about the data?*

   (2)   *Use this new information to revise your hypothesis about what the data represents.*

★★★ NEWS FLASH ★★★

A cleanup volunteer has just discovered a document that appears to have some information about the data you are researching. The document has been heavily damaged by floodwater and is difficult to interpret. The table on your answer sheet has the information that could be deciphered.

?   *d*   Read the table on the answer sheet and use the mapped data to complete the missing pieces. The definitions in the table belong to all six fields. Fill in the table on the answer sheet with the appropriate field name from the attribute table.

With this information, you are very close to being able to create a simple metadata document. You know what the data looks like because you've mapped it, and you have made educated guesses on the definitions for each field. An important piece of information still missing is how to interpret the legend for the numeric fields.

Each of the numeric themes represents a scale. The question is what order does the scale follow? Is 1 the lowest or the highest item in the scale? The ranking descriptions without their numeric value are found below for each of the numeric fields. The title for each table is its field name.

?   *e*   Complete the tables on your answer sheet by filling in the numeric values for each descriptor. Keep in mind that the descriptors may not be listed in numeric order and don't forget about zero. The first table has been filled in for you.

Congratulations! All of the details that you have discovered will go in the attribute information section of a metadata document. That section defines each of the fields in the database or attribute table, identifies what type of data it is, and provides the user with details on how types of data are ranked or identified.

Step 5   View metadata documentation

Now that you have had the opportunity to fill in the blanks in the definitions of the data, you will create a formal metadata document. Metadata means data about data. You will create a document that lists information about the data in the wwf_eco.shp theme that you investigated.

The metadata included with this book is in HTML format. Before you create your own metadata document, you will look at a sample document.

?   *a*   Save your ArcView project. Ask your teacher for instructions on how to rename the file and where to save it. Record this information on the answer sheet.

     b   Minimize ArcView.

     c   Navigate to the module 7 data directory (**C:\esri\mapworld\mod7\data\world**).

d   Double-click **continent.htm** to open the metadata document for continent.shp.

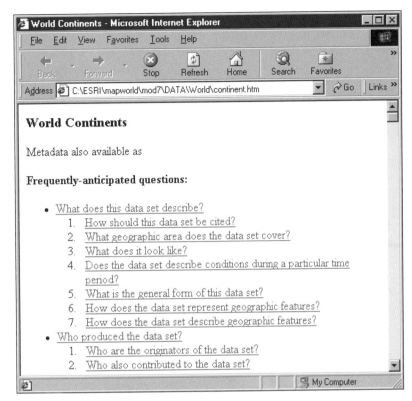

An HTML-formatted document opens in your Web browser.

e   Explore the links in continent.htm and answer the following questions:

   (1)   How would you describe this data set?

   (2)   Who made this data?

   (3)   What time period does the data cover?

   (4)   Which section explains whether or not you can reuse the data for a project or publish a work that uses the data?

f   Minimize continent.htm. You will need to refer to it when you are writing the metadata document for wwf_eco.shp.

Step 6   **Create a simple metadata document**

The metadata for continent.shp is very thorough and complete. For simplicity purposes, you will write a metadata document that addresses a few key areas.

a   Open a text editor or a word-processing document.

b   Title the document **Your Name's Metadata for Wwf_eco.shp**.

For example: Joe Smith's paper would be titled "Joe Smith's Metadata for Wwf_eco.shp."

c Type up the following information into the metadata document. Use continent.htm as a reference.

Identification—where did the data come from, who created it, and what does the data represent?

Attribute information—define each of the fields in the attribute table.

Metadata reference—who created the metadata and contact information.

> *Hint: You will need to toggle back and forth between the metadata document and the ArcView project while you are writing the document.*

d Ask your teacher if there are other items that you should include in the metadata document.

Two additional items could include:
- Spatial data organization—type of data and what scale the data is in.
- Spatial reference information—coordinate system and coverage information.

e Save this document with a unique name that matches up with your project name. For example, if your project was saved as **jsadv7.apr**, your metadata file could be saved as **jsadv7meta.doc**.

f Print your metadata document.

Step 7 Exit ArcView

Congratulations on a job well done! Your efforts will help the IWC get their data records back on track.

a Save your project one more time before you exit.

b Click Exit from the File menu. When asked if you want to save changes, click No.

NAME _____ DATE _____

Student answer sheet
Module 7
Human/Environment Interaction

Advanced investigation: Data Disaster

Step 2   Add the mystery data

h   What do you think this map is showing?

_____

_____

_____

Step 3   Evaluate the attribute table and map two themes

b   Write each field name and the type of data it contains (text, numeric, or string) in the table below.

FIELD NUMBER	FIELD NAME	FIELD TYPE
1		
2		
3		
4		
5		
6		

c   Which field does this theme map?

_____

m-1   Record the similarities and differences between these two themes.

_____

_____

_____

_____

m-2   Write a hypothesis on what you think this data represents.

_____

_____

_____

_____

o   Write the new name you gave the project and where you saved it.

_____   _____

(Name of project.   **(Navigation path to where project is saved.**
For example: kzeregion5.apr)   **For example: C:\student\kze\mapworld)**

Step 4   Map the data and analyze it

c-1   After mapping out all the data fields, what additional conclusions can you make about the data?

_____

_____

_____

_____

c-2   Use this new information to revise your hypothesis about what the data represents.

_____

_____

_____

_____

d   Read the table below and use the mapped data to complete the missing pieces. The definitions in the table belong to all six fields. Fill in the table with the appropriate field name from the attribute table.

FIELD NAME	DEFINITION
	The final assessment of the ecoregion as the estimated threat to the ecoregion over the next 20 years.
	Descriptive name for the _____ that are relatively large areas of land or water that share a large majority of their species, dynamics, and environmental conditions.
	The biological distinctiveness index. It is based on the species richness, endemism, rareness, and so forth.
	Represent global terrestrial and freshwater ecoregions defined as relatively large areas of land or water in the world that share a large majority of their species, dynamics, and environmental conditions.
	The major habitat for the area.
	Degree of threat to the ecoregion. Some examples include logging, conversion to agriculture/ urbanization, and so on.

e   Complete the following tables by filling in the numeric values for each descriptor. Keep in mind that the descriptors may not be listed in numeric order and don't forget about zero. The first table has been filled in for you.

BDI	
1	Globally outstanding in biological distinctiveness
2	Regionally outstanding in biological distinctiveness
3	Bioregionally outstanding in biological distinctiveness
4	Locally important in biological distinctiveness
0	Not assessed or unknown

THREAT	
	Relatively stable in degree of threat
	Relatively intact in degree of threat
	Vulnerable in degree of threat
	Not assessed or unknown
	Endangered in degree of threat
	Critical in degree of threat

FINAL	
	Not assessed or unknown
	Critical in estimated threat over the next 20 years
	Endangered in estimated threat over the next 20 years
	Relatively intact in estimated threat over the next 20 years
	Vulnerable in estimated threat over the next 20 years
	Relatively stable in estimated threat over the next 20 years

Step 5   View metadata documentation

a   Save your ArcView project. Ask your teacher for instructions on how to rename the file and where to save it.

(Name of project. For example: kzeregion5.apr)	(Navigation path to where project is saved. For example: C:\student\kze\mapworld)

e-1   How would you describe this data set?

e-2   Who made this data?

e-3   What time period does the data cover?

_____

e-4   Which section explains whether or not you can reuse the data for a project or publish a work that uses the data?

_____

Step 6   Create a simple metadata document

e   Save this document with a unique name that matches up with your project name. For example, if your project was saved as jsadv7.apr, your metadata file could be saved as jsadv5meta.doc.

(Name of project. For example: jsadv7.apr)	(Navigation path to where project is saved. For example: C:\student\js\mapworld)

Module 1   ArcView: The Basics

Part 1: Introducing the software

Step 6   Work with themes

f   Which themes are not visible on the map but are turned on in the table of contents? Rivers, Lakes

h   What happened on your map? The rivers show up on the map.

i-1   What happened on your map? The lakes show up on the map.

i-2   What would happen if you dragged the Lakes under World Countries Population again?
       The lakes would disappear again; or, the lakes will be covered up by the World Countries Population theme.

Step 7   Identify a country and record country data

h-1   What is the second listing in the column? Cntry_name

h-2   What is the third listing in the column? Curr_type

h-3   What is the final listing in this column? Color_map

i-1   What do you guess the field entitled "Sqmi" stands for? Square miles

i-2   What is the number to the right of the field "Sqmi"? 3,648,398.75

Step 8   Compare the Identify Results data with the table data

a   Which row in this table has the attributes for the United States? The third row

b-1   Where are these field names displayed in the table? Across the top of the table; or, as column headings in the table

b-2   How many square miles of land are in the United States? 9,449,362.00

b-3   Give a brief explanation of the relationship between the Identify Results window and the table.
       The Identify Results window has the same items of information that are in the table. The Identify Results window
       has information about only one country while the table has all the countries. (Students may not pick up on the
       second part just yet.)

Step 10   Explore city data on the world map

h   Use the Identify tool to find out the name and country of any two cities you choose.
       Answers will vary. Possible answers include:

CITY NAME	COUNTRY WHERE THE CITY IS LOCATED
Zaragoza	Spain
Hamburg	Germany

Step 11   Explore Europe with an attribute table

b   What is the name of the table you opened? Attributes of World Countries Population

d   What happens to the map when you click these rows in the table? The four countries all turn yellow.

e-1   What happens to Russia and the other countries that were highlighted? The yellow disappears.

MODULE 1
ArcView

MODULE 2
global

MODULE 2
regional

MODULE 3
global

MODULE 3
regional

MODULE 3
advanced

MODULE 4
global

MODULE 4
regional

MODULE 4
advanced

MODULE 5
global

MODULE 5
regional

MODULE 5
advanced

MODULE 6
global

MODULE 6
regional

MODULE 7
global

MODULE 7
regional

MODULE 7
advanced

e-2   Did you see Iceland turn yellow on the map? If not, why not?

Most students will not be able to see Iceland turn yellow because it isn't in the part of the world they are looking at. (Depending on how much students zoomed the map, Iceland may be displayed underneath the table or out of the view extent. Some students may see Iceland if they didn't zoom in enough, if they have a large screen, or if they moved their table.)

f   Why can you see Iceland now when you couldn't see it in the previous step?

The instructions had the students zoom in too far for them to be able to see Iceland. Once they zoomed to the full extent of the view, they were able to see Iceland.

Step 12   Practice using the Identify tool

b   What do you see on your map?   South America only

d-1   What country is it?   Brazil

d-2   What is this country's total population?   172,860,000

g-1   What city is it?   Manaus

g-2   What population range is this city in?   1,000,000–5,000,000

j   Name these two large cities.   São Paulo, Rio de Janeiro

Step 13   Practice using the Zoom Out tool

c-1   What does your map look like?   It's small or reduced in size.

c-2   Which zoom button could you use to return your map to full size?

Zoom to Full Extent (the button that looks like three sheets of paper)

Step 14   Continue to explore the world map

h-1   How many tourists arrive in Sudan each year?   30

h-2   How many people live in Sudan?   35,080,000

h-3   Does this seem like a low or high number of tourists for this population?

It is a very low number for that population. Students should wonder why this is such a low number. This is a good example of how geographic analysis can lead to further questions.

i   Do you see Qatar on your map?   No

k-1   How many people live in Qatar?   744,000

k-2   How many cell phones do they have?   43,476

k-3   How many people are there for every cell phone in Qatar?   17.1 people per cell phone

k-4   What large country is directly west of Qatar?   Saudi Arabia

n-1   What boot-shaped country do you see on the map?   Italy

n-2   What is the population of that country?   57,634,000

n-3   How many cell phones does that country have?   11,737,904

n-4   How many people are there for every cell phone in this country?   4.09 people per cell phone

o-1   What is the population of Japan?   126,550,000

o-2   How many cell phones does Japan have? 38,253,892

o-3   How many people are there for every cell phone in Japan? 3.3 people per cell phone

q   What happened on your map? The map gets smaller and Japan stays in the middle of the map.

t   What happened to Qatar? It turned back to its original color (the yellow disappeared).

Step 15   Get help from the Help tool
c-1   What does it say you must do to make more than one theme active at a time?
Hold down the Shift key when you click on the themes.

Part 2. The geographic inquiry model

Step 4   Ask a geographic question and develop a hypothesis
a   What makes this a geographic question?
Answers will vary. Students should recognize that the question involves the distribution of something (phone lines) in different places (countries).

b   Hypothesis:
Answers will vary. This is the student's best guess about what they think the answer is to the geographic question.

Step 5   Add a theme to your view
a   What other attribute of countries do you need in order to investigate your hypothesis? Number of phone lines

e   What is the name of the theme that has been added to your table of contents? Phone_line.shp

Step 6   Explore the World Phone Lines map
b-1   What color in the legend indicates countries with the fewest phone lines? Light orange or white

b-2   What color indicates countries with the most phone lines? Dark orange or brown

b-3   What color indicates countries with no data available for this theme? Gray

c   What other theme in your view has a graduated color legend? World Countries Population

d-1   Which country has the most phone lines? United States

d-2   On which continent are most of the countries with the fewest phone lines? Africa

d-3   Which two countries have the largest populations? China and India

d-4   Name three countries that are in the same population class (color) as the Untied States.
Answers will vary. Examples include: Brazil, Russia, Nigeria, Indonesia, Japan, Pakistan, Bangladesh.

d-5   Name which of the three countries, if any, are in the same phone line class (color) as the United States.
None of the countries

h   What two fields might help in answering the geographic question?
Tot_pop (total population)
Phone_lns (phone lines)

MODULE 1
ArcView

MODULE 2
global

MODULE 2
regional

MODULE 3
global

MODULE 3
regional

MODULE 3
advanced

MODULE 4
global

MODULE 4
regional

MODULE 4
advanced

MODULE 5
global

MODULE 5
regional

MODULE 5
advanced

MODULE 6
global

MODULE 6
regional

MODULE 7
global

MODULE 7
regional

MODULE 7
advanced

Step 7   Research and record phone line and population data

c   Use your Identify tool and Find button to locate the countries in the table below. Record the population and phone lines for each country in the appropriate columns.

COUNTRY NAME	COUNTRY POPULATION	PHONE LINES	NUMBER OF PEOPLE PER PHONE LINE
China	1,261,832,000	70,310,000	17.95
India	1,014,004,000	17,802,000	56.96
U.S.	275,563,000	172,452,000	1.60
Indonesia	224,784,000	4,982,000	45.12
Brazil	172,860,000	17,039,000	10.14
Russia	146,001,000	26,875,000	5.43
Pakistan	141,554,000	2,557,000	55.36
Japan	126,550,000	60,381,000	2.10

Step 10   Analyze results of your research

a   Use the data in your table to list the countries on the right according to the number of people per phone line. Then draw lines connecting the same country in each list.

RANKED BY POPULATION (HIGHEST TO LOWEST)	RANKED BY NUMBER OF PEOPLE PER PHONE LINE (LOWEST NUMBER OF PEOPLE PER PHONE LINE TO HIGHEST)
China	United States
India	Japan
United States	Russia
Indonesia	Brazil
Brazil	China
Russia	Indonesia
Pakistan	Pakistan
Japan	India

b-1   Which country has the fewest people who have to share a phone line? United States

How many people is that? 1.60

b-2   How does the country in question 1 rank in population size with the other seven countries in your table?
It's the third largest in population.

b-3   Which country has the most people who have to share a phone line? India

How many people is that? 56.96

b-4   How does the country in question 3 rank in population size with the other seven countries in your table?
It's the second largest.

b-5   What is the population of Japan? 126,550,000

How many people have to share a phone line in Japan? 2.10

MODULE 1
ArcView

MODULE 2
global

MODULE 2
regional

MODULE 3
global

MODULE 3
regional

MODULE 3
advanced

MODULE 4
global

MODULE 4
regional

MODULE 4
advanced

MODULE 5
global

MODULE 5
regional

MODULE 5
advanced

MODULE 6
global

MODULE 6
regional

MODULE 7
global

MODULE 7
regional

MODULE 7
advanced

b-6   What country has the second-most phone lines after the United States? China

How does the number of people per phone line in this country compare with the seven other countries in your table?
It is fifth.

b-7   Russia and Pakistan have about the same number of people. Why do you suppose these two countries have such a different number of people who have to share a phone line? What factors do you think contribute to this disparity?
Answers will vary. Examples: The countries have different cultures, a different economic base of government (GDP), and different economics of citizens.

b-8   What do you think the answer to the geographic question is?
The number of phone lines does not necessarily increase at the same rate as an increase in population.

b-9   Write your initial hypothesis here: Answers will vary but should match step 4b.

How does your hypothesis compare with your answer to the geographic question? Answers will vary.

Step 11   Develop a plan of action

a   Use the information in your table to describe the current phone line situation in your chosen country.

Answers will vary. Possible answers include:

China has the second-highest number of phone lines of any country in the world, but its extremely large population (over one billion people) means that nearly 18 people must share a phone line.

Approximately 10 Brazilians must share a phone line. Brazil ranks fourth among countries that have shared phone lines, after the United States, Japan, and Russia. Brazil is considerably ahead of the fifth-ranked country, China, where nearly 18 people must share a phone line.

In Indonesia, 56.96 people must share a phone line. Indonesia falls far behind other countries of similar size such as the United States (1.6 people per phone line) and Brazil (10.14 people per phone line).

The United States has more than 172 million phone lines—more than any other country in the world. This country also leads the world in access to phone lines because only 1.6 people have to share a phone line.

b   Do you think that increasing the number of phone lines operating in your chosen country would improve the quality of life there? Why or why not?

Answers will vary.

c   List three concerns you have about increasing the number of phone lines in your chosen country.

Answers will vary. Students may list issues such as (1) difficulty of building phone lines in rural areas or among the many islands of Indonesia, (2) cost of building additional phone lines, or (3) a preference for expanding more modern cell phone infrastructure.

d   List two new geographic questions that you would like to investigate to help you develop a sound plan.

Answers will vary but should include a geographic component.

Module 2   Global perspective: The Earth Moves

Step 3   Look at earthquake location data

b-1   Do earthquakes occur in the places you predicted? List the regions you predicted correctly for earthquake locations.

Answers will vary depending on where students predicted earthquakes.

b-2   What patterns do you see in the map?

Answers will vary, but they should indicate that earthquakes largely occur on the western coast of North and South America, along the eastern coast of Asia, and along the islands of the Pacific Rim. The pattern follows the Ring of Fire. They may also note a pattern of earthquakes down the center of the Atlantic Ocean and a string in the southern Atlantic, from South America eastward through to the Indian Ocean. Another pattern that is evident is a string that runs east and west through the south-central part of Asia and the southern part of Europe.

Step 4   Sort and analyze earthquake magnitudes

h   How do the 15 selected locations compare to your original paper map?

Answers will vary based on their original predictions. If they selected any spots around the Ring of Fire, then their predictions were fairly close to reality.

Step 5   Look at volcano data

a-1   How do the volcano locations compare with your original predictions? List the regions of volcanic activity you predicted correctly.

Answers will vary based on their original predictions. If they selected any spots around the Ring of Fire, then their predictions were fairly close to reality.

a-2   What patterns do you see in the volcano points and how do they compare with the earthquake patterns?

The earthquake and volcano points line up so that they are similar in their patterns with the exception of the volcanoes along the eastern side of Africa.

Step 6   Select all active volcanoes

g-1   Does this data provide any patterns that were not evident before? Identify those patterns.

The majority of the world volcanoes on the map are active, particularly in the islands of the Pacific Rim.

g-2   Create a hypothesis as to why volcanoes and earthquakes happen where they do.

Student answers will vary. Their hypothesis should begin to allude to the idea of plate tectonics and the fact that movement at plate boundaries causes disruptions on the earth's surface.

Step 7   Identify the active volcanoes on different continents

d   Zoom in to the continent of your choice. Use the Identify tool to find the name, elevation, activity level, and country location of three volcanoes. Write that information in the space below.

Student answers will vary. Here is a correct example: On Take, 3,063 m, Solfatara Stage, Japan; Banahao, 2,177 m, Active, Philippines; and Ibu, 1,340, Active, Indonesia.

Step 8   Add the plate boundaries theme

h   Compare the actual plate boundaries to the ones you drew on your paper map. Record all similarities and differences.

Answers will vary based on students' original hypotheses on plate boundary locations. If they drew the boundaries to follow the patterns of earthquakes and volcanoes, then they are on target.

Step 9   Add a theme and an image file

h   Use the Identify tool to find out the names of all major landforms formed at plate boundaries. Write them below
      and label them on your paper map.

NAME OF LANDFORM	FORMED BY
Mid-Atlantic Ridge	The separation of South American and African plates
Aleutian Trench Aleutian Islands	The boundary between the North American and Pacific plates
Rocky Mountain Range Cascade Mountain Range Sierra Nevada Mountain Range Baja California	The boundary of the eastern edge of the Pacific plate and the western edge of the North American plate
Sierra Madre Occidental Sierra Madre del Sur	The boundary of the North American, Pacific, and Coca plates
Andes Mountain Range Peru–Chile Trench	The boundary of Nazca and the South American plate
Alps Mountain Range Atlas Mountains	The boundary of the African and Eurasian plates
East Africa Rift Valley	The boundary of the African, Arabian, and Somali plates
Himalayas Tibetan Plateau	The boundary of the Indo-Australian, Indo-Chinese, Eurasian, and Amur plates
Mariana Trench	The boundary of the Philippine and the Pacific plates

Note: The above list covers landforms identified in feature.shp in the ArcView project. Students can find additional
landforms by consulting an atlas or physical map of the world.

Step 10   Identify major cities at high and low risk for seismic activity

b   Find the names of specific cities that are high-risk or low-risk for a seismic event. Write those names in the table.

HIGH RISK	LOW RISK
1 Tokyo, Japan	1 Kazan, Russia
2 Reykjavik, Iceland	2 Tombouctou, Mali
3 San Francisco, United States	3 Kansas City, United States
4 Quito, Ecuador	4 Minsk, Belarus
5 Managua, Nicaragua	5 Godhavn, Greenland

MODULE 1
ArcView

MODULE 2
global

MODULE 2
regional

MODULE 3
global

MODULE 3
regional

MODULE 3
advanced

MODULE 4
global

MODULE 4
regional

MODULE 4
advanced

MODULE 5
global

MODULE 5
regional

MODULE 5
advanced

MODULE 6
global

MODULE 6
regional

MODULE 7
global

MODULE 7
regional

MODULE 7
advanced

Module 2   Regional case study: Life on the Edge

Step 2   Open the region2.apr file and look at cities data

f   Use the Identify tool to locate one city within each country listed below and record that city's population.

Student answers will vary. This is an example of a table with correct information:

CITY NAME	COUNTRY NAME	CITY POPULATION
Kunming	China	1,280,000
Delhi	India	7,200,000
Tokyo	Japan	23,620,000

Step 3   Look at population density

d   Use the Identify tool to locate two world cities in East Asia in areas where the population density is greater than 200 people per square kilometer. Record below.

Student answers will vary. This is an example of a table with correct information:

WORLD CITIES THAT HAVE A POPULATION DENSITY GREATER THAN 200 PEOPLE PER SQUARE KILOMETER
Nanjing, China
Calcutta, India

Step 4   Look at earthquake magnitudes

a-1   Where did the largest earthquakes occur?

Student answers will vary. They should include a statement that large earthquakes tend to fall along the islands of the Pacific Rim, including some that occur under the ocean.

a-2   Did large earthquakes occur near densely populated areas? Where?

No. For the most part, the largest earthquakes occurred underwater.

Step 5   Measure the distance between active volcanoes and nearby major cities

g-1   Are there many active volcanoes located close to highly populated areas? What is the closest distance you found? Record the name of the volcano and the city, and their distance apart.

Yes, there are many active volcanoes near highly populated areas. There are several possible answers for close distances. For example, Manado, Indonesia, is only 4 miles from Mahawu, an active volcano.

g-2   What patterns do you see in the volcano points and how do they compare with the earthquake patterns? (Hint: You should turn the Earthquake Magnitude theme on and off.)

The patterns of earthquake and volcanic activity are virtually the same, especially along the islands of the Pacific Rim (the western edge of the Ring of Fire).

Module 3   Global perspective: Running Hot and Cold

Step 3   Observe annual world temperatures

b   Are the temperatures listed as degrees Fahrenheit or degrees Celsius? Fahrenheit

c   Write three observations about the pattern of temperatures displayed on the map.

Student answers will vary. Possible observations include the following:

The warmest temperatures are clustered halfway between the North and South poles.

Temperatures get steadily colder as you go from the equator toward the North Pole.

There are many cities with cold temperatures in the Northern Hemisphere, but none in the Southern Hemisphere.

Step 4   Label the latitude zones

p   Use the Identify tool to find information on cities and complete the table on your answer sheet.

ZONE	TYPICAL TEMPERATURE RANGE	EXAMPLE CITY (IT REFLECTS TYPICAL TEMPERATURES OF THAT ZONE)	ANOMALIES (CITIES THAT DO NOT FIT THE PATTERN OF THEIR ZONE)
Tropical	65°–85° F	Any of the 28 cities colored red or orange and that are between the Tropic of Capricorn and the Tropic of Cancer.	Quito (51°–60°) and La Paz (41°–50°)
North Temperate Zone	31°–64° F	Any of the cities colored purple, blue, or green and that are between the Tropic of Cancer and the Arctic Circle.	Students should look for cities that seem to differ from the others around them or from other cities at the same latitude. Examples include Lhasa or Ankara.
South Temperate Zone	54°–64° F	Any of the cities colored green or orange and that are between the Tropic of Capricorn and the Antarctic Circle. (Buenos Aires, Cape Town, Johannesburg, Sydney, Melbourne, Wellington, Auckland.)	None

p-1   Why do you think there aren't any cities in the North or South polar zones?

The temperatures are probably too cold to support major cities.

p-2   How is the North Temperate Zone different from the South Temperate Zone?

Student answers will vary. Possible observations include the following:

There is more land area in the North Temperate Zone.

There are cities with average temperatures below 51° in the North Temperate Zone, but none in the South Temperate Zone.

Step 5   Observe climate distribution

d-1   Complete the table.

LATITUDE ZONES	CHARACTERISTIC CLIMATE(S)
Tropical zones	Tropical Wet, Tropical Wet and Dry, and Dry Some areas of Arid, Semiarid, Humid Subtropical, Highlands
Temperate zones	Humid Subtropical, Humid Continental, Marine, Mediterranean, Subarctic Some areas of Arid, Semiarid, Highlands, Tundra
Polar zones	Subarctic, Tundra, Ice Cap

d-2   Which zone has the greatest number of climates?   North Temperate Zone

e   Give an example of a city in each of the following climate zones.

Student answers will vary. This list represents sample answers:

Arid	Khartoum
Tropical Wet	Kisangani
Tropical Wet and Dry	Bamako
Humid Subtropical	Atlanta
Mediterranean	Rome
Marine	Paris
Humid Continental	Warsaw
Subarctic	Irkutsk
Highland	Lhasa

Step 6   Observe monthly temperature patterns in the Northern Hemisphere

d-1   What does the chart show now?

Average monthly temperatures in Miami

d-2   What city is highlighted in the view?   Miami

e-1   What does the chart show now?   Average monthly temperatures in Miami and Boston

e-2   What city or cities are highlighted in the view?   Miami and Boston

f   Complete the table below.

CITIES	COLDEST MONTH	LOWEST TEMPERATURE (°F)	HOTTEST MONTH	HIGHEST TEMPERATURE (°F)	TEMPERATURE RANGE OVER 12 MONTHS
Boston	January	28°	July	73°	45°
Miami	January, February	68°	August	83°	15°

g-1   What is the name of the city?   Quebec

g-2   How does its monthly temperature pattern differ from Boston's?

The overall pattern is the same, but winter temperatures are colder and summer temperatures are slightly cooler. The annual temperature range is slightly greater: –55°. Summer is slightly shorter and winter slightly longer in the more northern city.

h-1   What is the name of the city?   Kingston

h-2   How does its monthly temperature pattern differ from Miami's?

Kingston has a smaller temperature range (5°) than Miami. Both cities are warm year-round, but Miami shows more seasonal variation. They have identical high temperatures, but Kingston's lows are not as cool as those in Miami.

i   List the name of each of the cities displayed in the chart and complete the information in the table below.

CITY	LATITUDE	COLDEST MONTH	LOWEST TEMPERATURE (°F)	HOTTEST MONTH	HIGHEST TEMPERATURE (°F)	TEMPERATURE RANGE OVER 12 MONTHS
Quebec	45.5°	January	15°	July	70°	55°
Boston	42°	January	28°	July	73°	45°
Miami	25.5°	January, February	68°	August	83°	15°
Kingston	18°	January, February	78°	July, August	83°	5°

j   Based on the information displayed in the chart, the map, and the table on your answer sheet, state a hypothesis about how the monthly temperature patterns change as latitude increases.

Student answers will vary. Answers should include the following points:

As latitude increases, the range of temperatures over the year increases.

The lower latitudes have less seasonal variation and tend to be warm year-round.

Temperatures get steadily colder as latitude increases.

January and February are the coldest months and July and August are the hottest months. (Note: this is a correct observation based on the four cities students are observing in this problem. A later step will focus on the difference between the Northern and Southern hemispheres.)

Step 7   Test your hypothesis

d-1   Complete the table below.

Student answers will vary.

CITY	LATITUDE
Stockholm	59°
Berlin	53°
Warsaw	53°
Prague	50°
Vienna	48°
Budapest	48°
Athens	38°
Rome	42°

d-2   Do the cities you selected confirm or dispute your hypothesis? Explain.

Student answers will vary depending on their hypothesis. These cities show a similar pattern to that observed in North America.

Step 8   Analyze temperature patterns in the Southern Hemisphere

c   Complete the table.

CITY	LATITUDE	COLDEST MONTH	LOWEST TEMPERATURE (°F)	HOTTEST MONTH	HIGHEST TEMPERATURE (°F)	TEMPERATURE RANGE OVER 12 MONTHS
Darwin	−13°	July	78°	January, February	84°	6°
Brisbane	−28°	July	59°	January, February	77°	18°
Sydney	−34°	July	53°	January, February	72°	19°
Melbourne	−38°	July	48°	February	68°	20°

d   Compare the monthly temperature patterns in the Southern Hemisphere to those in the Northern Hemisphere.

Patterns in the Southern Hemisphere mirror those in the Northern Hemisphere. Winter temperatures are not as cold in the Southern Hemisphere because none of the cities has a latitude greater than −38°. The major difference is that the warmest and coldest months are reversed.

Formulate a hypothesis about the relationship between monthly temperature patterns and increases in latitude.

Student answers will vary.

Step 9   Test your hypothesis on how latitude affects monthly temperature patterns in the Southern Hemisphere

b-1   Complete the table below.

CITY	LATITUDE
Cape Town	−34°
Johannesburg	−26°
Gabarone	−25°
Luanda	−9°

b-2   Based on your observations, do the cities you selected confirm or dispute your hypothesis about how latitude affects monthly temperature patterns in the Southern Hemisphere? Explain.

Student answers will vary. The patterns are the same as those seen in Australia.

Step 10   Investigate the ocean's influence on temperature

a-1   In which Canadian city would you experience the coldest winter temperatures? Winnipeg

a-2   In which Canadian city would you experience the warmest winter temperatures? Vancouver

a-3   Looking at the map, why do you think the warmest city has temperatures that are so much warmer than the others in the winter?

Vancouver is the only Canadian city (on this map) located on the coast. The proximity to the ocean has a steadying effect on the air temperature in Vancouver throughout the year. Therefore, the fluctuation between summer and winter temperatures is not as large as with inland cities at the same latitude.

h-1   Complete the table.

CITY	LATITUDE
London	51°
Amsterdam	52°
Berlin	52°
Warsaw	52°
Kiev	50°

h-2   What do these cities have in common as to their location on the earth?
All the cities are in the Northern Hemisphere, on the continent of Europe, and at approximately 50° north latitude.

h-3   Which cities have the mildest temperatures? London and Amsterdam

h-4   What happens to the winter temperatures as you move from London to Kiev?
Winter temperatures get steadily colder as you move east and inland.

h-5   Why do you think some cities have milder temperatures than the others?
Students should note that these cities are the warmest near the ocean (London on an island, Amsterdam on the coast).

i   Based on your observations of Canada and Western Europe, state a hypothesis about the influence of proximity to the ocean (or distance from it) on patterns of temperature.
Student answers will vary. They should note that those cities closest to the ocean have milder temperatures than cities at the same or similar latitudes located on the interior of continental landmasses.

Step 11   Investigate the impact of elevation on temperature patterns

d-1   Complete the table below.

CITY	LATITUDE
Kisangani	1°
Libreville	0°
Quito	1°
Singapore	0°

d-2   What do these cities have in common as to their location on the earth? All are located very close to the equator.

d-3   What temperature pattern do these four cities have in common?
All five cities show very little range in monthly temperatures throughout the year (6° or less).

d-4   How is Quito different from the other three?
Its temperatures are significantly cooler than the other three cities.

d-5   Since all these cities are located on or very near the equator, what other factor could explain the difference in their temperature patterns?
Student answers will vary. They should not predict that Quito's close proximity to the ocean causes its cooler temperatures, because they just learned that proximity to the ocean causes milder temperatures.

f   Analyze the attribute table and complete the table below.

CITY	ELEVATION (FEET)
Kisangani	1,361
Libreville	32
Quito	9,226
Singapore	104

h   Based on your observation of temperatures along the equator and the information in the table above, state a hypothesis about the influence of elevation on patterns of temperature.

Student answers will vary. Students should note that cities at significantly higher elevations have cooler temperatures than other cities at a similar latitude.

Step 12   Revisit your initial ideas

f   Rank the 13 cities from lowest to highest according to their average January temperatures.

1. Irkutsk
2. Minneapolis
3. Helsinki
4. Lhasa
5. Vancouver
6. London
7. Tunis

8. Quito
9. Wellington
10. Miami
11. Khartoum
12. Buenos Aires
13. Singapore

j   Rank the 13 cities from highest to lowest according to their average July temperatures.

1. Khartoum
2. Miami
3. Singapore
4. Tunis
5. Minneapolis
6. London
7. Irkutsk

8. Vancouver
9. Helsinki
10. Lhasa
11. Quito
12. Buenos Aires
13. Wellington

l   Put a check mark (✔) next to those answers that you predicted correctly.

Module 3   Regional case study: Seasonal Differences

Step 3   Observe patterns of rainfall

a-1   Which month gets the most rainfall in Bombay? July

a-2   Which months appear to get little or no rainfall in Bombay? December–April

a-3   Approximately how much rainfall does Bombay get each year (in inches)? 83

a-4   Write a sentence summarizing the overall pattern of rainfall in Bombay in an average year.
       Bombay gets more than 80 inches of rain per year in a concentrated period from June to September.

b-1   How did this change the map? The new city is selected. It turns yellow on the map.

b-2   How did this change the charts? Both charts now reflect data for the new city.

c   Analyze the charts and fill in the Mangalore section of the table on your answer sheet.

CITY	MONTHS WITH RAINFALL	HIGHEST MONTHLY RAINFALL (INCHES)	TOTAL ANNUAL RAINFALL (INCHES)
Mangalore	April–November	40	135
Bombay			
Ahmadabad			

d-1   Complete the rest of the table in step c above.

CITY	MONTHS WITH RAINFALL	HIGHEST MONTHLY RAINFALL (INCHES)	TOTAL ANNUAL RAINFALL (INCHES)
Mangalore	April–November	40	135
Bombay	May–October	26	84
Ahmadabad	June–September	13	32

d-2   As you move northward along the subcontinent's west coast, how does the pattern of rainfall change?
       The rainy season gets shorter. It starts later in the year and ends earlier. The monthly and yearly rainfall totals decline.

d-3   Although the monthly rainfall amounts differ, what similarities do you see among the overall rainfall patterns of these three cities?
       In all three cities, the rainy seasons and dry seasons are at the same time of year. In each city, July has the highest rainfall total of any month, and the period from December through March is dry.

Step 4   Compare coastal and inland cities

a   Complete the table on your answer sheet.

CITY	MONTHS WITH RAINFALL	HIGHEST MONTHLY RAINFALL (INCHES)	TOTAL ANNUAL RAINFALL (INCHES)
Bangalore	April–December	7	36

b   How does the rainfall pattern of Bangalore compare with that of Mangalore?
       Similarities: The two cities have a rainy season between April and December.
       Differences: Mangalore gets approximately four times as much rain as Bangalore.

d   What is the distance between the two cities? Approximately 170 miles

e   How can this data help you explain the differences between patterns of rainfall in inland Bangalore and coastal Mangalore?

Mangalore is on the coast while Bangalore is on the interior (Deccan) plateau. A narrow coastal mountain range (the Western Ghats) separates the two cities. The significant difference in total and monthly rainfall results from the orographic effect produced by the Western Ghats. Moist monsoon winds are forced to rise to go over these mountains as they come ashore. Condensing in the cooler upper atmosphere, most of the monsoon's moisture falls on the windward side of the mountains, leaving the inland side much drier.

Step 5   Compare eastern and western Asian cities

a-1   Analyze the charts and complete the table on your answer sheet.

CITY	MONTHS WITH RAINFALL	HIGHEST MONTHLY RAINFALL (INCHES)	TOTAL ANNUAL RAINFALL (INCHES)
Kabul	December–May	3	11
Herat	December–April	2	10

a-2   Describe the pattern of rainfall in these two cities.

Both of these cities are extremely dry. What little rainfall they do receive falls in the early months of the year.

a-3   How do you think Afghanistan's rainfall pattern will affect the way of life in that country?

There is not enough rainfall to support agriculture. They will have to rely on activities such as nomadic herding or extractive industry if any appropriate resources exist.

b-1   Analyze the charts and complete the table on your answer sheet.

CITY	MONTHS WITH RAINFALL	HIGHEST MONTHLY RAINFALL (INCHES)	TOTAL ANNUAL RAINFALL (INCHES)
Calcutta	February–November	13	64
Dhaka	February–November	16	79

b-2   Describe the pattern of rainfall in these two cities.

These two cities have significant annual rainfall total with a distinct rainy season that lasts longer than the rainy season on the southwest coast. The dry season lasts from November to February. The majority of the rain falls between May and October.

c   What is happening to the patterns of rainfall as you move from west to east across South Asia?

The amount of annual rainfall increases as you move eastward and the length of the rainy season gets longer.

Step 6   Observe yearly precipitation

g-1   Which regions within South Asia get the least rainfall? The northwest (Afghanistan and Pakistan)

g-2   Which regions within South Asia get the most rainfall? The southwest coast and the northeast

g-3   In step 5c you were comparing Calcutta, Herat, New Delhi, and Dhaka. Does the map of yearly rainfall that is on your screen now reflect the observation you made at that time? Explain.

Student answers will vary, but essentially, students should observe that precipitation does increase as you move from west to east across South Asia.

h   What relationships do you see between South Asia's patterns of yearly rainfall and its physical features?

The region's heaviest rainfall is on the windward side of the Western Ghats and the Himalayas. Orographic lift is responsible for these areas of heavy rainfall. Cities on the Deccan Plateau, on the subcontinent's interior, get significantly less rainfall because they lie in the rain shadow of the mountains.

Step 7   Explore the monsoon's impact on agriculture and population density

b-1   Which regions or countries of South Asia are suitable for agriculture and which are not? Explain.

Student answers will vary. The western section of South Asia (Afghanistan, Pakistan, and western India) does not get enough rainfall to support agriculture. Additionally, much of Afghanistan and Pakistan is in the mountains, making agriculture unlikely there. Most of the remainder of the subcontinent is suitable for farming because it gets sufficient rainfall and is either a plain or plateau.

b-2   In which regions of South Asia do you expect to see the lowest population density? Explain.

Student answers will vary. Students should expect the dry mountainous west to have the lowest population density because the region cannot produce enough food to support a large population.

b-3   In which regions of South Asia do you expect to see the highest population density? Explain.

Student answers will vary. Students should recognize the importance of rivers to agriculture (alluvial flood plain, fertile deltas, and a steady source of water) and predict a high population density there.

j-1   Does the Agriculture theme reflect the predictions you made in step 7b? Explain.

Answers will vary depending on their answer in step 7b. However, the data does illustrate lack of farming in the dry mountainous regions.

j-2   Why are grazing, herding, and oasis agriculture the major activities in Afghanistan?

Mountainous terrain and scarce rainfall make these the only viable economic activities for most people.

j-3   What do you know about rice cultivation that would help explain its distribution on the agriculture map?

Students familiar with rice cultivation will note that this is a crop that is often grown in flooded fields (wet rice cultivation) and requires a lot of water. Therefore, it makes sense that rice is cultivated in areas with the highest rainfall.

j-4   Is there any aspect of the agriculture map that surprised you? Explain.

Student answers will vary. Some students may be surprised about the agricultural activity in Pakistan since the area is so dry.

p-1   Does the People/sq km theme reflect the predictions you made in step 7b? Explain.

Student answers will vary depending on their prediction.

p-2   Why is Afghanistan's population density so low?

Its harsh conditions make this an area that cannot support a large population.

p-3   Since most of Pakistan gets little to no rainfall, how do you explain the areas of high population density in that country?

The Indus River provides a rich alluvial flood plain and a year-round supply of water for irrigation.

p-4   What is the relationship between population density and patterns of precipitation in South Asia?

Overall population density is highest where rainfall amounts are conducive to agriculture. The notable variation to this pattern is the high population density along the rivers—particularly in the west. The rich soil and dependable source of water on the Indo-Gangetic Plain enable agriculture to support dense populations in spite of insufficient rainfall in some areas or at some times of year.

p-5   What is the relationship between population density and physical features in South Asia?

Population density is lowest in mountainous areas of Afghanistan and Pakistan and highest on the Indo-Gangetic Plain.

Module 3   Advanced investigation: Sibling Rivalry

Step 4   Start ArcView and add the georeferenced images

h   Save your project and record its location.

Student answers will vary depending on how they named their project and where they chose to save it.

Step 5   Analyze characteristics of El Niño and La Niña

c   Record your observations of precipitation characteristics for the same time period on your answer sheet.

YEAR / SO	TEMPERATURE CHARACTERISTICS	PRECIPITATION CHARACTERISTICS
1997 / El Niño	Ranges above normal from 1–5 degrees Celsius	500 mm above normal over much of the Pacific
1999 / La Niña	Ranges below normal from 1–3 degrees Celsius	300 mm below normal over Pacific west of South America

f   Synthesize the information you've recorded and develop a definition of El Niño and La Niña. Write these definitions on your answer sheet.

Student answers will vary.

Step 6   Are El Niño and La Niña equal and opposite?

a   Complete the table on your answer sheet by filling in your observations for El Niño and La Niña's effect on the different regions of the world.

WORLD REGION	TEMPERATURE		PRECIPITATION		OTHER		OTHER	
	DEC. 1997 EL NIÑO	MAR. 1999 LA NIÑA	DEC. 1997 EL NIÑO	MAR. 1999 LA NIÑA	DEC. 1997 EL NIÑO	MAR. 1999 LA NIÑA	DEC. 1997 EL NIÑO	MAR. 1999 LA NIÑA
North America	Ranges above normal from 1–2° C.	1–1.5° above normal in north and 1–1.5° C below normal in south.	Parts of southeast 25–50 mm above normal with scattered areas 10–20 mm below normal.	Predominantly normal with small areas in north 10–20 mm above and 10–50 mm below in south.				
South America	Northern half of continent ranges above normal from 1–5° C. Southern parts are 0–1° C below normal.	Predominantly normal.	10–500 mm below normal in north with small areas 10–50 mm above normal.	10–50 mm below normal in east and 10–250 mm above normal in central part of the region.				
Europe	Central Europe 1° C above normal.	Normal.	South and west 10–50 mm above normal.	Normal.				

Step 6   Are El Niño and La Niña equal and opposite? (continued)

a   (continued)

WORLD REGION	TEMPERATURE		PRECIPITATION		OTHER		OTHER	
	DEC. 1997 EL NIÑO	MAR. 1999 LA NIÑA	DEC. 1997 EL NIÑO	MAR. 1999 LA NIÑA	DEC. 1997 EL NIÑO	MAR. 1999 LA NIÑA	DEC. 1997 EL NIÑO	MAR. 1999 LA NIÑA
Africa	1–2° C above normal.	Central Africa 1–1.5° C above normal.	Mixed central and south 10–400 mm below normal. Small areas 10–100 mm above normal.	South 10–100 mm above normal. Small areas 10–250 mm below normal.				
Asia	1–2° C below normal in north and 1–2° C above normal in south.	1–2° C above normal in the South. Small scattered areas 1° C below normal.	Southeast 10–400 mm above normal. Small areas 10–20 mm below normal.	Small areas 10–30 mm above normal. Much of south 10–50 mm above normal.				
Oceania	1–3° C above normal in most areas. Some spots slightly below normal.	Predominantly normal except western Australia 1–2° C above normal. Small sections of east 1° C below normal.	10–500 mm below normal with isolated areas 250 mm above normal.	Western areas 10–500 mm below normal. Eastern areas 10–500 mm above normal.				
Pacific Ocean	Ranges above normal from 1–5° C.	Ranges below normal from 1–3° C.	500 mm above normal over much of the Pacific.	300 mm below normal over Pacific west of South America.				

b   Based on the data you recorded in the previous question, is La Niña equal and opposite to El Niño? Explain your answer.

No. El Niño is a much longer-lasting event and produces a larger difference in precipitation and temperature than La Niña. It is true that they are opposite. Where El Niño is associated with an increase in water temperature (off the coast of South America), La Niña is associated with a decrease in water temperature. However, the increase in water temperature during El Niño is greater than the decrease in water temperature during La Niña. Therefore, they are not equal effects.

c-1   Was one year better than the other for you and your community?

Student answers will vary depending on which region of the world they are living in or addressing.

c-2   If in one year you received greater than normal rainfall, did your town have problems with flooding?

Student answers will vary depending on which region of the world they are living in or addressing.

c-3   If weather was unseasonably warm and mild, did outdoor activities such as amusement parks have greater turnout?

Student answers will vary depending on which region of the world they are living in or addressing.

c-4   How did these things affect the local economy?

Student answers will vary depending on which region of the world they are living in or addressing.

Module 4   Global perspective: The March of Time

Step 3   Look at cities in 100 C.E.
b-1   Where are they located on the earth's surface?

Many of the cities are approximately 30 degrees north latitude. All of the cities are in the northern or eastern hemispheres.

b-2   Where are they located in relation to each other?

Five of the cities are located in the lands surrounding the Mediterranean Sea. All but three of the cities are in Asia. None of the cities is located in North or South America or Australia.

b-3   Where are they located in relation to physical features?   All of the cities are located near rivers or near the coast.

c   What are possible explanations for the patterns you see on this map?

Answers will vary. Possible answers include the influence of climate, the extent of the Roman Empire, trade, suitability for agriculture, and so on.

Step 4   Find historic cities and identify modern cities and countries
g, i   Use the Find tool and Identify tool to complete the information in the table on the answer sheet.

HISTORIC CITY NAME	MODERN CITY NAME	MODERN COUNTRY
Carthage	Tunis	Tunisia
Antioch	Antioch	Turkey
Peshawar	Peshawar	Pakistan

Step 5   Find the largest city of 100 C.E. and label it
a   What's your best guess at how many people lived in the world's largest city in 100 C.E.?   Answers will vary.

g-1   What was the largest city in 100 C.E.?   Rome

g-2   What was the population of the world's largest city in 100 C.E.?   450,000

Step 6   Look at cities in 1000 C.E. and label the most populous city
c-1   What notable changes can you see from 100 C.E. to 1000 C.E.?

The Mediterranean Sea is no longer the site of half the world's largest cities.

c-2   What similarities can you see between 100 C.E. to 1000 C.E.?

Cities still cluster around 30 degrees north latitude.
All of the cities are under 1,000,000 population.
All but two of the cities are in Asia.
None of the top 10 cities is located in the Americas or Australia.

h-1   What was the largest city in 1000 C.E.?   Cordova

h-2   What was the population of the world's largest city in 1000 C.E.?   450,000

Step 7   Observe and compare other historic periods

b   Explore the map to complete the table on the answer sheet.

YEAR C.E.	LARGEST CITY	POPULATION OF LARGEST CITY	MAJOR DIFFERENCES FROM PREVIOUS TIME PERIOD
100	Rome	450,000	Not applicable
1000	Cordova	450,000	Mediterranean Sea is no longer a center of urban development.
1500	Beijing	672,000	Four of the 10 largest cities are now in China.
1800	Beijing	1,100,000	A city exceeds 1,000,000 for the first time. Europe now has three of the largest cities. Japan now has three of the largest cities.
1900	London	6,480,000	Nine of the 10 largest cities are now in Europe (six) and North America (three). Five of the European cities are in Western Europe. Only one of the cities is in Asia. All 10 cities are over 1,000,000. For the first time, major cities move into more northern middle latitudes. The size of the largest city is six times what it was 100 years earlier.
1950	New York	12,463,000	First time a South American city is on the Top Ten list. First city in Southern Hemisphere is listed. Europe drops back to three of the top 10 cities with only two in Western Europe.
2000	Mexico City	26,300,000	Half of cities are over 15,000,000. Western Europe has only one of the top 10 cities.

Step 8   State a hypothesis

b   Complete the left column by stating which periods in history are associated with the greatest changes. In the right column, state possible explanations for the changes that you see.

TIME PERIODS OF SIGNIFICANT CHANGE	EXPLANATION FOR CHANGE
1800 through 1900	The Industrial Revolution caused rapid growth of European and North American cities. It also led to the immigration of millions of people to the United States from Europe.
1950 through 2000	The size of the world's largest cities mushroomed because of rapid economic growth in the developed world and the loss of agricultural jobs in the developing world (sending people to cities in hopes of finding a job).

Step 9   Investigate cities in the present time

a-1   How many of your original guesses are among the cities in Top Ten Cities, 2000? Answers will vary.

a-2   Which cities did you successfully guess? Answers will vary.

k-1   What are the rankings of the cities you chose that are not in the top 10? Answers will vary.

k-2   How far are they from the top 10? Answers will vary.

477

Module 4   Regional case study: Growing Pains

Step 3   Compare birth rate and death rate data

a-1   Which world region or regions have the highest birth rates? Sub-Saharan Africa, Southwest Asia

a-2   Which world region or regions have the lowest birth rates?

North America (Canada and the United States), Australia, Europe (with Asian Russia)

b-1   Which world region or regions have the highest death rates? Sub-Saharan Africa

b-2   Which world region or regions have the lowest death rates?

Mexico, Central America, Australia, China, and the Middle East

c-1   If the overall rate of growth is based on the formula BR – DR = NI, which world regions do you think are growing the fastest? Slowest?

Fastest: Most sub-Saharan African countries

Slowest: Many European countries

h   Choose two European countries and two African countries and record their birth and death rates in the table on the answer sheet.

Answers will vary. The answers provided below are examples of correct responses:

COUNTRY AND CONTINENT NAME	BIRTH RATE/1,000	DEATH RATE/1,000
Niger (Africa)	51.45	23.17
Spain (Europe)	9.22	9.03
Hungary (Europe)	9.26	13.34
Ethiopia (Africa)	45.13	17.63
Chad (Africa)	48.81	15.71

i   List three questions that these two maps raise in your mind.

Answers will vary.

Step 4   Add a natural increase theme

c-1   What is happening to the population in the countries that are red?

Their death rates exceed their birth rates—over time these populations will decline unless migration into the country makes up for the net loss from natural increase.

c-2   Which world region is growing the fastest? Sub-Saharan Africa

c-3   Which world region is growing the slowest? Europe

c-4   Think about what it would mean for a country to have a population that is growing rapidly or one that is growing slowly. Which of these two possibilities (fast growth or slow growth) do you think would cause more problems within the country? On the answer sheet, briefly list some of the problems you would expect to see.

Most problems would occur in countries with fast growth.

Problems there would include:

Answers will vary, but students should recognize that a country with rapid growth will have a difficult time keeping up with the constantly increasing need for education, health care, social infrastructure, resources, and jobs.

Step 5   Look at standard of living indicators for Europe and Africa

d   Complete the chart on the answer sheet.

INDICATOR	COMPARE SUB-SAHARAN AFRICA AND EUROPE	WHAT DOES THIS "INDICATE" ABOUT THE STANDARD OF LIVING IN THESE REGIONS?
Population > 60 years	Africa: Most countries have 3.15–5.7% in this age group. Europe: Most countries have 17.13–23.91% in this age group.	Africa: Low percent indicates many people die prematurely and do not reach old age. Low standard of living. Europe: High percent indicates more people live into their sixties and beyond because of good health care, sanitation, adequate food supply, and so forth. High standard of living.
GDP per capita	Africa: Most countries have the lowest level of GDP. Europe: Most countries have the highest or second-highest level of GDP.	This is not the same as average income—be sure that students do not make that assumption. A higher GDP per capita does indicate a wealthier country, and that means more money to spend on the infrastructure. High GDP means a high standard of living and enough capital to continue to grow and expand economically.
Infant mortality rate	Africa: All countries have 19.84 infant deaths/1,000 born or higher. Europe: All countries have fewer than 19.84 infant deaths/1,000 born.	High rate of infant mortality indicates a low standard of living. This statistic is typically used to evaluate the health conditions (sanitation, health care, food supply, disease, and so forth) in a country because newborns are much more susceptible to death from such health problems than adults or older children.
Life expectancy	Africa: There is a mix of low to high life expectancy, with most countries having a life expectancy of 37.24–57.42 years. Europe: Most countries have the highest life expectancy of 73.79– 83.46 years.	A higher life expectancy reveals a higher standard of living because it reflects prevailing conditions in a country at this time.
Literacy rate	Africa: Literacy rate varies from country to country, with no country at the highest level. Europe: All but one country is at the highest level of literacy.	Information on literacy, while not a perfect measure of education in a country, is probably the most easily available and valid for international comparisons. Low levels of literacy and education in general can impede the economic development of a country in the current rapidly changing, technology-driven world.
Percent of workforce in service sector	Africa: The service sector of the workforce varies from the lowest level to the highest level from country to country. Europe: The service sector of the workforce varies between the top two levels throughout Europe.	A higher percent of the workforce in the service sector indicates a higher standard of living. As a country becomes more developed economically, a larger percent of its workforce is employed in the service sector. The workforce of less developed countries is characterized by higher percent of workers in agriculture and industry.

Step 6   Add the net migration theme

a   In step 5 you compared standard of living indicators in Europe and sub-Saharan Africa. Based on your observations of those indicators, which region would you expect to have a negative net migration? A positive net migration?

Negative: Countries with high standards of living

Positive: Countries with low standards of living

Explain your answer:

Answers will vary, but students should recognize that there will be more out-migration from countries with low standards of living and more in-migration to countries with a higher standard of living.

h   Summarize the overall patterns of net migration in Europe and sub-Saharan Africa in the table on the answer sheet.

NET MIGRATION IN SUB-SAHARAN AFRICA	NET MIGRATION IN EUROPE
Generally, there is a tendency for out-migration from sub-Saharan Africa.	Generally, there is a tendency for in-migration to Europe with the exception of certain Eastern European countries.

i   What are possible political or social conditions or events that could explain any of the migration patterns you see on the map?

Possible answers include Balkan wars, reunified Germany, political unrest in Liberia and Rwanda, and so on.

Step 7   Draw conclusions

e   Based on your map investigations, write a hypothesis about how a country's rate of natural increase affects its standard of living and its net rate of migration.

Answers will vary, but students should note that natural increase has a direct effect on standard of living, and that standard of living creates push–pull factors that influence migration.

f   Illustrate your hypothesis with data from one European country and one sub-Saharan African country.

Answers will vary depending on the hypothesis that was created in step 7e. However, students should include data for each European and African country that includes natural increase, net migration, and other data sets that support their hypothesis.

EUROPE	DATA	AFRICA
	Country name	
	Natural increase	
	Net migration	

Step 8   Create a layout

j   What are the units of measurement?   Miles

Module 4   Advanced investigation: Generation Gaps

Step 2   Create new themes

2-1   What world regions have the highest rates of natural increase today? Africa and the Middle East

2-2   What world regions have the lowest rates of natural increase today? Europe and Russia

Step 3   Thematically map world age structure

i-1   What regions of the world have populations with a high percent below age 15?
Most African countries and some parts of the Middle East

i-2   What regions of the world have populations with a relatively low percent below age 15?
Europe and North America

i-3   Why do you think there is a much higher percent of children in some populations than in others?
Student answers will vary. There is a high percentage of children in growing populations because of high birth rates and sustained high fertility rates. As a result, each generation will be larger than the one before it.

i-4   Why are there far more people 60 years of age and older in some populations than in others?
Student answers will vary. Populations with high standards of living and good medical care tend to have large populations over 60 years old.

i-5   Based on the maps in Generation Gaps View 1, how do you think natural increase is related to age structure?
Where natural increase is high, the population less than 15 years of age is also high. Countries with the greatest percentage of population over age 60 have a low natural increase. A high rate of natural increase leads to an age structure with a large percent of children.

Step 7   Analyze census data for your county

d   Are there identifiable concentrations of children between birth and five years old in your county? How do you explain these concentrations?
Student answers will vary based on their chosen county.

e   Where is the greatest number of this age cohort found?
Student answers will vary based on their chosen county.

f-1   Do you have any colleges in your county? Any retirement communities? Are these institutions reflected in the census data?
Student answers will vary based on their chosen county.

f-2   Identify patterns of age distribution in your county and suggest explanations for those patterns.
Student answers will vary based on their chosen county.

Module 5   Global perspective: Crossing the Line

Step 3   Explore mountain ranges as physiographic boundaries

m   The Pyrenees Mountains are the border between which two countries? Spain and France

n   Complete the table:

COUNTRIES THAT HAVE MOUNTAIN RANGES AS POLITICAL BOUNDARIES	MOUNTAINS THAT FORM THE BOUNDARY
Italy and Switzerland	Alps
Italy and France	Alps
Poland and Czech Republic	Sudetan Mountains

Step 4   Explore bodies of water as physiographic boundaries

c   Record the names of three sets of countries that share a boundary that's a river.
Some examples include:

COUNTRIES THAT HAVE RIVERS AS BOUNDARIES	RIVER THAT FORMS THE BOUNDARY
France and Germany	Rhine
Germany and Poland	Oder
Romania and Bulgaria	Danube
Belarus	Dnieper

d   Look at the view and name three landlocked countries in Western Europe.
Some examples include: Switzerland, Austria, Czech Republic, Slovakia, Hungary, and Serbia.

Step 5   Explore geometric boundaries

e   Record three sets of countries in the table on the answer sheet.
Some examples include:

COUNTRIES THAT ARE SEPARATED BY GEOMETRIC BOUNDARIES		
Egypt and Libya	Niger and Algeria	Namibia and Botswana
Sudan and Chad	Libya and Sudan	Algeria and Mauritania
Libya and Chad	Algeria and Mali	Angola and Zambia

Step 6   Explore anthropographic boundaries based on language and religion

h   Determine the principal language groups in the regions listed on the answer sheet.
South America: Indo-European
Western Europe: Indo-European, Uralic, and Others

j   Locate three examples in the world where political boundaries coincide with anthropographic boundaries based on language.

Possible answers include:

ANTHROPOGRAPHIC BOUNDARIES BASED ON LANGUAGE COINCIDE WITH POLITICAL BOUNDARIES BETWEEN		
India and China	Kazakhstan and Russia	Haiti and Dominican Republic
Georgia and Russia	Finland and Norway/Sweden	Brazil and Uruguay
North Korea and China	Brazil and Paraguay	Venezuela and Guyana

n   Determine the principal religions in the following regions:

North America: Protestant, Roman Catholic, Mixed Christian, Mormon, Indigenous

Africa: Sunni Muslim, Indigenous, Roman Catholic, Mixed Christian, Protestant, Eastern, Orthodox

p   Locate three examples in the world where political boundaries coincide with anthropographic boundaries based on religion.

Possible answers include:

ANTHROPOGRAPHIC BOUNDARIES BASED ON RELIGION COINCIDE WITH POLITICAL BOUNDARIES BETWEEN		
India and China	Vietnam and Cambodia/Laos	Finland and Russia
India and Burma	Kazakhstan and Russia	Ireland and United Kingdom
India and Pakistan	Mongolia and Russia	Bangladesh and India
Thailand and Malaysia	Iran and Pakistan/Iraq/Turkmenistan	

Step 7   Review physiographic, geometric, and anthropographic boundaries

Find additional examples of physiographic, geometric, and anthropographic boundaries between countries and record your findings in the table on the answer sheet.

Answers will vary. See answers above for possible answers to each category.

Step 8   Explore the impact of boundary shape, cultural diversity, and access to natural resources

e   Locate another example of each type of country. Record them in the following table.

Some examples include:

TYPE OF COUNTRY	EXAMPLE	EXAMPLE 2
Elongated	Chile	Vietnam, Panama
Fragmented	Philippines	Indonesia, Japan
Circular/Hexagonal	France	Uruguay, Zimbabwe
Small/Compact	Bulgaria	Costa Rica, Belgium
Perforated	South Africa	Italy (Vatican City, San Marino), Malaysia (Brunei)
Prorupted	Namibia	Thailand, Afghanistan

f-1   By using language as an indicator of cultural uniformity, identify three countries that reflect cultural uniformity.

Answers will vary. Some examples include Japan, France, Argentina, Italy, and Hungary.

f-2   By using language as an indicator of cultural diversity, identify three countries that reflect cultural diversity.

Answers will vary. Some examples include Canada, Spain, Nigeria, Burkina Faso, Syria, Turkey, India, Switzerland, Myanmar (Burma), Sudan, Sri Lanka, and Namibia.

h   Use the ArcView tools and buttons you've learned in this investigation to find an example of a landlocked country on each of the following continents. For a continent that does not have a landlocked country, write "none."

Complete the table:

CONTINENT	LANDLOCKED COUNTRY
North America (including Central America)	None
South America	Bolivia and Paraguay
Africa	Mali, Burkina Faso, Niger, Chad, Central African Republic, Rwanda, Burundi, Uganda, Zambia, Zimbabwe, Botswana, Lesotho
Asia	Jordan, Afghanistan, Nepal, Bhutan, Laos, Mongolia, Kazakhstan, Uzbekistan, Turkmenistan, Kyrgyzstan, Tajikistan, Armenia, Azerbaijan

m-1   Name two Southeast Asian countries that do not have any oil and gas resources within their borders.

Cambodia, Laos, Vietnam

m-2   Name two Southeast Asian countries that have oil and gas resources within their borders.

Indonesia, Brunei, Thailand, Malaysia, Philippines, Myanmar (Burma)

Step 9   Explore boundary changes in the 1990s

e-1   Describe three political boundary changes you see between 1992 and 2000.

Answers will vary, but they should focus on the changes in Eastern Europe and the former USSR.

e-2   Name two countries that existed in 1992, but do not exist in 2000. USSR, Czechoslovakia, Yugoslavia

g   Select three countries from group A and three from group B and complete the table.

Answers will vary based on countries selected. Refer to global5.apr for specific data.

Module 5   Regional case study: A Line in the Sand

Step 3   Identify countries that border the Arabian Peninsula

d   Record the names of the countries on the map that border the Arabian Peninsula at the north.

Jordan, Iraq, Kuwait

Step 4   Investigate the physical characteristics of the Arabian Peninsula

a-1   Is any part of the Arabian Peninsula mountainous? Yes

a-2   If so, where are the mountains located?

The peninsula is mountainous along its west (Red Sea) coast. A second region of mountains can be seen on the northeast (Gulf of Oman) coast.

b-1   Are there any parts of the Arabian Peninsula that do not have any water at all? If so, where are these regions?

Yes. The south-central part of the peninsula has no permanent bodies of water or streams.

b-2   Do you see any relationship between landforms and the availability of water?

Mountains and areas of higher elevation have more surface water.

d   Describe the bodies of water:

The permanent bodies of water look like disconnected fragments of rivers and lakes.

g-1   How many millimeters equal 10 inches? 254 mm

g-2   Based on the amounts of rainfall displayed on the map, do you think there is much farming on the Arabian Peninsula? Explain.

No. Most of the Arabian Peninsula is so dry that agriculture wouldn't be possible without an alternative source of water such as a river. Egypt, for example, is just as dry, but the Nile River provides water for agriculture.

g-3   Approximately what percent of the Arabian Peninsula is desert?

Approximately 70–90 percent of the Arabian Peninsula is desert.

h   What is the approximate range of temperatures across the Arabian Peninsula during this period?

Sept.–Nov.: 15° C to 33° C; 59° F to 91° F

i-1   Which season is the hottest? Summer (June–August)

i-2   What is the approximate range of temperatures across the Arabian Peninsula during this period?

15° C to 37° C; 59° F to 100° F

j-1   What relationship do you see between the Arabian Peninsula's ecozones as displayed on this map and its patterns of landforms, precipitation, and temperature?

The limited zones of temperate grassland on the Arabian Peninsula are found in the mountains where there is more precipitation and milder temperatures.

MODULE 1
ArcView

MODULE 2
global

MODULE 2
regional

MODULE 3
global

MODULE 3
regional

MODULE 3
advanced

MODULE 4
global

MODULE 4
regional

MODULE 4
advanced

MODULE 5
global

MODULE 5
regional

MODULE 5
advanced

MODULE 6
global

MODULE 6
regional

MODULE 7
global

MODULE 7
regional

MODULE 7
advanced

j-2   Complete the table. List three observations for each physical characteristic.

PHYSICAL CHARACTERISTICS OF THE ARABIAN PENINSULA	
Landforms and bodies of water	The Arabian Peninsula has a narrow coastal plain along the Red Sea that is separated by a low mountain range from the rest of the peninsula. The highest mountains in this range are at the southern point of the peninsula where the Red Sea meets the Gulf of Aden. A second, smaller range of mountains is found on the northeastern point of the peninsula along the Gulf of Oman. There are few permanent bodies of water on the Arabian Peninsula. The region has many intermittent streams.
Climate	Most of the Arabian Peninsula is a desert because it gets less than 10 inches of rainfall per year. Winters are mild in terms of temperature, but in the summer, average temperatures are extremely high.
Ecozones	Most of the Arabian Peninsula has a desert ecosystem. The primary exception to this are the temperate grasslands. They can be found in the mountains and higher elevations of the western part of the peninsula, and in the mountains of the northeast.

j-3   In your opinion, which of the region's physical characteristics would be considered "valuable" in a boundary decision? Explain.

In a boundary decision, the valuable characteristics are grassland ecosystems, areas with greater than 500 mm (19.5 inches) of annual precipitation, and areas with access to permanent bodies of water or springs.

Step 5   Investigate the human characteristics of the Arabian Peninsula

d-1   What is the principal agricultural activity on the peninsula?

The principal agricultural activity of the Arabian Peninsula is nomadic herding.

d-2   Based on what you now know about the physical characteristics of the region, why do you think the agricultural activity is so limited?

There is not sufficient water for farming in most of the region. Livestock can be herded from place to place depending on the seasonal availability of water and pastureland.

e-1   How does Yemen compare to the rest of the Arabian Peninsula in population density?

Southwestern Yemen has the largest area of relatively high population density (greater than 50 people per sq. km.).

e-2   Describe the overall population density of the Arabian Peninsula.

Most of the Arabian Peninsula has fewer than 25 people per sq. km. and at least half of that area has fewer than one person per sq. km.

g   Speculate on the most frequent use of the water at these springs and water holes.

Student answers will vary. Because of the large amount of nomadic herding, a logical conclusion is that most of the springs and water holes are used to water livestock.

h   Use your answers from previous step 5 questions and analysis of the maps to complete the table. List three observations for each human characteristic.

HUMAN CHARACTERISTICS OF THE ARABIAN PENINSULA	
Agricultural activities	The principal agricultural activity of the Arabian Peninsula is nomadic herding. Farming exists on a very limited basis in desert oases, on irrigated lands along the Red Sea, and in the high elevations of southern Yemen.
Population density and distribution	The Arabian Peninsula is very sparsely populated overall. Most of the population centers are along the coast (with a few exceptions in Saudi Arabia). The largest concentration of people is found in southern Yemen.

If an international boundary were to be drawn across some part of the Arabian Peninsula, how would these characteristics influence the perception of certain regions as being more "valuable" than others?

Nomadic herders would place a high value on having access to sources of water and grazing land. The population density of a region is a direct reflection of the ability of land to support population. Those areas with very low population density would be least valuable and those with high population density would be most valuable.

Step 6   Locate and describe the Empty Quarter

b-1   Complete the table. List three observations in each column.

THE EMPTY QUARTER	
PHYSICAL CHARACTERISTICS	**HUMAN CHARACTERISTICS**
The Rub' al-Khali or Empty Quarter is a desert region with virtually no permanent bodies of water and less than 2 inches of rainfall per year.	The Rub' al-Khali or Empty Quarter has no agricultural activity at all. The region is virtually uninhabited—most of it has less than one person per sq. km. And about one third has no people at all. The lack of roads in this region indicates minimal human presence.

b-2   What difficulties would an area like this present if an international boundary must cross it?

International boundary difficulties include:

There's no one living in the area to enforce the boundaries.

The lack of permanent landmarks (due to shifting sand dunes) makes the line difficult to mark and see.

Nomadic herders would want easy access throughout the border area.

Step 7   Explore Saudi Arabia's southern boundaries

d-1   Are the boundaries what you expected them to be? Student answers will vary.

d-2   Which boundary remained unsettled?

The border between Saudi Arabia and Yemen was the only border that remained undefined.

i   What does the area between the green and purple lines represent?

It is claimed by both Saudi Arabia and Yemen—it is the disputed territory between these two countries.

j   What is the principal economic activity of the regions in dispute?

Nomadic herding. The disputed territory is an area of land that borders the Empty Quarter.

k   Describe the population distribution in the disputed territory.

The disputed territory is mostly uninhabited. Most of the disputed territory has fewer than one person per sq. km. The only area with a higher concentration of people is the western part of the territory with 1–25 people per sq. km.

Step 8   Draw the Saudi–Yemeni boundary

b-1   Does the red line go through any cities or towns? (Hint: you may need to zoom in again to answer the question.) If yes, approximately how many does the boundary pass through?

Student answers will vary. There are fewer than 10 villages that the boundary line actually passes through, but if you consider villages within a mile or two of the border (approximately 1.25 km.), there are many more.

b-2   How would you decide which side of the town to put the boundary on? Remember, this decision would determine whether the residents of that village would be citizens of Saudi Arabia or Yemen.

Student answers will vary. One way to decide where to put the boundary line is to survey villagers to find out whether they feel a closer affiliation with Yemen or Saudi Arabia. Such affiliations are based on long-standing tribal traditions.

MODULE 1
ArcView

MODULE 2
global

MODULE 2
regional

MODULE 3
global

MODULE 3
regional

MODULE 3
advanced

MODULE 4
global

MODULE 4
regional

MODULE 4
advanced

MODULE 5
global

MODULE 5
regional

MODULE 5
advanced

MODULE 6
global

MODULE 6
regional

MODULE 7
global

MODULE 7
regional

MODULE 7
advanced

g   Does the new line seem to favor Yemen or Saudi Arabia? Explain.

The new border seems to favor Yemen. It gained control of all the disputed territory and gained even more territory beyond the previous boundary.

h   Look at your map. What body of water does the maritime boundary traverse?   The Red Sea

t   How does the actual boundary established by the Treaty of Jeddah compare with the boundary you drew earlier?

Student answers will vary. In most cases, students will find that the Treaty of Jeddah gave more land to Yemen than they predicted.

u   Write three observations about the boundary line created by the Treaty of Jeddah.

Student answers will vary. Here are three possible responses:

The new boundary increased Yemen's territory.

Most of Yemen's new territory is land used by nomadic herders and desert.

The border settlement probably did not have a significant impact on Yemen's overall population as most of the new territory is uninhabited or very sparsely settled.

Step 9   Define the pastoral area

a   How many miles is 20 kilometers? (Hint: 1 kilometer = .6214 miles.)   20 km. = 12.428 miles

d-1   In which part of the Saudi–Yemeni border will the pastoral area be most significant? Explain.

The pastoral area will be most significant in the western part of the boundary region (corresponding to Yemen1.shp) because this is the part of the boundary where nomadic herding is the characteristic agricultural activity. The remainder of the new boundary is north of the nomadic herding areas.

d-2   Why do you think the Treaty of Jeddah created a pastoral area?

Student answers will vary. However, students should understand that the establishment of a pastoral area recognizes that nomadic herding is incompatible with fixed and finite boundaries. The pastoral area represents a compromise between the need to clearly define the boundary between Saudi Arabia and Yemen and the reality that the border area is populated by people whose nomadic traditions include territory on both sides of that boundary.

Module 5   Advanced investigation: Starting from Scratch

Step 3   Explore project themes

List other important factors that influence boundary decisions.

Historic events (wars and treaties), ethnicity, natural resources are some. Student answers will vary.

b   Use the Religion legend to determine three principal religions of South Asia. Record them here.

Hindu, Sunni Muslim, Buddhist

c   The boundary between which two religions corresponds to a physiographic boundary visible in the Satellite Image?

Hindu and Buddhist

d   Identify the principal language groups in South Asia. Indo-European, Dravidian, Sino-Tibetan

e   The boundary between which two language groups corresponds to a physiographic boundary visible in the Satellite Image?

Sino-Tibetan and Indo-European

Step 6   Draw the boundaries of new countries

a   Choose the continent you will be working on and record it here. Student answers will vary.

Module 6   Global perspective: The Wealth of Nations

Step 3   Evaluate the legends and patterns of the maps

a-1   What do the darkest colors represent?   A very high percentage in that field

a-2   What do the lightest colors represent?   A very low percentage in that field

b-1   What does the description "Very High" refer to in this map?
It refers to the percentage of the workforce that is in the agriculture field.

b-2   Where are the countries with a high percentage of agricultural workers generally located?   Africa and parts of Asia

b-3   Where are the countries with a low percentage of agricultural workers generally located?
North America and Australia

c-1   Where are the countries with a high percentage of service workers generally located?
North America, Australia, and Europe

c-2   Where are the countries with a low percentage of service workers generally located?   Central Africa and India

d   What patterns do you see on the map?   The lightest colors are in Africa, the darkest are in Russia and Argentina.

e   Using the workforce information in all three maps, in what part or parts of the world do you find the greatest number of developing countries?
In Africa and parts of the Middle East and Asia

Step 4   Analyze data on Bolivia and other countries

i-1   What percentage of workers in Bolivia are involved in agriculture?   38.9

i-2   What percentage of workers in Bolivia are involved in the industrial field?   20.8

i-3   What percentage of workers in Bolivia are involved in service?   40.3

i-4   Would you classify Bolivia as a developed or developing country? Explain.
Developing. Regardless of their answer, it is important for students to support their answer with the data.

k   Complete the table.

COUNTRY	AGRICULTURE	INDUSTRY	SERVICE	DEVELOPING	DEVELOPED
Bolivia	High	Low	High	X	
India	Very High	Low	Low	X	
New Zealand	Low	Moderate	Very High		X
South Korea	Low	Moderate	Very High		X
Portugal	Low	Moderate	High		X
Uruguay	Very Low	Moderate	Very High		X

Step 7   Analyze GDP per capita and energy use data

b-1   What level is the GDP per capita for Bolivia? Low

b-2   Based on this new information and the workforce data, should Bolivia be classified as a developing or developed country? Developing

Why?

Low GDP is consistent with a developing nation. Previous data on workforce statistics also indicate that Bolivia is a developing country.

c-1   What level is the Energy use for Bolivia? Very Low

c-2   Based on this new information and previous data, should Bolivia be classified as a developing or developed country?

Developing

c-3   Why does energy use increase when a country develops?

Energy is used more when countries build infrastructure and establish manufacturing plants than when the primary mode of economic production is agriculture. Industry- and service-oriented production consume more energy than agriculture.

d-1   Complete the table below.

COUNTRY	GDP PER CAPITA	ENERGY USE	DEVELOPED OR DEVELOPING	IS THIS A CHANGE FROM YOUR EARLIER CLASSIFICATION?
Bolivia	Low	Very Low	Developing	Student answers will vary.
India	Very low	Moderate	Developing	
New Zealand	High	Very Low	Developed	
South Korea	High	Low	Developing	
Portugal	Moderate	Low	Developed	
Uruguay	Moderate	Very Low	Developed	

d-2   Name one country from above that you earlier classified as developed and that has GDP and Energy Use data that indicates it's developing.

Possible answers include Portugal and Uruguay.

d-3   For the country you named in question 2, explain your current classification.

Students should note that although Portugal and Uruguay have a high percentage of workers in the service industry, they have an average GDP and low energy use. Both of these factors make Portugal and Uruguay lean toward a developing classification. Students should begin to realize that there are many factors that determine a country's classification as developed or developing and that as they examine different factors, their classifications may change.

MODULE 1
ArcView

MODULE 2
global

MODULE 2
regional

MODULE 3
global

MODULE 3
regional

MODULE 3
advanced

MODULE 4
global

MODULE 4
regional

MODULE 4
advanced

MODULE 5
global

MODULE 5
regional

MODULE 5
advanced

MODULE 6
global

MODULE 6
regional

MODULE 7
global

MODULE 7
regional

MODULE 7
advanced

Module 6   Regional case study: Share and Share Alike

Step 3   Examine the map and attribute table

c-1   Which years does the theme contain data for? 1991 to 2000

c-2   How many attributes are there for each year? Six

e-1   What is the name of the table? Attributes of NAFTA Countries

e-2   How many rows are there for each country on the map? One

f-1   What happened on the table? The row turned yellow (was selected).

f-2   What happened on the map? The country matching the table row turned yellow.

g   What was the value of goods and services exported from Canada to the United States in 2000? $230,838,000,000

Step 4   Link another table to the theme table

e-1   How many rows are there for each country? Four

e-2   What information is collected under the field title "item" for Canada?
Total Exports to United States
Trade Balance with United States
Total Exports to Mexico
Trade Balance with Mexico

e-3   Describe in general terms the information collected in the field titled "item."
Total Exports and Trade Balance for each of the other two NAFTA countries

e-4   How many years of data are represented in the table? 10

j   What happens in the two tables and the map?
The U.S. rows are highlighted (yellow) in both tables and the United States is highlighted (yellow) on the map.

l   What happens in the two tables and the map? Canada is highlighted on the map and in both tables.

m-1   What happens in the two tables and the map?
The Mexico row that was clicked in the NAFTA Trading Statistics table becomes highlighted but Canada remains selected in the attribute table and map.

m-2   What have you observed about the way the NAFTA Trading Statistics table is tied to the NAFTA Countries attribute table and map theme?
The link only goes in one direction. It goes from the attribute table to the NAFTA table but not the other way around.

Step 5   Examine export charts

b-1   Which country exported more goods and services to Canada—Mexico or the United States? United States

b-2   Why is the chart empty in the space for Canada? A country (Canada) doesn't export goods to itself.

d   What happened to the chart? Only the data for Mexico is displayed.

e-1   How many years of data are represented on the chart? 10

e-2   What year does the first bar on the left represent? 1991 e-3   Compare the numbers on the y-axis with those in the NAFTA Trading Statistics table. Are the numbers on the chart in thousands, millions, or billions of dollars?

Millions

e-4   How do you know?   The numbers are the same as those in the table, and we know the table is in millions.

e-5   Looking at the chart, how would you describe the trend of Mexican exports to Canada over the 10-year period?

They increased dramatically

e-6   What was the approximate value of Mexican exports to Canada in 1991? $2,300,000,000 ($2.3 billion)
In 2000? $8,000,000,000 ($8 billion)

e-7   Approximately how many times greater is the 2000 export figure than the 1991 export figure?

3.5 (or 4, depending on how the student has rounded)

g-1   How would you describe the trend of Mexican exports to the United States over the 10-year period?

They increased steadily and dramatically.

g-2   Approximately how many times greater is the 2000 export figure than the 1991 export figure?

4.6 (or 5, depending on how the student has rounded)

Step 6   Examine a trade balance chart

c-1   Did Mexico have a trade surplus or deficit with the United States for 1992?   Deficit

c-2   What was the approximate value of the trade balance for 1992? (Remember, the y-axis is in millions of dollars.)

$–5,000,000,000 ($–5 billion)

c-3   What was the first year that Mexico exported more to the United States than it imported from the United States?

1995

c-4   Describe the trend of Mexico's trade balance with the United States over the 10-year period.

It went from a deficit to a surplus that continued to grow until 2000.

d-1   Did Canada have a deficit trade balance with the United States anytime during the 10-year period?   No

d-2   In 1998, was Canada's trade balance with the United States greater, smaller, or about the same as Mexico's?

About the same

d-3   In 2000, was Canada's trade balance with the United States greater, smaller, or about the same as Mexico's?

Greater

d-4   Referring to the NAFTA Trading Statistics table, what was the exact value of Canada's trade balance with the United States in 2000?

$51,897,000,000 surplus

Step 8   Change the look of your chart

m   What are the minimum and maximum y-axis values?

Minimum: –25000 (or, –25,000,000,000 dollars)

Maximum: 10000 (or, 10,000,000,000 dollars)

p   What changes did this make on your chart?

The minimum value changed to –60000, the maximum value changed to 60000, and horizontal lines across the chart were added. Students may also note that the increments changed from 5000 to 20000.

Step 9   Evaluate the effectiveness of NAFTA

g   Use the charts to find the value of exports between each set of countries for the year 2000.

Student answers may vary slightly because they will be approximating the values from the chart.

DIRECTION OF EXPORT FLOW	VALUE OF EXPORTS (MILLION $)	TOTAL VOLUME BETWEEN PARTNERS (MILLION $)
United States to Mexico	115,000	255,000
Mexico to United States	140,000	
United States to Canada	180,000	410,000
Canada to United States	230,000	
Canada to Mexico	1,000	11,000
Mexico to Canada	10,000	

h   Add the export values together for each pair of countries (for example, United States to Mexico plus Mexico to United States) and write that number in the Total Volume Between Partners column.

Refer to totals in the table above.

i   Rank the trading partners by the overall volume of trade between the two countries. Use 1 for the partners trading the most and 3 for the partners trading the least.

United States–Mexico: 2   United States–Canada: 1   Canada–Mexico: 3

i-1   Do you think that NAFTA had a positive (+), negative (–), or neutral (n) effect on trade volume between each set of partner countries?

United States–Mexico: +   United States–Canada +   Canada–Mexico +

i-2   Do you think that any one of these three countries benefited more than the other two by NAFTA? If so, which country? Explain your answer.

Student answers will vary because they are asked to speculate. Trade volume has increased for all the trading relationships.

l-1   What country has a healthier trade balance with Canada—Mexico or the Untied States? Mexico

l-2   On what chart do you find a set of bars that looks like a mirror image of those for the U.S. trade balance with Canada?

Canada on the Trade Balance with U.S. chart

m-1   What country had the most dramatic change for the better after NAFTA came into being? (Remember, NAFTA went into effect in 1994.) Mexico

m-2   Estimate the U.S. trade deficit with Canada and Mexico for 1999 and 2000. Express the values in billions of dollars, using round numbers that you estimate from the charts.

	1999 ($ BILLION)	2000 ($ BILLION)
U.S. trade balance with Mexico	$ –23 billion	$ –2.5 billion
U.S. trade balance with Canada	$ –30 billion	$ –51 billion
Combined deficit (total)	$ –53 billion	$ –76 billion

m-3   Did the U.S. combined trade balance get better or worse between 1999 and 2000? Worse
By how much? $23 billion

Step 10   Prepare charts for the layout

i   What happens on your chart? The bars descend from the zero value line at the top of the chart.

Module 7   Global perspective: Water World

Part 1: A South Pole point of view

Step 3   Look at Antarctica
a   Do you think this map gives you a realistic representation of Antarctica? Explain your answer.

Student answers will vary. They should observe that the map of Antarctica is very skewed in size, shape, and determining distance.

c   Does this projection give you a better view of the region around the South Pole? No

d-1   Do any of these projections work well for viewing Antarctica?

No. None of the projections represent Antarctica in a realistic way.

d-2   Which projection do you think works best? Why?

Student answers will vary. The projection that displays Antarctica the most accurately is the Orthographic projection.

Part 2: Just Add Water

Step 6   Open the Water World view
e   What significant differences to you see between today's country outlines and the elevation map of 20,000 years ago?

Student answers will vary. Possible answers include: Alaska was connected to Russia, Florida was much larger, and Australia was connected to the islands of Indonesia.

Step 7   Analyze global sea levels if Antarctic ice sheets melted
a   Record your general observations of each theme in the table below.

Student answers will vary. Possible answers include:

SEA LEVEL	OBSERVATIONS
Today	Country outline matches up perfectly with the shorelines.
Plus 5 meters	There is no dramatic change. However, some coastal cities in the southern United States (Miami) will be under water.
Plus 50 meters	There is dramatic change. Most of Florida is under water, there is a large gap in the Amazon Basin in South America, and parts of Europe are gone.
Total Thaw (plus 73 meters)	Much of eastern Europe and western Asia are under water. Large portions of Australia, South America, and southeast United States are gone. Africa is the least affected.

Step 8   View changes in water levels

d-1   What kinds of changes do you see in the rivers and lakes? Provide a specific example.

Student answers will vary. In South America, a large lake appears in the north central area. This lake is the result of the increased water level of the Amazon River. The Parana River in Argentina and the Amazon River Basin are significantly shorter in length.

d-2   With a sea level increase of 50 meters, what kinds of consequences do you foresee for the major river ecosystems of South America? Provide a specific example.

Student answers will vary. The Amazon Basin has the possibility of flooding the rain forests of central South America.

d-3   There are several locations around the globe that are on the interior of landmasses and are below sea level. One of them is in South America. Hypothesize how these low-lying areas were formed.

Student answers will vary. One explanation is that plate boundaries are drifting apart at that location. Another is that the land surrounding the Amazon Basin could have risen up due to tectonic activity.

Step 9   View changes in political boundaries

g   Record your results in the table below.

REGION	COUNTRIES/AREAS AFFECTED	POSSIBLE CONSEQUENCES
Middle East	Iraq	Boundary disputes over lost land.
Asia	Cambodia	Cambodia is almost completely under water. There will be a large migration of the population moving to neighboring countries.
Europe	Netherlands	Netherlands would be completely submerged. As a result, a large migration of the population would move to neighboring countries.
Africa	Guinea-Bissau	Guinea-Bissau would be completely submerged. As a result, a large migration of the population would move to neighboring countries. There might be a climate change to the drier areas of the continent.
Oceania	Australia	There would be a loss of major cities and economic areas. These people would move inland.
North America	United States	There would be a loss of most major southern ports: Houston, New Orleans, Miami—and most of Florida. These people would move inland.
Latin America	Brazil	The loss of much of the Brazilian rain forest could have negative repercussions on the global environment.

Step 10   Look for additional data to explore

Based on your previous observations, list other possible themes or data you would like to analyze to study the effect of this phenomenon. In your final assessment, you will have the opportunity to explore many other data sets; this list will help to guide you in further explorations.

Student answers will vary. For a complete list of available data, refer to the assessment data sheet.

Module 7   Regional case study: In the Eye of the Storm

Part 1: The calm before the storm

Step 3   Focus on the capital cities of Central America

f   Record the capital cities for each of the countries in Central America in the Populated Places column in the table.
Answers will vary. The table on the following page presents a sampling of possible answers.

Step 4   Focus on Central America prior to Hurricane Mitch

b   Record this data in the Populated Places and Transportation Network columns on the answer sheet on the next page.
Answers will vary. The table on the following page presents a sampling of possible answers.

m   Analyze the precipitation for each country and record the data in the Average Precipitation column on the answer sheet on the next page.
Answers will vary. The table on the following page presents a sampling of possible answers.

p-1   Which country has the most area devoted to agriculture? El Salvador

p-2   Which country has the most area covered by mountains? Honduras

p-3   Which country has the most extensive transportation network? El Salvador

Central America Prior to Hurricane Mitch

COUNTRY	POPULATED PLACES	TRANSPORTATION NETWORK	MAJOR EXPORTS	LAND USE	AVERAGE PRECIPITATION	PHYSICAL LANDMARKS
Belize	Capital: Belmopan Distribution: Throughout the country, but concentrated around the capital	Roads: Sparse road network Railways: None Airports: 1 civilian	Bananas	Primarily forest with some irrigated land and little cropland	Primarily 2,000–2,800 mm	Maya Mountains
Guatemala	Capital: Guatemala Distribution: Concentrated near southern coast	Roads: Well devl. network, esp. along southern coast Railways: Primarily in south Airports: 1 civilian, 7 total	Coffee	About 1/2 forest mixed with irrigated land, cropland, and forested wetlands along the south coast	Mixed from 1,000–5,600 mm	Sierra De Santa Cruz Sierra De Los Cuchumatanes Sierra Madre
Honduras	Capital: Tegucigalpa Distribution: Concentrated near the capital; bare in the east	Roads: Well devl. along eastern half of country Railways: Along northern coast Airports: 12 total, 1 civilian	Coffee	1/4 forest, some cropland, some grazing land, 1/2 nonirrigated land	Mixed from 600 mm in center of country to 4,000 mm along the coast	11 mountains Montana De Botaderos Montana
El Salvador	Capital: San Salvador Distribution: Concentrated around the capital	Roads: Well developed Railways: Well devl. network Airports: 4 total	Coffee	Primarily cropland with some grazing, forested wetlands along the coast, little forest	Primarily 2,800–4,000 mm	Volcan De Santa Ana Cordillera De Celague Volcan De San Miguel
Nicaragua	Capital: Managua Distribution: Coast communities and concentration near capital; bare patches in central	Roads: Well devl. in most of country, esp. west coast Railways: Sparse; near capital Airports: 7 total	Coffee	One-third forest, some crop-land, grazing land, some forested wetland along the east coast	Mixed from 600 mm in the northwest to 5,600 mm along the east coast	3 mountains, 2 volcanoes Volcan Cosiguina Cordillera Dariense
Costa Rica	Capital: San Jose Distribution: Concentrated near Pacific coast and capital	Roads: Along west coast Railways: Through center of country and capital Airports: 13 total, 1 civilian	Garments	Little agriculture: mostly nonirrigated land, small amounts of cropland and forest	4,000–5,600 mm along most of coastline with the exception of 1,400 mm in the northwest	4 mountains, 2 volcanoes Volcan Barva Cerro Chirripo
Panama	Capital: Panama Distribution: Concentrated in southern coast and capital; bare in southeast region	Roads: Southwest coast, west of capital Railways: Sparse Airports: 16 total, 1 civilian	Bananas	No agricultural places identified	Primarily a range from 4,000–8,000 mm	6 mountains, 1 volcano Cerro Santiago Volcan De Chiriqui

Part 2: The storm

Step 2   Track Hurricane Mitch

f-1   At what time was Tropical Storm Mitch at this location? 10/22/21Z

f-2   What does the "Z" mean in the time?
Zulu. Instead of being local time, Zulu refers to the time in Greenwich, England.

f-3   What was Mitch's wind speed at this location? 40 mph

h-1   What are the latitude and longitude coordinates for Hurricane Mitch at this location?
14.3 latitude, −77.7 longitude

h-2   At what time was Hurricane Mitch at this location? 10/24/09Z

h-3   What was Mitch's wind speed at this location? 80 mph

j-1   At what time was Hurricane Mitch at this location? 10/27/21Z

j-2   What was Mitch's wind speed at this location? 135 mph

q   Write down the times for each event in the appropriate spot and determine the difference. The time is written in this format: Month/Day/Hour (of 24).
Hurricane - 5: 10/26/12Z
Tropical_Storm: 10/24/03Z
Time Difference: 00/02/09 (it took 2 days and 9 hours)

r   Examine the attribute table further and identify the maximum wind speed. 155 mph

s-1   What was the wind speed at this location? 135 mph

s-2   Did the speed of the wind increase or decrease when the hurricane hit land? Decrease

s-3   Why do you think the wind speed changed?
Student answers will vary. Most students should understand that the warm, humid air that's prevalent over the ocean is what fuels a hurricane. When a hurricane hits land, it loses that fuel and often becomes less intense. The result is decreased wind speed.

Step 3   Measure the size of the storm

h   What is the diameter of the eye of Hurricane Mitch? 25 miles

i, o   Note: Answers in this table are approximate values. Student values will differ depending on the precise location of each measurement taken.

	DIAMETER OF EYE	DIAMETER OF STORM	DISTANCE OF EYE TO COASTLINE OF HONDURAS	HOW HAS THE STORM CHANGED FROM THE PREVIOUS IMAGE?
Mitch2sat.tif	25 miles	850 miles	110 miles	Not applicable.
Mitch3sat.tif	19 miles	995 miles	50 miles	Storm appears more intense and enlarged. It's closer to the coastline.
Mitch4sat.tif	0 (not visible)	835 miles	0 (on shore)	The eye is not visible, clouds are much thicker, but the spiral shape is still visible.
Mitch5sat.tif	0 (not visible0	875 miles	0 (on shore)	There is still a large amount of clouds, but the spiral shape is gone.

Step 4   Analyze rainfall from Hurricane Mitch

c-1   What pattern do you notice in the amount of rainfall within the storm?

The greatest amount of precipitation is on the southwest arm of the storm.

c-2   Is this a pattern you expected to find? Why or why not?

Student answers will vary depending on their familiarity with hurricanes.

e-1   At the Mitch4 location, what was the highest range of rainfall measured?   24–29 inches

e-2   Which country received the majority of this heavy rain?   Nicaragua

e-3   Describe the difference between the rainfall patterns on October 30 and October 31, 1998.

The rainfall pattern on October 30 was centered heavily over the western coast of Nicaragua and southern El Salvador, with other bands extending due north and one off the eastern coast of Nicaragua and Honduras. On October 31, the main rain center was much larger in area, but less intense in rainfall. The outside bands appear to have merged with the main rain band from October 30.

e-4   What kind of damage do you expect to find with this type of storm? What aspects of the region will be most affected? Elaborate on your answer using your table, Central America Prior to Hurricane Mitch, as a resource.

Student answers will vary. They should mention the possibility of flooding, potential problems with landslides, damaged utility lines, and power outages due to high winds. Nicaragua, Honduras, and El Salvador were severely affected by the rainfall and wind.

Module 7   Advanced investigation: Data Disaster

Step 2   Add the mystery data

h   What do you think this map is showing?

Student answers will vary. They should observe that the data addresses the environment in some way. The legend displays the names of different biomes.

Step 3   Evaluate the attribute table and map two themes

b   Write each field name and the type of data it contains (text, numeric, or string) in the table below.

FIELD NUMBER	FIELD NAME	FIELD TYPE
1	Shape	String
2	Ecoregion	String
3	Mht_name	String
4	BDI	Numeric
5	Threat	Numeric
6	Final	Numeric

c   Which field does this theme map? Mht_name

m-1   Record the similarities and differences between these two themes.

The Ecoregion field contains much more detail than the Mht_name field.

m-2   Write a hypothesis about what this data represents.

Student answers will vary, but they should include some of the following information:

The numeric fields appear to be some sort of scale. It's unclear which value is considered most critical—1 or 5.

Zero is more than likely a "no data" field.

There is some sort of relationship between the numeric fields and the text/string fields.

The numeric field represents the status of the various ecoregions.

Step 4   Map the data and analyze it

c-1   After mapping out all the data fields, what additional conclusions can you make about the data?

Possible answers include:

The numeric fields appear to be some sort of scale.

Zero is likely to be a "no data" field.

There is a relationship between the numeric and string fields.

c-2   Use this new information to revise your hypothesis about what the data represents.

Student answers will vary and depend on their response to the previous question.

d   Read the table below and use the mapped data to complete the missing pieces. The definitions in the table belong to all six fields. Fill in the table with the appropriate field name from the attribute table.

FIELD NAME	DEFINITION
Final	The final assessment of the ecoregion as the estimated threat to the ecoregion over the next 20 years.
Ecoregion	Descriptive name for the _____ that are relatively large areas of land or water that share a large majority of their species, dynamics, and environmental conditions.
BDI	The biological distinctiveness index. It is based on the species richness, endemism, rareness, and so forth.
Shape	Represent global terrestrial and freshwater ecoregions defined as relatively large areas of land or water in the world that share a large majority of their species, dynamics, and environmental conditions.
Mht_name	The major habitat for the area.
Threat	Degree of threat to the ecoregion. Some examples include logging, conversion to agriculture/urbanization, and so on.

e   Complete the tables below by filling in the numeric values for each descriptor. Keep in mind that the descriptors may not be listed in numeric order and don't forget about zero.

THREAT	
5	Relatively stable in degree of threat
4	Relatively intact in degree of threat
3	Vulnerable in degree of threat
0	Not assessed or unknown
2	Endangered in degree of threat
1	Critical in degree of threat

FINAL	
0	Not assessed or unknown
1	Critical in estimated threat over the next 20 years
2	Endangered in estimated threat over the next 20 years
4	Relatively intact in estimated threat over the next 20 years
3	Vulnerable in estimated threat over the next 20 years
5	Relatively stable in estimated threat over the next 20 years

MODULE 1
ArcView

MODULE 2
global

MODULE 2
regional

MODULE 3
global

MODULE 3
regional

MODULE 3
advanced

MODULE 4
global

MODULE 4
regional

MODULE 4
advanced

MODULE 5
global

MODULE 5
regional

MODULE 5
advanced

MODULE 6
global

MODULE 6
regional

MODULE 7
global

MODULE 7
regional

MODULE 7
advanced

Step 5   View metadata documentation

e-1   How would you describe this data set? It represents the boundaries for the continents of the world.

e-2   Who made this data? Environmental Systems Research Institute, Inc. (ESRI)

e-3   What time period does the data cover? 1996

e-4   Which section explains whether or not you can reuse the data for a project or publish a work that uses the data?
The question "Are there legal restrictions on access or use of the data?" displays the use constraints of the data. It includes information on redistribution rights and the data license agreement for continent.shp.

Step 6   Create a simple metadata document

c   Type up the following information into the metadata document. Use continent.htm as a reference.
Student answers will vary. You can view the actual metadata document for Wwf_eco.shp from within the module 7 data directory. The file is wwf_eco.htm.

Bibliography

ArcAtlas: Our earth [Computer Software]. (1996). Redlands, Calif.: ESRI.

Bureau of Economic Analysis. (2001). *Bureau of Economic Analysis regional economic accounts.* Retrieved from www.bea.doc.gov/bea/regional/docs/regional_overview.htm

Chandler, T. (1989). *Four thousand years of urban growth: An historical census.* Lewiston, N.Y.: Edwin Mellen Press.

Council of the National Seismic System. (2000). *CNSS worldwide earthquake catalog, 2000* [Data file]. Available from Northern California Earthquake Data Center Web site, quake.geo.berkeley.edu/cnss/catalog-search.html

CountryWatch. (2000). Various demographic data sets [Data file]. Available from Country Watch Web site, www.countrywatch.com

DeBlij, H. J., and P. O. Muller. (1992). *Geography: Regions and concepts.* New York: John Wiley & Sons, Inc.

ESRI data & maps 1999 [Data CD]. (1999). Redlands, Calif.: ESRI.

ESRI data & maps media kit (Version 8) [Data CD]. (2000). Redlands, Calif.: ESRI.

ESRI Schools & Libraries Program. (2000). Mapping GLOBE visualizations. Retrieved October 29, 2001, from gis.esri.com/industries/k-12/arclessons/arclessons.cfm

Fellmann, J., A. Getis, and J. Getis. (1992). *Human geography: Landscapes of human activities,* 3d ed. Dubuque, Iowa: Wm. C. Brown Publishers.

Ferrigno, J. G., and R. S. Williams (eds.). (1999). Estimated present-day area and volume of glaciers and maximum sea level rise potential. In *Satellite image atlas of glaciers of the world* (chap. A, U.S. Geological Survey Professional Paper 1386-A). Retrieved September 6, 2001, from pubs.usgs.gov/factsheet/fs133-99//gl_vol.html

Food and Agricultural Organization of the United Nations. (n.d.). *FAO homepage.* Retrieved July 2001, from www.fao.org

Getting to know ArcView GIS, 3d ed. (1999). Redlands, Calif.: ESRI.

GIS for schools and libraries (Version 5) [Computer Software]. (1999). Redlands, Calif.: ESRI.

The GLOBE Program. (n.d.) *GLOBE visualizations.* Retrieved from viz.globe.gov/viz-bin/home.cgi?l=en&b=g&rg=n

Guiney, J. L. and M. B. Lawrence. (1999, January 28). *Preliminary report: Hurricane Mitch 22 October–05 November 1998.* National Hurricane Center. Retrieved September 17, 2001, from www.nhc.noaa.gov/1998mitch.html

Hardwick, S. W., and D. G. Holtgrieve. (1996). *Geography for educators: Standards, themes, and concepts* (82–112). Upper Saddle River, N.J.: Prentice Hall.

Instituto Nacional de Estadistica, Geographica e Informatica, Mexico. (2001). Export and import data [Data file]. Available from Instituto Nacional de Estadistica, Geographica e Informatica, Mexico, Web site, www.inegi.gob.mx

International Society for Technology in Education. (1998). *National educational technology standards for all students.* Eugene, Oreg.: Author.

Kennedy, H. (ed.), et al. (2001). *The ESRI Press dictionary of GIS terminology.* Redlands, Calif.: ESRI Press.

Kious, J., and R. I. Tilling. (1996). *This dynamic earth: the story of plate tectonics.* Retrieved June 7, 2001, from pubs.usgs.gov/publications/text/dynamic.html

Mcfalls, J. A., Jr. (1998). Population: A lively introduction. *Population Reference Bureau Population Bulletin* 53(3).

National Aeronautics and Space Administration. (n.d.). *NASA visible earth: Cryosphere.* Retrieved September 6, 2001, from visibleearth.nasa.gov/Cryosphere/Sea_Ice/Icebergs.html

National Council for Teachers of Mathematics. (2000). *Principles and standards for school mathematics.* Reston, Va.: Author.

National Geographic Society. (1994). *The National Geography Standards, 1994.* Washington, D.C.: The Geography Education Standards Project.

National Oceanic and Atmospheric Administration. (2001). *The south geographic pole* (Image ID: Corp1566, NOAA Corps Collection) [Data file]. Available from NOAA Photo Library Web site, www.photolib.noaa.gov/corps/corp1566.htm

National Research Council—National Academy of Sciences. (1995). *National science education standards.* Washington, D.C.: National Academy Press.

National Snow and Ice Data Center. (n.d.). *Antarctic ice shelves and icebergs.* Retrieved September 6, 2001, from nsidc.org/iceshelves

Population Reference Bureau. (n.d.). *PRB home page.* Retrieved July 2001, from www.prb.org

Population Reference Bureau. (n.d.). *World population: Towards the next century.* Washington, D.C.: Author.

Schofield, Richard. Abridged version of speech given March 31, 1999, published in *British–Yemeni Society Journal* (July 2000) and retrieved from www.al-bab.com.

Sheets, B., and J. Williams. (2001). *Hurricane watch: Forecasting the deadliest storms on earth.* New York: Vintage Books.

United Nations. (1985). Estimates and Projections of Urban, Rural and City Populations (ST/ESA/Ser.R/50).

U.S. Geological Survey. (2001). Global GIS database: Digital atlas of Central and South America (Digital Data Series DDS-62-A) [Computer software and data]. U.S.: Author.

U.S. Geological Survey. (2001). Global GIS database: Digital atlas of the Middle East [Computer software and data]. Unpublished, Author.

U.S. Geological Survey. (n.d.). *USGS earthquake hazards program.* Retrieved June 7, 2001, from earthquake.usgs.gov

U.S. Geological Survey. (n.d.). *USGS Hurricane Mitch program.* Retrieved September 10, 2001, from mitchnts1.cr.usgs.gov

U.S. Geological Survey. (2000, February). *USGS TerraWeb: Antarctica.* Retrieved September 6, 2001, from terraweb.wr.usgs.gov/projects/Antarctica

U.S. Geological Survey. (n.d.). *USGS volcano hazards program.* Retrieved June 7, 2001, from volcanoes.usgs.gov

U.S. Census Bureau. (2001). Export and import data [Data file]. Available from U.S. Census Bureau Foreign Trade Web site, www.cenus.gov/foreign-trade/www

Warnings from the ice. (1998, April). Retrieved September 6, 2001, from www.pbs.org/wgbh/nova/warnings

Whitaker, B. (2000, July 1). Commentary on the border treaty. *YEMEN Gateway.* Retrieved from www.al-bab.com/yemen/pol/border000629.htm

Whitaker, B. (2000, June 12). Translation of the Treaty of Jeddah. Retrieved from www.al-bab.com/yemen/pol/int5.htm

White, Frank. (1998, September). *Overview Effect: Space Exploration and Human Evolution,* 2d ed. Reston, Va.: American Institute of Aeronautics and Astronautics.

Williams, J. (2001). Ice shelves float on the sea. *USA Today.com.* Retrieved September 6, 2001, from www.usatoday.com/weather/antarc/aiceshlf.htm

Williams, J. (2001). Warming effect on sea level unsure. Edited summary of Intergovernmental Panel on Climate Change's book *Climate Change 2001.* USA Today.com. Retrieved September 6, 2001, from www.usatoday.com/weather/antarc/iceipcc.htm

World Climate.com. (n.d.). Various world climate data sets [Data file]. Available from World Climate.com Web site, www.worldclimate.com

World geography: Building a global perspective. (1998). Upper Saddle River, NJ: Prentice Hall.

World Wildlife Fund. (n.d.). *Endangered spaces: The global 200.* Retrieved September 19, 2001, from www.worldwildlife.org/global200

License Agreement

Permitted Uses:

- Licensee may install the Software, Data, and Related Materials or portions of the Data onto permanent storage device(s) and thereafter use, copy, reproduce, and distribute the Software, Data, and Related Materials, including Documentation, in quantities sufficient to meet Licensee's own internal needs, to be used solely in conjunction with the exercises and context of this book.

- Licensee may make only one (1) copy of the original Software, Data, and Related Materials for archival purposes during the term of this Agreement.

- Licensee may modify the Data and merge other data sets with the Data for Licensee's own internal use. The portions of the Data merged with other data sets will continue to be subject to the terms and conditions of this Agreement.

- Licensee may use, copy, alter, modify, merge, reproduce, and/or create derivative works of the online Documentation for Licensee's own internal use. The portions of the online Documentation merged with other software, data, hard copy, and/or digital materials shall continue to be subject to the terms and conditions of this Agreement and shall provide the copyright attribution notice acknowledging ESRI and its Licensor(s) proprietary rights in the online Documentation as instructed in the accompanying metadata file.

- Licensee may use, copy, reproduce, and/or redistribute the Data or any derived portion(s) of the Data in published hard-copy and/or in static, electronic (i.e., .gif, .jpeg, etc.) formats, provided that Licensee affixes a legend statement acknowledging the appropriate Data Publisher as the source of the portion(s) of the Data displayed, printed, or plotted.

- Licensee may customize the Software using any (i) macro or scripting language, (ii) open application programming interface (API), or (iii) source or object code libraries, but only to the extent that such customization is described in the Documentation.

- Licensee may use the Data only as described in the Distribution Rights section of the help files or metadata files delivered with the Software, Data, and Related Materials.

Uses Not Permitted:

- Licensee shall not sell; rent; lease; sublicense; lend; assign; time-share; or act as a service bureau or Application Service Provider (ASP) that allows third party access to the Software, Data, and Related Materials except as provided herein; or transfer, in whole or in part, access to prior or present versions of the Software, Data, or Related Materials, any updates, or Licensee's rights under this Agreement.

- Licensee shall not redistribute the Software, in whole or in part, including, but not limited to, extensions, components, or DLLs without the prior written approval of ESRI as set forth in an appropriate redistribution license agreement.

- Licensee shall not reverse engineer, decompile, or disassemble the Software, Data, or Related Materials, except to the extent that such activity is expressly permitted by applicable law notwithstanding this restriction in order to protect ESRI and its Licensor(s) trade secrets and proprietary information contained in the Software, Data, or Related Materials.

- Licensee shall not make any attempt to circumvent the technological measure(s) (e.g., License Manager, etc.) that controls access to or use of the Software, Data, and Related Materials, except to the extent that such activity is expressly permitted by applicable law notwithstanding this restriction.

- Licensee shall not use the Software to transfer or exchange any material where such transfer or exchange is prohibited by copyright or any other law.
- Licensee shall not remove or obscure any ESRI or its Licensor(s) patent, copyright, trademark, or proprietary rights notices contained in or affixed to the Software, Data, or Related Materials.

Term: The license granted by this Agreement is for a period of one (1) year and shall commence upon installation of the Software. In the event Licensee licenses a copy of ESRI Software or the K-12 Bundle, Licensee may continue to use the Data and Related Materials after termination of this Agreement. The Agreement shall automatically terminate without notice if Licensee fails to comply with any provision of this Agreement. Licensee shall then return the Software, Data, and Related Materials to ESRI. The parties hereby agree that all provisions that operate to protect the rights of ESRI and its Licensor(s) shall remain in force should breach occur.

Limited Warranty: ESRI warrants that the media upon which the Software, Data, and Related Materials are provided will be free from defects in materials and workmanship under normal use and service for a period of ninety (90) days from the date of receipt. The Software, Data, and Related Materials are excluded from the warranty, and Licensee expressly acknowledges that the Data contain some nonconformities, defects, errors, or omissions. ESRI and its Licensor(s) do not warrant that the Data will meet Licensee's needs or expectations, that the use of the Data will be uninterrupted, or that all nonconformities, defects, or errors can or will be corrected. ESRI and its Licensor(s) are not inviting reliance on these data, and Licensee should always verify actual data.

EXCEPT FOR THE LIMITED WARRANTIES SET FORTH ABOVE, THE SOFTWARE, DATA, AND RELATED MATERIALS CONTAINED THEREIN ARE PROVIDED "AS IS," WITHOUT WARRANTY OF ANY KIND, EITHER EXPRESS OR IMPLIED, INCLUDING, BUT NOT LIMITED TO, THE IMPLIED WARRANTIES OF MERCHANTABILITY AND FITNESS FOR A PARTICULAR PURPOSE.

Exclusive Remedy and Limitation of Liability: ESRI and/or its Licensor(s) entire liability and Licensee's exclusive remedy shall be to terminate the Agreement upon Licensee returning the Software, Data, and Related Materials to ESRI with a copy of Licensee's invoice/receipt and ESRI returning the license fees paid to Licensee.

IN NO EVENT SHALL DATA PUBLISHERS AND/OR THEIR LICENSOR(S) BE LIABLE FOR COSTS OF PROCUREMENT OF SUBSTITUTE GOODS OR SERVICES, LOST PROFITS, LOST SALES OR BUSINESS EXPENDITURES, INVESTMENTS, OR COMMITMENTS IN CONNECTION WITH ANY BUSINESS, LOSS OF ANY GOODWILL, OR FOR ANY INDIRECT, SPECIAL, INCIDENTAL, EXEMPLARY, OR CONSEQUENTIAL DAMAGES ARISING OUT OF THIS AGREEMENT OR USE OF THE SOFTWARE, DATA, AND RELATED MATERIALS, HOWEVER CAUSED, ON ANY THEORY OF LIABILITY, AND WHETHER OR NOT DATA PUBLISHERS AND/OR THEIR LICENSOR(S) HAVE BEEN ADVISED OF THE POSSIBILITY OF SUCH DAMAGE. THESE LIMITATIONS SHALL APPLY NOTWITHSTANDING ANY FAILURE OF ESSENTIAL PURPOSE OF ANY EXCLUSIVE REMEDY.

No Implied Waivers: No failure or delay by ESRI and/or its Licensor(s) in enforcing any right or remedy under this Agreement shall be construed as a waiver of any future or other exercise of such right or remedy by ESRI and/or its Licensor(s).

U.S. Government Restricted/Limited Rights: Any software, Documentation, and/or data delivered hereunder is subject to the terms of the License Agreement. In no event shall the U.S. Government acquire greater than RESTRICTED/LIMITED RIGHTS. At a minimum, use, duplication, or disclosure by the U.S. Government is subject to restrictions as set forth in FAR §52.227-14 Alternates I, II, and III (JUN 1987); FAR §52.227-19 (JUN 1987) and/or FAR §12.211/12.212 (Commercial Technical Data/Computer Software); and DFARS §252.227-7015 (NOV 1995) (Technical Data) and/or DFARS §227.7202 (Computer Software), as applicable. Contractor/Manufacturer is ESRI, 380 New York Street, Redlands, CA 92373-8100 USA.

Export Control Regulations: Licensee expressly acknowledges and agrees that Licensee shall not export, reexport, or provide the Software, Data, or Related Materials, in whole or in part, to (i) any country to which the United States has embargoed goods; (ii) any person on the U.S. Treasury Department's list of Specially Designated Nationals; (iii) any person or entity on the U.S. Commerce Department's Table of Denial Orders; or (iv) any person or entity where such export, reexport, or provision violates any U.S. export control law or regulation. Licensee shall not export the Software, Data, and/or Related Materials or any underlying information or technology to any facility in violation of these or other applicable laws and regulations. Licensee represents and warrants that it is not a national, resident, located in or under the control of, or acting on behalf of any person, entity, or country subject to such U.S. export controls.

Severability: If any provision(s) of this Agreement shall be held to be invalid, illegal, or unenforceable by a court or other tribunal of competent jurisdiction, the validity, legality, and enforceability of the remaining provisions shall not in any way be affected or impaired thereby.

Governing Law: This Agreement shall be construed and enforced in accordance with and be governed by the laws of the United States of America and the State of California without reference to conflict of laws principles.

Entire Agreement: The parties agree that this constitutes the sole and entire agreement of the parties as to the matter set forth herein and supersedes any previous agreements, understandings, and arrangements between the parties relating hereto.

Installing the data and software

Mapping Our World: GIS Lessons for Educators includes two CDs: one with education and GIS resources for teachers, and one with ArcView 3.x software for Windows and Macintosh platforms and the exercise data. ArcView 3.2a software for Windows is supported on Windows 95®, Windows 98®, Windows NT®, Windows 2000® and Windows XP®. ArcView 3.0a software for Macintosh is supported on OS 7.5 through OS 8.1 (it has not been tested on later operating systems). The software takes up about 100 megabytes of hard disk space, the data about 300 megabytes. The included ArcView 3.x software must be installed on individual computers.

Please read the important information provided below before installing the software:

The ArcView software included on this CD–ROM is intended for educational purposes only. If you already have a licensed copy of ArcView 3.x installed on your computer (or accessible through a network), we recommend that you install the included software anyway. The projects in the book will work on any version of ArcView 3.x. Once installed, the software will run for 365 days. The software cannot be reinstalled nor can the time limit be extended. It is recommended that you uninstall this software when it expires. You do not need to uninstall it before installing other versions of ArcView 3.x or ArcView 8; all can be used simultaneously.

The installation includes three parts and the entire process will take about 10 minutes.

1 Data installation

2 ArcView 3.2a Education Edition software installation

3 Extension installation

Installing the data on a Windows platform

Follow the steps below to install the exercise data. Do not copy the files directly from the CD to your hard drive.

1 Put the software and data CD in your computer's CD–ROM drive. The CD auto-runs and the Welcome appears. Read it.

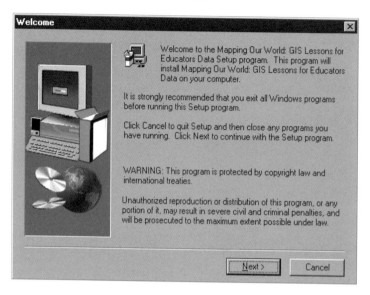

2 Click Next. Read the Important Installation Information.

3 Click Next. For the destination location, either accept the default installation folder or navigate to the drive where you want to install the data. If you choose a new drive, be sure to inform your students that the data is stored in a different location. Exercises in the book reference the default location of C:\esri\mapworld.

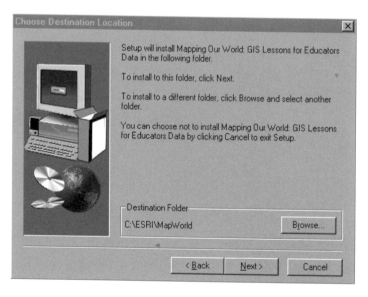

4 When asked to Confirm New Folder, click Yes. Setup automatically installs the data on your computer in a folder called MapWorld. When the data installation is complete, you will install the ArcView GIS Education Edition software that's included on the CD.

Installing the software

5 The ArcView 3.2a Education Edition Setup Welcome screen appears. Read it and click Next.

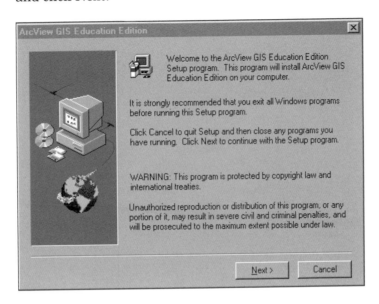

6 The ArcView GIS Education Edition License Agreement appears. Read it and click Yes.

7 Information about the ArcView GIS Education Edition appears. Read it and click Next.

8 Accept the default destination folder or browse to select a new one. Click Next.

9 Click Next to select the default program folder name.

10 Click Install to start copying ArcView GIS Education Edition files.

Note: Windows NT and Windows 2000 users may get a failure to open and a failure to create error at this step. If you get an error, click OK and ArcView will be installed correctly.

11 Click No to restart your computer later. Click Finish.

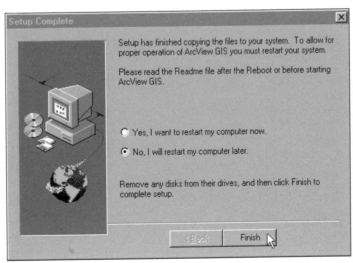

12 Setup installs the book's extension and finishes.

13 Restart your computer.

ESRI provides no technical support for this complimentary CD. If you have questions or encounter problems during the installation process, refer to the complete installation guide included on the CD or go to *Mapping Our World: GIS Lessons for Educators* Web site at www.esri.com/mappingourworld.

Uninstalling the data and software

To uninstall the exercise data or software from your computer, open your operating system's control panel and click the Add/Remove Programs icon. In the Add/Remove Programs Properties dialog, select whichever of the following entries you want to uninstall and follow the prompts to remove them:

ESRI Mapping Our World: GIS Lessons for Educators Data

ArcView GIS Education Edition

Installing the data and software

Mapping Our World: GIS Lessons for Educators includes two CDs: one with educational and GIS resources for teachers, and a second disc with the exercise data, ArcView 3.2a software for Windows, and ArcView 3.0a software for Macintosh. This guide will help you install the exercise data and ArcView software for Macintosh, including the Extension (an additional file that enhances the ArcView software). The procedure you will follow is:

1 Read the installation guide

2 Install the data

3 Install the ArcView 3.0a software for Macintosh

4 Install the Extension

The ArcView 3.0a software for Macintosh is supported on OS 7.5 through OS 8.1. It has not been tested on later operating systems.

The software takes up approximately 60 megabytes of hard disk space; the data about 300 megabytes. The included ArcView 3.0a software for Macintosh must be installed on individual computers. When you install the data and the software, the entire process will take about 15 minutes per computer.

If you already have a licensed copy of ArcView 3.0a installed on your Macintosh computer, do not install the ArcView software. Use your licensed software to do the exercises in this book. However, you must install the data and the book's Extension on your computer as described below.

Installing the data on the Macintosh platform

Follow the steps below to install the exercise data. Do not copy the files directly from the CD to your hard drive.

1 Put the Software/Data CD in your computer's CD drive. You will see the Mac Install Guide file and three applications:

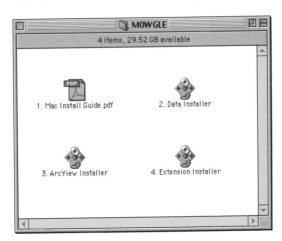

2 The Mac Install Guide.pdf file contains instructions for installing the three applications. It is a digital version of the information you are reading now. Read through the entire document before continuing.

3 Close any applications that you have running. Your computer will be restarted during this installation process and unsaved information in open applications will be lost.

4 Double-click Data Installer to begin the installation.

5 An information screen appears; click Continue.

6 Read the installation guidelines. When finished, click Continue.

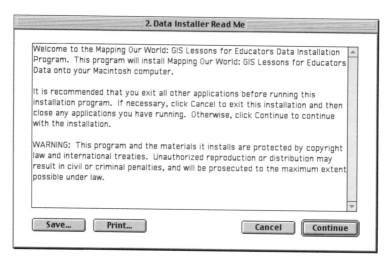

7 You will need 300 MB of free hard drive space to install the data. If you have at least 300 MB free, click Install.

8 By default, the data will be installed into a folder named "ESRI" which will be created for you automatically. You can accept the default installation folder or navigate to the location where you want to install the data.

Note: If you choose a new location, be sure to inform your students that the data is stored in a different location. Exercises in the book reference the default location of C:\esri\mapworld.

Click Choose to begin copying the files to your computer.

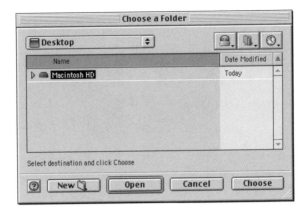

9 A status bar displays the installation's progress. A message appears when installation is complete. Click OK to exit the installation. Now you will install the ArcView 3.0a software and the book's Extension (an additional file to enhance the ArcView software). If you already have a licensed copy of ArcView 3.0a for Macintosh, proceed to the section "Installing the Extension on a Macintosh platform."

Installing ArcView 3.0a software on the Macintosh platform

Please read the important information provided below before installing the software.

The ArcView software included on this CD–ROM is intended for educational purposes only. The software is licensed to you for 365 days. After this time, the software should be uninstalled. If you wish to continue using the materials after this time, please contact ESRI to purchase a permanent license (1-800-447-9778 or www.esri.com).

1 Double-click ArcView Installer to begin the installation. A welcome screen appears.

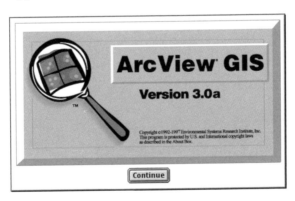

2 Click Continue.

3 Read the important installation information. When finished, click Continue.

4 Click Install.

5 By default, the software will be installed into a folder named "ESRI." You can accept the default installation folder or navigate to the location where you want to install the data. Click Install to install the software.

Note: If you choose a new location, be sure to inform your students that the program is found in a different location than the book describes.

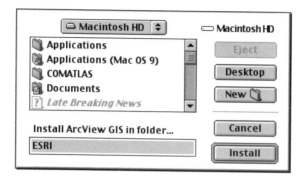

6 A message appears indicating that your computer must be restarted after the installation. Click Yes.

7 A status bar displays the installation's progress. A message appears when installation is complete. Click Restart to restart your computer.

Installing the Extension on the Macintosh platform

1 Double-click Extension Installer to begin the installation.

2 Click Continue.

3 Read the installation guidelines. When you are finished, click Continue.

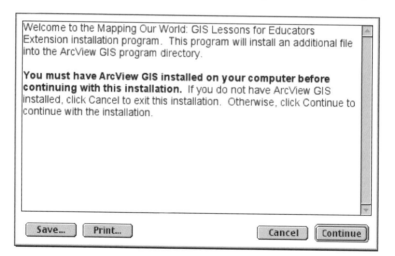

4 Select the ESRI folder.

Note: The Extension must be installed into the folder where the ArcView software is installed to function properly. If you chose a different location when installing the ArcView software, navigate to that location instead of the location shown below.

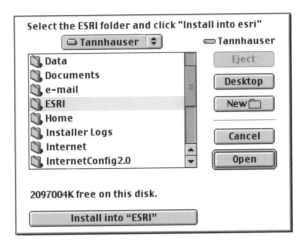

5 A message appears when the installation is completed. Click OK to exit the installation. You now are ready to use the materials. Refer to the text for additional information about working through the exercises.

Tips for optimizing performance

1 Increase the memory available to ArcView. The recommended minimum setting is 16 MB (16,384K). Better performance will be obtained by making more memory available to ArcView (e.g., 32 MB or 64 MB). However, your ideal setting will depend on the amount of memory installed on your computer.

2 Close other applications to free up memory.

Registration information

When you first start ArcView 3.0a, you will be prompted for registration information. Enter your name and organization, and enter the following code for the registration number: 708301184411

Uninstalling the data and software

To uninstall the exercise data and software from your computer, drag the ESRI folder to the trash.

Macintosh Technical Guide

In this book, all the illustrations show how ArcView software looks under the Microsoft Windows operating systems. The appearance and operation of ArcView is nearly identical on both Windows and Macintosh platforms. However, there are some cosmetic differences and a few procedural differences. This document describes the main differences that Macintosh users will encounter. When a Mac user's screen differs from the screen illustrated in this book for a particular step, reference to the information below will clarify and reconcile the difference.

1 **No application window.** The Windows software uses an application window that contains all the smaller windows. On a Macintosh, each window (the Project window and the View window as seen below) seems independent on the desktop.

Macintosh

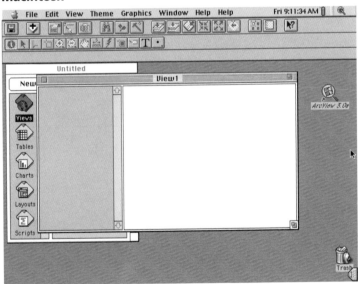

Windows

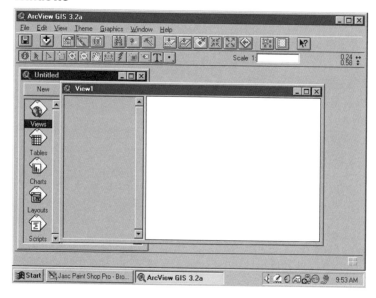

2 **Location of status bar.** ArcView software provides a status bar that explains the function of the tools and buttons. In the Macintosh environment, the status bar is located under the toolbar, near the top of the screen. The illustrations in the book show the status bar at the bottom of the ArcView application window.

Macintosh

Windows

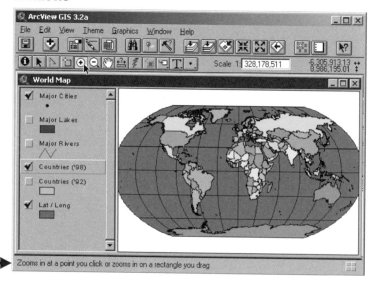

3 **Working in the active window.** If you want to perform an action in a window, you first must make it the active or front-most window and then click again in the window to perform the action. Thus, it may seem that you need to double-click an item, even though the instructions indicate a single click. It is helpful to know which is the active window at each step of the exercise. On a Macintosh, the active window will have lines/strips across the title bar and the nonactive windows will be blank (white).

Active window   **Nonactive window**

On Windows, a single click in a nonactive window will bring that window to the front and perform the desired action, while on Macintosh a single click will only bring that window to the front. For example, in a Windows environment, if a table is behind a view and you click on the table, two things happen: the table comes to the front and the record that you click is selected. In a Macintosh environment, only one thing happens: the table comes to the front.

4 **Adding themes.** When adding themes, the window that appears is cosmetically different between Macintosh and Windows. On a Macintosh, you will find the folders/directory on the left side and the data sources on the right side of the Add Theme window. Illustrations in the book will show the list of data sources on the left side and the folders/directory structure on the right side of the Add Theme window. Also, in Macintosh, you click Add or Add All to add data. In Windows, you click OK to add the data.

Macintosh

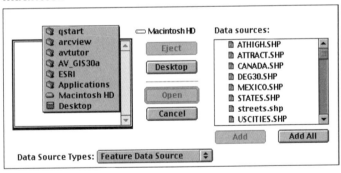

Windows

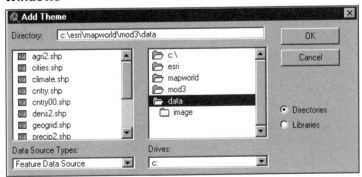

5 **Adding themes.** In the Add Theme window, Macintosh users should use Apple-click to select multiple, nonadjacent items without choosing the items in between. In the Windows version, Shift-click selects multiple, nonadjacent items without affecting intermediate items.

6 **Maximizing and closing windows.** On Macintosh, you close a window by clicking the button in the upper left corner of the window. To maximize or unmaximize a window, click the button in the upper right corner. On the Windows platform, you can minimize, maximize, or close a window by clicking one of the three buttons located in the upper right corner of each window.

7 **Resizing windows.** On a Macintosh, you can click the bottom right corner of most windows to perform custom resizing of windows. On the Windows platform, you can click any edges or corners of most windows to perform custom resizing.

8 **Help menu.** In Macintosh, two "Help" menus are visible. The left one is for the ArcView software, the right one is for the Macintosh operating system. In Windows, one "Help" menu is visible, which is for the ArcView software (see graphic under item 4 above).

9 **Pathways and folders.** When looking at a theme's properties, Macintosh indicates a change in folder using a colon (:) character, while Windows uses a backslash (\) character.

Macintosh navigation pathway
MacHD:esri:mapworld:mod1:module1.apr

Windows navigation pathway
C:\esri\mapworld\mod1\module1.apr

10 **Graphic export file formats.** When exporting a map to a graphic file, Macintosh users can export to a variety of formats, including the PICT format (compatible with most other Macintosh applications), while Windows users can export to BMP, WMF, and JPEG formats.

National Science and Technology Standards

STANDARD	SCIENCE STANDARDS							TECHNOLOGY STANDARDS					
	A	B	C	D	E	F	G	1	2	3	4	5	6
Module 1 *Global*	MH					MH		✔	✔	✔		✔	✔
Module 2 *Global* *Regional* *Advanced investigation*	MH MH MH	MH MH MH		M M M		MH MH MH		✔ ✔ ✔	✔ ✔ ✔	✔ ✔ ✔	 ✔	✔ ✔ ✔	✔ ✔ ✔
Module 3 *Global* *Regional* *Advanced investigation*	MH MH	 M	M 	 MH	H MH	MH MH MH	H M	✔ ✔ ✔	✔ ✔ ✔	✔ ✔ ✔	 ✔	✔ ✔ ✔	✔ ✔ ✔
Module 4 *Global* *Regional* *Advanced investigation*	 MH MH		 M 			MH MH 		✔ ✔ ✔	✔ ✔ ✔	✔ ✔ ✔	 ✔	✔ ✔ ✔	✔ ✔ ✔
Module 5 *Global* *Regional* *Advanced investigation*	MH MH 					MH MH 		✔ ✔ ✔	✔ ✔ ✔	✔ ✔ ✔		✔ ✔ ✔	✔ ✔ ✔
Module 6 *Global* *Regional* *Advanced investigation*	M M M					M M M		✔ ✔ ✔	✔ ✔ ✔	✔ ✔ ✔	 ✔	✔ ✔ ✔	✔ ✔ ✔
Module 7 *Global* *Regional* *Advanced investigation*	MH 		 M	H 	MH 	MH MH 		✔ ✔ ✔	✔ ✔ ✔	✔ ✔ ✔	 ✔	✔ ✔ ✔	✔ ✔ ✔

M = Middle school   H = High school

attribute   A piece of information that describes a geographic feature on a GIS map. The attributes of an earthquake would include the date it occurred, its latitude and longitude, depth, and magnitude.

Shape	Date	Lat	Lon	Depth	Mag
Point	20000101	36.8740	69.9470	54.30	5.10
Point	20000101	-60.7220	153.6700	10.00	6.00
Point	20000101	46.8880	-78.9300	18.00	5.20
Point	20000101	23.1120	143.6440	33.00	5.20
Point	20000102	-17.9430	-178.4760	582.30	5.50
Point	20000102	51.4470	-175.5580	33.00	5.80
Point	20000102	-20.7710	-174.2360	33.00	5.80
Point	20000102	51.4990	-175.5200	33.00	5.20
Point	20000103	9.6340	126.2800	67.90	5.20
Point	20000104	-20.7160	-177.6130	336.40	5.40
Point	20000105	-20.7710	-174.2140	33.00	5.10
Point	20000105	-20.9640	-174.0970	33.00	5.60
Point	20000105	-11.3710	165.3780	33.00	6.10
Point	20000105	-11.4620	165.4130	33.00	5.50

Attributes of Earthquakes

attribute table   Contains all of the attributes for like features on a GIS map. The attribute table for Top 10 Cities, 1950, C.E. includes data on each of the ten cities listed.

Modern City	Modern Country	Historic City	Rank1950	Pop_1950
Moscow	Russia	Mocow	6	5100000
London	United Kingdom	London	2	8860000
Paris	France	Paris	4	5900000
Chicago	United States	Chicago	8	4906000
New York	United States	New York	1	12463000
Tokyo	Japan	Edo	3	7000000
Shanghai	China	Shanghai	5	5406000
Calcutta	India	Calcutta	10	4800000
Buenos Aires	Argentina	Buenow Aires	7	5000000
Essen	Germany	Ruhr	9	4900000

Attributes of Top 10 Cities, 1950 C.E.

axis   The vertical (y-axis) or horizontal line (x-axis) in a chart or graph on which measurements can be illustrated and coordinated with each other. Each axis in a GIS chart can be made visible or invisible, labeled, and moved.

chart A graphic representation of tabular data. It's another word for graph in ArcView.

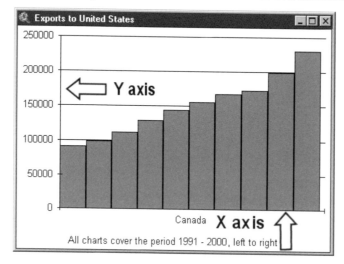

color palette The window that allows you to change the color of geographic features, lines, and text on your GIS map.

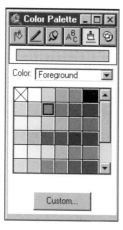

comma-delimited text file A data table in text form where the values are separated by commas. Spreadsheet programs use the commas to determine where a new cell stops and starts.

data Any collection of related facts, from raw numbers and measurements to analyzed and organized sets of information. In a GIS, data is stored in attribute tables and provides all the information about the geographical features on the GIS map.

data directory   A folder or area on the hard drive of your computer or your network's computer that is available for storage of GIS projects and data that you create.

Shape	Cntry_name	Curr_type	Sqkm	Sqmi	Tot_pop
Polygon	Afghanistan	Afghani	642115.94	247920.95	25889000
Polygon	Albania	Lek	28615.98	11048.63	3490000
Polygon	Algeria	Dinar	2320957.00	896121.50	31194000
Polygon	American Samoa	US Dollar	164.37	63.46	
Polygon	Andorra	Peseta	506.31	195.49	67000
Polygon	Angola	New Kwanza	1252148.63	483454.72	10145000
Polygon	Anguilla	EC Dollar	91.57	35.36	
Polygon	Antarctica		12303052.00	4750207.50	
Polygon	Antigua & Barbuda	EC Dollar	538.66	207.98	66000
Polygon	Argentina	Peso	2781207.00	1073824.00	36955000
Polygon	Armenia	Dram	29636.09	11442.49	3344000
Polygon	Aruba	Florin	200.35	77.35	
Polygon	Australia	Australia Dollar	7701651.00	2973607.50	19165000
Polygon	Austria	Schilling	83708.39	32319.81	8131000
Polygon	Azerbaijan	Manat	164129.88	63370.55	7748000

(Window title: Attributes of Country Outlines)

decimal degrees   Degrees of latitude and longitude expressed in decimals instead of in degrees. Latitude and longitude is most typically displayed in degrees, minutes, and seconds. Decimal degrees converts the degrees, minutes, and seconds into a decimal number using the mathematical formula of:

decimal degrees = degrees + minutes/60 + seconds/3,600
73° 59′ 15″ longitude is equal to 73.9875 decimal degrees

feature   The geographic objects on a map represented by points, lines, and polygons.

feature, line   A line on a map that represents a geographic object too narrow to show as a polygon. A line feature might represent a road, river, or a phone line.

feature, point   A point on a map that represents a geographic object too small to show as a line or polygon. A point feature might represent a tree, or a phone line, or even a city viewed from a satellite.

feature, polygon   An area on a map that represents a geographic object too large to show as a point or a line. A polygon feature might represent a lake, or a city viewed from an airplane, or a whole continent viewed from a satellite.

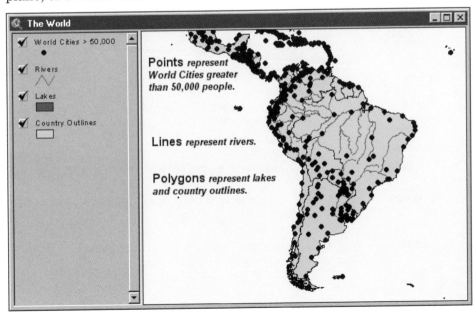

field   The vertical column in an attribute table that contains attributes or information about geographic features on a GIS map.

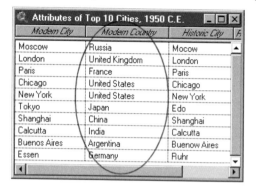

field name   The vertical column heading in an attribute table. ArcView abbreviates field headings to eight characters, but you are able to edit them. In the graphic below, Modern City, Modern Country, and Historic City are field names.

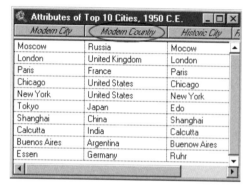

georeference   To assign coordinates from a reference system, such as latitude/longitude, to the page coordinates of an image or map.

graduated color legend   A map that uses a range of colors to show a sequence of numeric values. For example, on the population density map the more people per square kilometer the darker the purple color gets.

Identify tool   Allows the user to display the attributes of features in the current view.

image   A graphic representation of data such as a photograph, scanned picture, or a satellite photograph.

JPEG   Joint Photographic Experts Group, a way of creating and saving an image in a digital format that combines similar elements (pixels, for instance) as it compresses, sacrificing clarity and sharpness for small size and electronic transportability. (Compare TIFF.)

label   Text placed next to a geographic feature on a map to describe or identify it.

layout   An on-screen document where map elements such as the title, legend, and scale bar are arranged for printing.

legend   A list of symbols on a GIS map that contains a sample of each symbol as well as text that identifies the symbol.

line feature   See feature.

map projection   The process that transforms the locations of geographic features on the earth's curved surface to locations on a flat two-dimensional surface. Map projections are made using complex mathematical formulas that are part of ArcView's automatic functions.

metadata   Information that describes the data in an attribute table. Information in metadata may include the names of the authors or compilers of the data, the date it was created, and its projection, scale, and resolution.

pan   To move your map up, down, or sideways with the Pan tool.

point feature   See feature.

polygon feature   See feature.

polyline   A sequence of points, each pair of which can be connected by a straight line or arc.

project (.apr)   In ArcView, a file that organizes the views, tables, charts, layouts, and scripts used for geographic mapmaking and analysis.

projection   See map projection.

project window   The first window you see when you open an ArcView project. It displays and holds the different kinds of documents an ArcView project can contain. The project window has icons for Views, Tables, Charts, Layouts, and Scripts documents.

record	A row in a database or attribute table that contains all of the attributes values of a single entity.

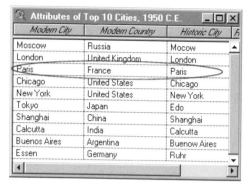

selected feature	In ArcView, a geographic feature can be clicked on using the Select tool so that various functions can be performed on that feature. When a geographic feature is selected it will be highlighted yellow.
shapefile (.shp)	ArcView's format for storing the location, shape, and attribute information of geographic features.
sort, ascending	To arrange information in an attribute table in order from lowest to highest numerically or alphabetically.
sort, descending	To arrange information in an attribute table in order from highest to lowest numerically or alphabetically.
symbol window	The window in ArcView that contains the palettes to change the way attributes and text look on the GIS map. The symbol window contains the Fill, Pen, Marker, Font, and Color palettes and is located in the Window pull-down menu.
table of contents	The area on a view window that holds the legends for the themes in an ArcView project.
theme	In ArcView, a set of geographic features of the same type along with their attribute tables. For example: "all cities in the world with populations greater than 1,000,000," or "all the countries in Latin America."
active, theme	Many actions that are performed in a view only work on active themes.

theme, turn on Turning on a theme allows the theme to display in the view. Place a check mark in the box next to the theme name in the table of contents.

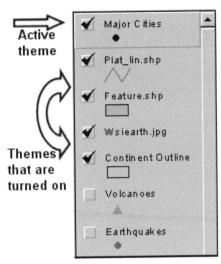

TIFF Tagged Image File Format, a way of creating and saving an image that compresses all the image's digital information without losing any of it. (Compare JPEG.)

vertex The junction of lines that form an angle.

view (window) The area of an ArcView project that displays interactive GIS maps as well as the table of contents.

zoom To display a larger or smaller region of a GIS map or image.